Methods in Molecular Biology

Series Editor
John M. Walker
School of Life Sciences
University of Hertfordshire
Hatfield, Hertfordshire, AL10 9AB, UK

For further volumes:
http://www.springer.com/series/7651

Patch-Clamp Methods and Protocols

Second Edition

Edited by

Marzia Martina

Department of Translational Bioscience, National Research Council of Canada, Ottawa, ON, Canada

Stefano Taverna

Department of Neuroscience & Brain Technologies, Italian Institute of Technology, Genoa, Italy

Humana Press

Editors
Marzia Martina
Department of Translational Bioscience
National Research Council of Canada
Ottawa, ON, Canada

Stefano Taverna
Department of Neuroscience & Brain Technologies
Italian Institute of Technology
Genoa, Italy

ISSN 1064-3745 ISSN 1940-6029 (electronic)
ISBN 978-1-4939-1095-3 ISBN 978-1-4939-1096-0 (eBook)
DOI 10.1007/978-1-4939-1096-0
Springer New York Heidelberg Dordrecht London

Library of Congress Control Number: 2014942964

Printed on acid-free paper

Humana Press is a brand of Springer
Springer is part of Springer Science+Business Media (www.springer.com)

Preface

"Patch-Clamp Methods and Protocols" was first published in 2007, but the rate of progress in this field has been so compelling that a second edition in 2014 is fully merited.

Since its development by Sakmann and Neher in the late 1970s, patch-clamp has been widely considered the gold standard method to record ion channel activity. This technique has become an indispensable means for conducting pharmacological, physiological, and biophysical research aimed at understanding crucial aspects of cellular and network function. Despite being so historically entrenched, patch-clamp development has not stood still. In fact, quite the opposite has occurred, with the pace of progress continuing to accelerate in both technology and theory. Significant resources have been devoted to develop new technologies to further improve the quality and throughput of the data obtained with patch-clamp. This second edition builds on the first by presenting a comprehensive collection of these new techniques, including the development of automated, high-throughput screening systems for pharmacological evaluation, the use of various patch-clamp configurations together with novel molecular biological and imaging methodologies, and enhanced stimulation protocols and perfusion systems. Biophysical analysis tools have also been improved along with modeling strategies to simulate electrical and synaptic activity.

As in the 2007 edition, the structure of the new "Patch-Clamp Methods and Protocols" includes three main parts: Pharmacology, Physiology, and Biophysics. A historic outlook on the development of the concept of bioelectricity and of various electrophysiology approaches employing patch-clamp is presented at the beginning of the book. This chapter is intended to provide a perspective to the following sections to underline the enormous progress that continues to be gained in the field.

The Pharmacology section presents chapters describing methods to reproduce more physiologically relevant conditions for drug application and screening technologies. Two chapters describing automated planar patch-clamp systems with improved throughput and increased quality of the electrophysiological recordings are also presented. This section closes with a review of the main materials and fabrication protocols for the production of patch-clamp chips and presents the concept of a patch-clamp chip array providing high resolution recordings from individual cells at multiple sites in a network of communicating neurons.

The Physiology section has been integrated with a wide variety of recently developed applications such as optogenetic stimulation, whole-cell recordings in freely moving animals, two-photon imaging combined with neurotransmitter uncaging, and optical control of neuronal activity at the single-receptor level.

The Biophysics section includes novel methods to create custom microelectrodes designed for reducing the access resistance and improving the rate of molecular diffusion through the pipette, the implementation of customized real-time dynamic-clamp protocols, the analysis of ion-selectivity properties of pore-forming peptides, and an analytical overview of single-channel kinetics.

The goal of this volume is to provide a useful technical and methodological guide to diverse audiences of electrophysiologists ranging from students to experienced investigators.

Genoa, Italy *Stefano Taverna*
Ottawa, ON, Canada *Marzia Martina*

Contents

Contributors

BRUNO ALLARD • *Centre de Génétique et de Physiologie Moléculaire et Cellulaire, Université Claude Bernard Lyon 1, UMR CNRS 5534, Villeurbanne, France*

MARCO AQUILA • *Department of Bioscience, University of Milan, Milan, Italy*

ANDREA BARBERIS • *Department of Neuroscience and Brain Technologies, Italian Institute of Technology, Genoa, Italy*

JEAN-CLAUDE BÉÏQUE • *Department of Cellular and Molecular Medicine, Heart and Stroke Foundation Partnership for Stroke Recovery and Centre for Neural Dynamics, University of Ottawa, Ottawa, ON, Canada*

MASCIA BENEDUSI • *Section of Biophysics and Physiology, Department of Biology and Evolution, University of Ferrara, Ferrara, Italy*

GÉZA BERECKI • *Health Innovations Research Institute, RMIT University, Melbourne, VIC, Australia*

NEELIMA BHARGAVA • *NanoScience Technology Center, University of Central Florida, Orlando, FL, USA*

ALESSANDRO BOSCA • *Department of Neuroscience and Brain Technologies, Italian Institute of Technology, Genoa, Italy*

MICHAEL BRECHT • *Bernstein Center for Computational Neuroscience Berlin, Humboldt University of Berlin, Berlin, Germany; Cluster of Excellence NeuroCure, Charité-Universitätsmedizin Berlin, Berlin, Germany*

ROBERT J. BUTERA • *School of Electrical and Computer Engineering, Georgia Institute of Technology, Atlanta, GA, USA; Department of Biomedical Engineering, Georgia Institute of Technology, Atlanta, GA, USA*

THIERRY CENS • *Centre de Recherche de Biochimie Macromoleculaire, UMR 5237 -CNRS, Montpellier, France*

PIERRE CHARNET • *Centre de Recherche de Biochimie Macromoleculaire, UMR 5237 -CNRS, Montpellier, France*

QIAN CHEN • *McGovern Institute for Brain Research and Department of Brain and Cognitive Sciences, MIT, Cambridge, MA, USA*

DAVID J. CHRISTINI • *Greenberg Division of Cardiology and Institute for Computational Biomedicine, Weill Cornell Medical College, New York, NY, USA*

CLAUDE COLLET • *Centre de Génétique et de Physiologie Moléculaire et Cellulaire, Université Claude Bernard Lyon 1, UMR CNRS 5534, Villeurbanne, France; CNRS UMR 5297, Interdisciplinary Institute for Neuroscience, University of Bordeaux, Bordeaux Cedex, France*

DENISE COOK • *Department of Cellular and Molecular Medicine, University of Ottawa, Ottawa, ON, Canada*

TANYA L. DAIGLE • *Department of Cell Biology, Duke University Medical Center, Durham, NC, Cambridge, MA, USA; McGovern Institute for Brain Research and Department of Brain and Cognitive Sciences, MIT, Cambridge, MA, USA*

MAINAK DAS • *NanoScience Technology Center, University of Central Florida, Orlando, FL, USA*

ALAN D. DORVAL II • *Department of Bioengineering, Brain Institute, University of Utah, Salt Lake City, UT, USA*
JÉRÔME EPSZTEIN • *Institut National de la Santé et de la Recherche Médicale (INSERM U - 901), Marseille, France; Aix-Marseille University, Marseille, France; Institut de Neurobiologie de la Méditérranée (INMED), Marseille, France*
ANNA FASOLI • *Section of Biophysics and Physiology, Department of Biology and Evolution, University of Ferrara, Ferrara, Italy*
GUOPING FENG • *McGovern Institute for Brain Research and Department of Brain & Cognitive Sciences, MIT, Cambridge, MA, USA*
KATE E. GILLING • *Charité-Universitätsmedizin, Berlin, Germany*
ANTONI C.G. VAN GINNEKEN • *Department of Anatomy, Embryology and Physiology, Academic Medical Center, University of Amsterdam, Amsterdam, The Netherlands*
NATHAN G. GLASGOW • *Department of Neuroscience and Center for Neuroscience, University of Pittsburgh, Pittsburgh, PA, USA*
ROMAIN GUINAMARD • *Groupe Signalisation, Electrophysiologie et Imagerie des Lésions d'Ischémie-Reperfusion Myocardique, Normandie Université, Caen Cedex, France*
JAMES J. HICKMAN • *NanoScience Technology Center, University of Central Florida, Orlando, FL, USA*
THOMAS HOF • *Groupe Signalisation, Electrophysiologie et Imagerie des Lésions d'Ischémie-Reperfusion Myocardique, Normandie Université, Caen Cedex, France*
VINCENT JACQUEMOND • *Centre de Génétique et de Physiologie Moléculaire et Cellulaire, Université Claude Bernard Lyon 1, UMR CNRS 5534, Villeurbanne, France*
JON W. JOHNSON • *Department of Neuroscience and Center for Neuroscience, University of Pittsburgh, Pittsburgh, PA, USA*
JUNG-FONG KANG • *NanoScience Technology Center, University of Central Florida, Orlando, FL, USA*
ALBERT K. LEE • *Janelia Research Campus, Howard Hughes Medical Institute, Ashburn, VA, USA*
KEVIN F.H. LEE • *Department of Cellular and Molecular Medicine, University of Ottawa, Ottawa, ON, Canada*
ROMAIN LEFEBVRE • *Centre de Génétique et de Physiologie Moléculaire et Cellulaire, Université Claude Bernard Lyon 1, UMR CNRS 5534, Villeurbanne, France*
HENRY MARKRAM • *Brain and Mind Institute, EPFL, Lausanne, Switzerland*
MARZIA MARTINA • *Department of Translational Bioscience, National Research Council of Canada, Ottawa, ON, Canada*
PETER MOLNAR • *NanoScience Technology Center, University of Central Florida, Orlando, FL, USA; Faculty of Natural Sciences, University of West Hungary, Szombathely, Hungary*
ÉMILIE MULLER-CHRÉTIEN • *Hospices Civils de Lyon, Hôpital des Charpennes, Centre d'Investigation Clinique, Villeurbanne, France*
FRANCIS A. ORTEGA • *Greenberg Division of Cardiology, Weill Cornell Medical College, New York, NY, USA*
VLADIMIR PARPURA • *Department of Neurobiology, Center for Glial Biology in Medicine, Atomic Force Microscopy & Nanotechnology Laboratories, Civitan International Research Center, Evelyn F. McKnight Brain Institute, University of Alabama, Birmingham, AL, USA; Department of Biotechnology, University of Rijeka, Rijeka, Croatia*

CHRIS G. PARSONS • *Pharmacology, Non-Clinical Science, Merz Pharmaceuticals GmbH, Frankfurt am Main, Germany*
ENRICA MARIA PETRINI • *Department of Neuroscience and Brain Technologies, Italian Institute of Technology, Genoa, Italy*
JOSHUA L. PLOTKIN • *Department of Physiology, Feinberg School of Medicine, Northwestern University, Chicago, IL, USA*
LIUDMILA POLONCHUK • *Roche Pharmaceutical Research and Early Development, Innovation Center Basel, F. Hoffmann-La Roche Ltd., Basel, Switzerland*
SANDRINE POUVREAU • *Centre de Génétique et de Physiologie Moléculaire et Cellulaire, Université Claude Bernard Lyon 1, UMR CNRS 5534, Villeurbanne, France; UR406 Abeilles et Environnement, INRA, Avignon Cedex, France*
CHRISTOPHE PY • *Electronics Team, Information and Communication Technologies, National Research Council Canada, Ottawa, ON, Canada*
FENG QIN • *Department of Physiology and Biophysics, State University of New York at Buffalo, Buffalo, NY, USA*
GIORGIO RISPOLI • *Section of Biophysics and Physiology, Department of Biology and Evolution, University of Ferrara, Ferrara, Italy*
GIOVANNI RUSSO • *Department of Neuroscience and Brain Technologies, Italian Institute of Technology, Genoa, Italy*
LAURENT SALLÉ • *Groupe Signalisation, Electrophysiologie et Imagerie des Lésions d'Ischémie-Reperfusion Myocardique, Normandie Université, Caen Cedex, France*
CARY SOARES • *Department of Cellular and Molecular Medicine, University of Ottawa, Ottawa, ON, Canada*
SONJA STOELZLE-FEIX • *Nanion Technologies GmbH, Munich, Germany*
D. JAMES SURMEIER • *Department of Physiology, Feinberg School of Medicine, Northwestern University, Chicago, IL, USA*
STEFANO TAVERNA • *Department of Neuroscience & Brain Technologies, Italian Institute of Technology, Genoa, Italy*
JONATHAN T. TING • *Human Cell Types Department, Allen Institute for Brain Science, Seattle, WA, USA*
MARIA TOLEDO-RODRIGUEZ • *School of Life Sciences, University of Nottingham, Queens Medical Centre, Nottingham, UK*
ARIE O. VERKERK • *Department of Anatomy, Embryology and Physiology, Academic Medical Center, University of Amsterdam, Amsterdam, The Netherlands*
ALEXEI VERKHRATSKY • *Faculty of Life Sciences, The University of Manchester, Manchester, UK; Achucarro Center for Neuroscience, IKERBASQUE, Basque Foundation for Science, Bilbao, Spain; Department of Neurosciences, University of the Basque Country UPV/EHU, Leioa, Spain*
JOHN A. WHITE • *Department of Bioengineering, Brain Institute, University of Utah, Salt Lake City, UT, USA*
RONALD WILDERS • *Department of Anatomy, Embryology and Physiology, Academic Medical Center, University of Amsterdam, Amsterdam, The Netherlands*

Chapter 1

History of Electrophysiology and the Patch Clamp

Alexei Verkhratsky and Vladimir Parpura

Abstract

We provide a historic outlook on the development of the concept of bioelectricity, with emphasis on the neuromuscular junction as a model that revolutionized our thinking of the nerve, nervous, and muscle tissue excitability. We abridge some crucial experiments in defining the electrical excitability of biological cells. We also provide an insight into developments of tools and methods, which gradually yielded a contemporary "palette" of electrophysiology approaches, including the patch clamp. Pioneering steps in this journey, ranging from Galvani's experiments using the Leyden jar to those of Neher and Sakmann using a gigaseal patch-clamp approach, are pictorially illustrated. This chapter is meant to be a perspective to the following sections in this volume dedicated to patch-clamp methods and protocols.

Key words Action potentials, Bioelectricity, Ionic theory, Intracellular perfusion, Microelectrodes, Single cell electrophysiology

1 Bioelectricity: The First Notions of Electrical Nature of Excitation

Ideas of nature of excitation and nerve conduction has ancient roots, as indeed already Galenic ventriculo-pneumatic doctrine postulated the existence of "psychic pneuma," an extremely light fluid filling the brain ventricles and hollow nerves and passing the signals from sensory inputs to the brain and then back to effector organs. René Descartes suggested a role for the "animal spirit" represented by "a very fine flame," that through the system of valves can move between nerves and the brain and between nerves and muscles, thus carrying nerve impulses and initiating body responses. Sir Isaac Newton (Fig. 1) was most likely the first to suggest the electric nature of nerve signals, indicating that "electric bodies operate to greater distances…and all sensation is excited, and the members of animal bodies move at the command of the will, namely, by the vibrations of this spirit, mutually propagated along the solid filaments of the nerves, from the outward organs of sense to the brain, and from the brain into the muscles. But these are things that cannot be explained in few words, nor are we furnished

Marzia Martina and Stefano Taverna (eds.), *Patch-Clamp Methods and Protocols*, Methods in Molecular Biology, vol. 1183,
DOI 10.1007/978-1-4939-1096-0_1, © Springer Science+Business Media New York 2014

Fig. 1 The makers of electrophysiology

with that sufficiency of experiments which is required to an accurate determination and demonstration of the laws by which this electric and elastic spirit operates." [1]. Stephen Halles (who incidentally was one of the fathers of hemodynamics and blood pressure measurements; see [2]), a Newton's Royal Society fellow, derived to a similar conclusion that nerve signals can be conveyed by some energy acting "along the surfaces [of nerves] like electrical powers" [3]. The electrical nature of excitability was supported by several physiologists throughout Europe, including Tommaso Laghi, Giuseppe Gardini, and Pierre Bertholon de Saint-Lazare, the latter inventing the term *animalis electricitas* or animal electricity.

Thus, by the middle of eighteenth century the neuroelectrical theory was born, and this theory was almost instantly subjected to a severe criticism led by Albrech von Haller, who developed an alternative doctrine of "irritability" which assumed that every muscle has intrinsic contractive capability and nerve provided a mere trigger that was not associated with any electrical phenomena [4], see also [5] for detailed historical narration.

First experimental corroboration of bioelectricity was obtained in 1770s, when John Walsh described electric sparks and electric shocks generated by live animals such as electric eel and electric ray; these observations having a fundamental importance rapidly circulated through European learned societies [6].

2 Luigi Galvani Sets Up the Electrophysiology and Predicts Ion Channels

Luigi Galvani (Fig. 1) begun his electrophysiological experiments in 1780s in Bologna, where he was a professor of anatomy and a paid member (*accademico benedettino*) of the Academy of Sciences (of which he was also elected as a president), in collaboration with his wife Lucia Galeazzi (who was a daughter of Galvani's anatomy lecturer Domenico Maria Gusmano Galeazzi) (Fig. 1) and later with his nephew, Giovanni Aldini (Fig. 1) (for a detailed account of Galvani's work see [5–8]). In these experiments Galvani used a frog neuromuscular preparation, which rapidly became a classic in experimental physiology; for many years this preparation was called "*rane preparate alla maniera di Galvani.*" The frog neuromuscular preparation in sensu stricto was, however, developed a century earlier by Dutch microscopist and natural scientist Jan Swammerdam (Fig. 1), who used the frog leg, from which "one of the largest muscles be separated from the thigh of a Frog, and, together with its adherent nerve, prepared in such a manner as to remain unhurt." In this preparation mechanical stimulation of the nerve triggered muscle contraction, and, moreover, Swammerdam even made some arrangements for monitoring this mechanical activity by inserting the muscle into a glass tube and attaching needles to each of the muscle ends so that contractions can be followed up through movements of needles [9–11]. It is unlikely that Galvani read Swammerdam's works, however.

Galvani conducted his electrophysiological experiments for almost a decade before publishing the first results in 1792 in his famous book, *De Viribus Electricitatis in Motu Musculari*, which appeared as the seventh volume of the *Commentarii* of the Institute of Sciences of Bologna [12]. In this book he described the discovery of the electrical excitation of the nerve–muscle preparation using an electrical machine, i.e., the Leyden (or Leiden) jar, invented independently by the dean of the Kamin Cathedral in Pomerania Ewald Jürgen Georg von Kleist and a professor of physics and mathematics at the University of Leyden Peiter van Mussechenbroek. This gizmo today would be referred to as a capacitor. When a spark from the Leyden jar was delivered to the frog preparation, it triggered vigorous muscle contractions (Fig. 2). In these first experiments Galvani also found the saturation of muscle contraction with an increase in intensity of stimulation and described the refractory phenomenon, by showing that repeated

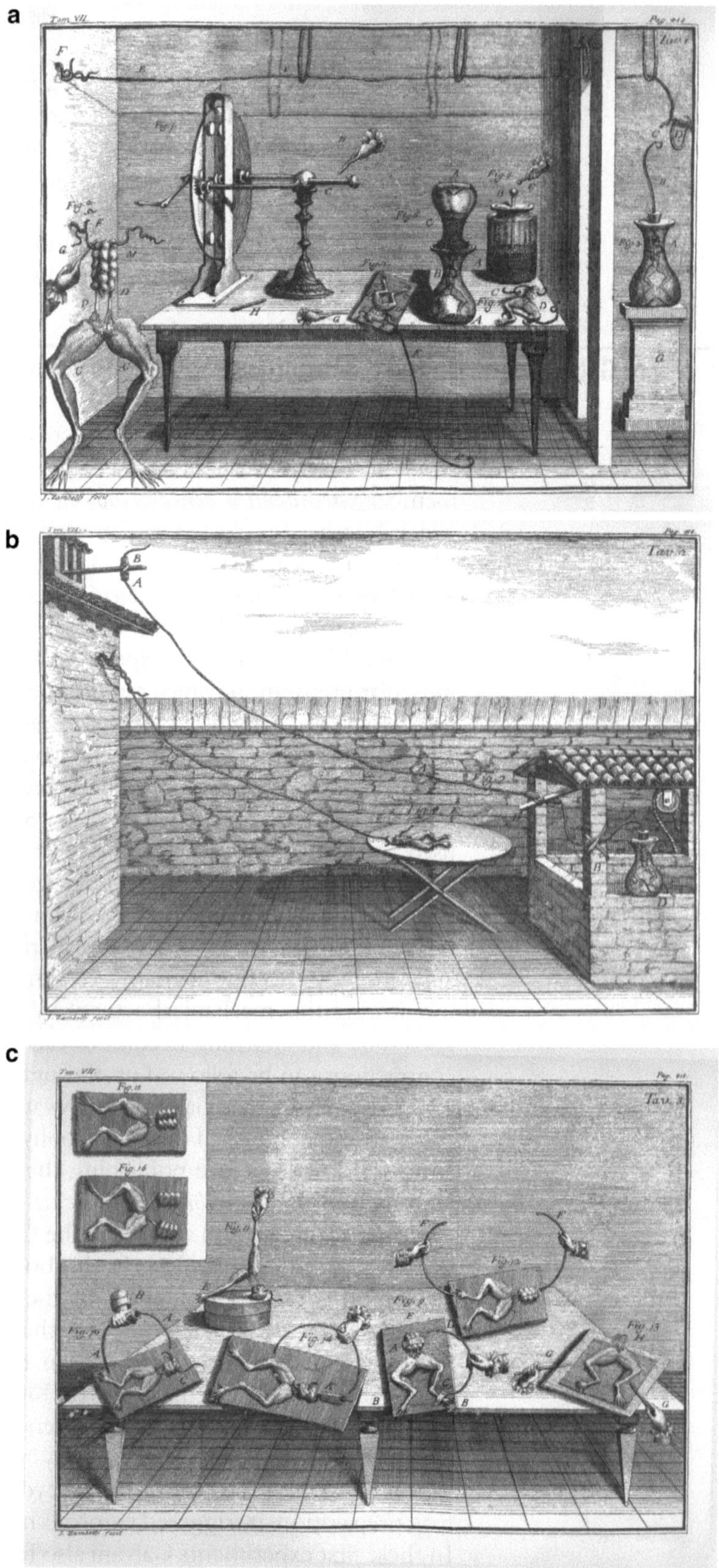

Fig. 2 Experiments of Luigi Galvani (scans of engravings were kindly provided by Prof. Marco Piccolino, University of Ferrara, Italy). These are plates 1–3 from the

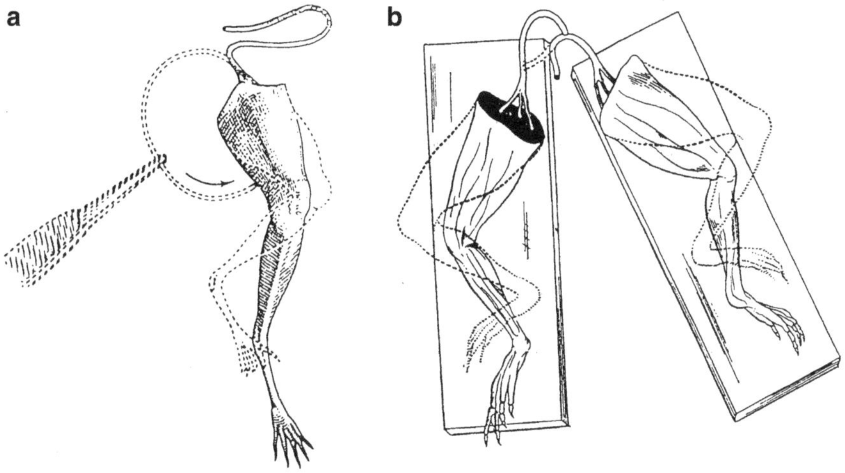

Fig. 3 Galvani experiments of the contraction without metals [13, 14]. (**a**) The 1794 experiment: when the surface of section of the nerve touches the muscle the leg contracts; (**b**) the 1797 experiment: when the surface of section of the right sciatic nerve touches the intact surface of the left sciatic nerve, both legs contract

stimulation leads to disappearance of contractions, which can be restored after a period of rest. Galvani also described a spectacular experiment when the muscle contraction was initiated by atmospheric electricity. For this purpose the nerves of the frog preparation were connected to long wires directed toward the sky and "... in correspondence of four thunders, contractions not small occurred in all muscles of the limbs, and, as a consequence, not small hops and movements of the limbs. These occurred just at the moment of the lightning; they occurred well before the thunders when they were produced as a consequence of these ones" [12] (Fig. 2). Several years later Galvani made another fundamentally important experiment when he used two frog legs with long sciatic nerves attached (Fig. 3). When the nerve of the first preparation was in contact with the nerve or muscle of the second, contraction occurred in both preparations. This was, in essence, the first demonstration of a propagating bioelectricity [13].

Galvani also produced coherent theory of animal electricity. He realized that the nerve conducts through the internal core that is insulated from the external milieu by a nonconductive sheath, and thus predicting an existence of non-conductive cell membrane. The electricity, according to Galvani results from accumulation of positive and negative charges on external and internal surfaces of the muscle or the nerve fiber, which he compared to the internal and external plates of the Leyden jar. The electrical current flow,

Fig. 2 (continued) *De Viribus Electricitatis in Motu Musculari* [12] which show frog preparations and the electric machine, i.e., the Leyden jar (**a**), experiments with atmospheric electricity (**b**) and experiments with metallic arcs (**c**)

which occurs during excitation, required a specific pathway, and Galvani contemplated the existence of water-filled channels, which penetrate the surface of the fibers, and allow for electrical excitability. Once more, comparing the biological tissue to a Leyden jar, he wrote: "...let one plaster then this conductor with some insulating substance, as wax... let one make small holes in some part of the plastering that concerns the conductor. Then let one moist with water or with some other conductive fluid all the plastering, having care that the fluid penetrate in the above mentioned holes, and come in contact with the conductor itself. Certainly, in this case, there is communication through this fluid between the internal and the external surface of the jar" [14] quoted from [15]. This was a portend of conductive water-filled channels that penetrate non-conductive outer membrane and thus allow for flow of charges underlying biological electricity.

Galvani's ideas were not universally acknowledged by the scientific society and he himself bitterly complained that "*I am attacked by two very opposite sects—the scientists and the know-nothings. Both laugh at me—calling me 'the frogs' dancing-master.' Yet I know that I have discovered one of the greatest forces in nature*" [13]. Nonetheless, the Galvanism became a cultural phenomenon and Galvani's experiments were repeated in saloons and in cabinets of scholars. After Galvani's death (which was precipitated by French occupation of Bologna in 1797; Galvani refused to swear allegiance to a new Cispaline republic and was stripped from all his positions and titles), Giovanni Aldini become the propagator of animal electricity, publishing books [16, 17] and making public demonstration of most illustrious nature when he galvanized (electrically stimulated) body parts of freshly executed criminals. One of the most famous of Aldini's demonstrations took place at the Royal College of Surgeons in London. On the day of January 17, 1803 George Foster, who drowned his wife and child in Paddington canal, was hanged and (in full accord to the sentence) his body was subjected to postmortem experimentation. Aldini took a pair of conducting rods connected to a powerful battery, and touched body parts with them. When the rods were applied to Forster's mouth and ear, "the jaw began to quiver, the adjoining muscles were horribly contorted, and the left eye actually opened." When one rod was moved to touch the rectum, the whole body convulsed: indeed, the movements were "so much increased as almost to give an appearance of re-animation" [18].

3 Electrophysiology Develops: From First Instrumental Recordings to Membrane Theory of Bioelectrical Potentials

Instrumental era of electrophysiology began with experiments of Leopoldo Nobili, an artillery officer by education, in Modena, whose service in Russian campaign earned him a *legion d'honneur* from Napoleon. Using an electromagnetic ("astatic") galvanometer, which he developed in 1825, Nobili made the very first recording of electrical activity from the neuromuscular preparation [19]. He measured a current associated with muscle contraction which he called *corrente di rana* (frog's current) or *corrente propria* ("intrinsic current"). Incidentally Nobili's interpretation was completely at odds with bioelectricity as he was convinced that he was measuring a thermoelectrical current resulting from unequal cooling of the two ends of the preparation. Another decade passed before Carlo Matteucci (Fig. 1) (who held a chair in Physics at University of Pisa) clearly demonstrated the biological nature of electric signals. He developed what he called a "rheoscopic frog," the cut nerve and an attached frog thigh muscle. The decisive experimental design was built around a pile of frog thighs arranged in a series—the more the thighs connected, the larger the galvanometer needle deflection [20]. Further advances were made by Emil Heinrich du Bois-Reymond (Fig. 1) at Humbold University of Berlin, Prussia, who directly measured what we now know as the action potential from both the muscle and nerve and he was also the first to realize that excitation decreases (or even completely eliminates) the potential difference between the intact surface and the cut portion of the tissue—and therefore he called the excitatory electrical response the "*negative Schwankung*" (negative fluctuation) [21].

In 1840s Johannes Peter Müller (who was a chair of anatomy and physiology at Humboldt University in Berlin) developed a concept of electrical signal propagation though the nerve. He believed, however, that the speed of this propagation must be immense and hence could not be instrumentally measured. This obstacle however, was proven wrong by his pupil Herman Ludwig Ferdinand von Helmholtz, who measured the speed of nerve impulse propagation, by measuring the delay (which he called "*le temps perdu*"—the lost time) between the application of an electrical stimulus and the muscle contraction [22, 23]. To do so, Helmholtz used a technique developed earlier by Sorbonne professor Claude Servais Mathias Marie Roland Pouillet, who found that galvanometer excursions induced by brief pulses of current were proportional to the pulse duration (this principle was used in military practice for determining the speed of cannon balls; see [24] for further details). The next step was made by Julius Bernstein (Fig. 1) who invented the "differential rheotome" which allowed for extremely fast recordings (the sampling rate of this rheotome was in the range of several tens of microseconds—see [25] for a detailed account). Bernstein was able to record the resting and action potential; he estimated the resting potential of ~−60 mV, and he found that the action potential was not simple negation of the resting electronegativity but the potential deflection actually crossed the "zero potential" line causing the "sign reversal," which clearly reflected the action potential overshoot (Fig. 4; [26, 27]).

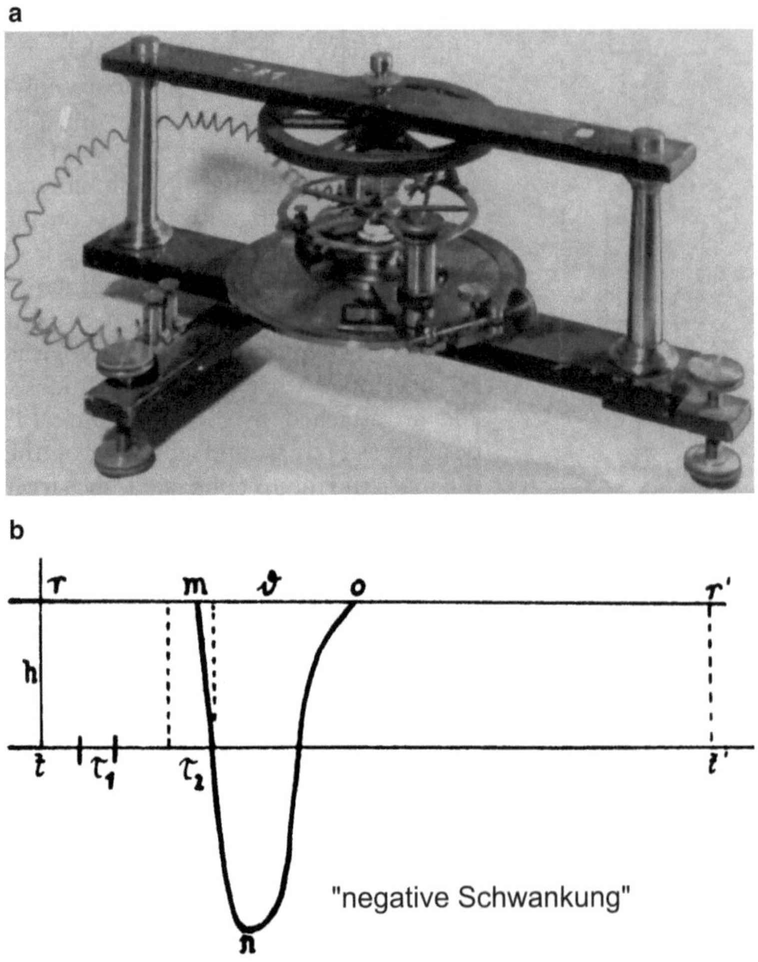

Fig. 4 The first recording of the action potential from a nerve were made by Julius Bernstein [26, 27]; Reprinted from [47] with permission

These experimental advances allowed Bernstein to develop a membrane theory of excitation; he presumed that it is selective K^+ permeability of the membrane, which is responsible for the generation of the resting membrane potential [28] (incidentally, Bernstein adapted electrolytic theory of Walther Herman Nernst to biological systems being influenced by his student Vassily Tschagovetz). Bernstein ideas were further developed by Charles Ernst Overton, who demonstrated that Na^+ ions are required for producing the potential overshoot, and suggested that the excitation process results from the exchange of Na^+ and K^+ [29]. Overton also was the first to propose a "*lipoidal membrane*" model of the plasmalemma, based on the fact that lipid-soluble dyes enter cells substantially easier than the water-soluble ones [30]. This theory was subsequently developed by Evert Gorter and François Grendel who proposed the lipid bilayer model [31]. This model finally was perfected by James Frederic Danielli and Hugh Dawson [32] who regarded plasmalemma as the bilayer lipid membrane associated with numerous proteins and which is penetrated by narrow water-filled pores, which allow for the passage of lipid-insoluble molecules, including ions.

4 Voltage-Clamp and Ionic Theory of Membrane Excitability

Formalization of the ionic theory of membrane potential and membrane excitability resulted from a new methodology that allowed for direct measurements of ion-carried transmembrane currents. This was prompted by the discovery of the giant squid axon by John Z. Young in 1936 [33], which became a standard preparation for nerve physiology. Using the squid axon Kenneth S. Cole and Howard J. Curtis, measured, with extracellular electrodes, changes in the impedance and revealed the rapid decrease in membrane resistance during the development of the action potential which was indicative of transmembrane current generation. Subsequently, Cole and Curtis [34] and independently Alan L. Hodgkin and Andrew F. Huxley (Fig. 1) [35] fabricated mini-electrodes which can be inserted into the axon (which diameter was ~1 mm) and preformed first direct recordings of resting (~−50 mV) and action potentials. Several years later the voltage-clamp technique that allowed for direct recordings of membrane currents at rest and during excitation was developed by Cole [36] and by George Marmont [37]. This technique was then adopted by Hodgkin and Huxley, who, in collaborating with Sir Bernhard Katz (Fig. 1), described ion currents underlying axonal activity and demonstrated that membrane excitability is determined by passive ion fluxes across biological membranes according to their electrochemical gradients [38–44] (Fig. 5). The ion theory also postulated that the specific membrane permeabilities to ions define the excitable characteristics of the given cell type. The Hodgkin–Huxley theory also assumed the existence of specific pathways for transmembrane ion fluxes that could be either aqueous pores (i.e., ion channels) or carriers. Incidentally, the theory of ionic/electric basis for excitability was not immediately and universally accepted and an alternative views associating nerve impulse propagation with chemical agents, such as for example acetylcholine, persisted well into 1970s [45, 46].

5 The Patch Clamp

The way to the patch-clamp technique was quite long and was paved by the efforts of many researchers and laboratories; numerous techniques were developed and subsequently disappeared,

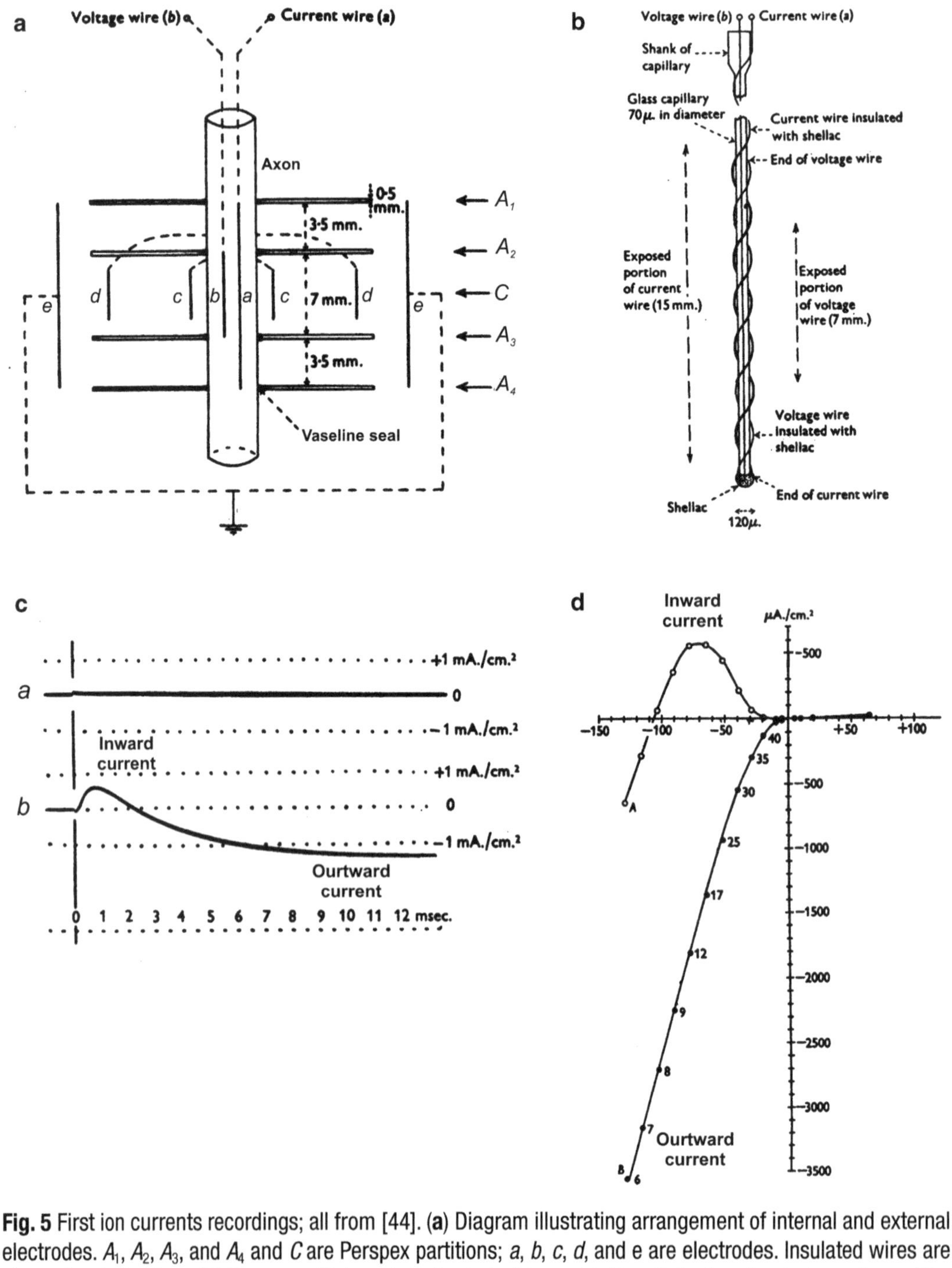

Fig. 5 First ion currents recordings; all from [44]. (**a**) Diagram illustrating arrangement of internal and external electrodes. A_1, A_2, A_3, and A_4 and *C* are Perspex partitions; *a*, *b*, *c*, *d*, and e are electrodes. Insulated wires are shown by *dotted lines*. (**b**) Diagram of internal electrode (not to scale). The pitch of each spiral was 5 mm. The exposed portions of the wires are shown by *heavy lines*. (**c**) Records of membrane current under a voltage clamp. At zero time the membrane potential was increased by 65 mV (record **a**) or decreased by 65 mV (record **b**); this level was then maintained constant throughout the record. Inward current is shown as an upward deflexion. axon diameter, 585 μm. Temperature, 3.8 °C. (**d**) I-V plots of inward and outward currents obtained in c

although all of them left a mark and were indispensible for the perfection of the method [47].

The very first direct recordings of unitary currents associated with openings of ion channels in artificial lipid bilayers were made by Paul Müller and Donald O. Rudin in 1963 [48, 49]. In these experiments lipid membranes were exposed to antibiotics (e.g., gramicidin A) or certain proteins, which inserted themselves into the membranes and behaved like a voltage-dependent aqueous ion-selective pores; their openings were manifested by step-like, discrete events of transmembrane currents [50, 51].

The beginning of single cell electrophysiology is directly associated with the introduction of glass microelectrodes with submicron tips suitable for low-traumatizing penetration of individual cells. The microelectrodes for intracellular recordings of the membrane potential of the frog sartorius muscle were developed by Gilbert N. Ling and Ralf W. Gerard in 1949 [52]; soon microelectrodes were adapted for almost all conceivable cell types from unicellular organisms to neurones in live brain [53]. Many cells tolerated long-lasting impalement with several microelectrodes that allowed for application of voltage clamp sometimes even in combination with ionophoresis or intracellular injections.

Extracellular electrodes also produced from glass had a longer history being for the first time used in 1919 when Frederick H. Pratt and John P. Eisenberger used a fine-pointed capillary pore electrode with outer diameter ~4–8 μm for focal stimulation of single skeletal muscle fibers [54]. In 1960s Alfred Strickholm developed liquid-filled micropipettes with the tip diameter of several micrometers that he used for extracellular recordings from muscle cells. These electrodes were pressed against the cell membrane, thus providing a degree of electrical isolation of the membrane patch underneath. Using these pipettes it was possible to measure the impedance of frog muscles and even monitor currents, flowing through the small membrane part of membrane isolated by the tip of the extracellular pipette [55, 56]. Similar technique of extracellular glass microelectrodes with micrometer diameter tips was developed by Karl Frank and Ladislav Tauc in 1963 for probing of heterogeneous distribution of sodium channels in molluscan neurones [57]. In 1969 Erwin Neher (Fig. 1) and Hans Dieter Lux used another version of an extracellular recording technique to monitor membrane currents from the somatic membrane of subesophageal ganglion neurones of *Helix pomatia* snails [58]. In these experiments, Neher and Lux pulled micropipettes from asymmetrical double-barrelled capillaries to obtain tips of about 100–150 μm in diameter; the pipette tip was subsequently fire-polished (Fig. 6). Here, for the first time a gentle suction (2–10 mmHg) was applied to the pipette interior, which helped the approach to the neuronal membrane in the ganglia (normally covered by glial cells) and improved the shunt resistance between the pipette wall and the cell membrane.

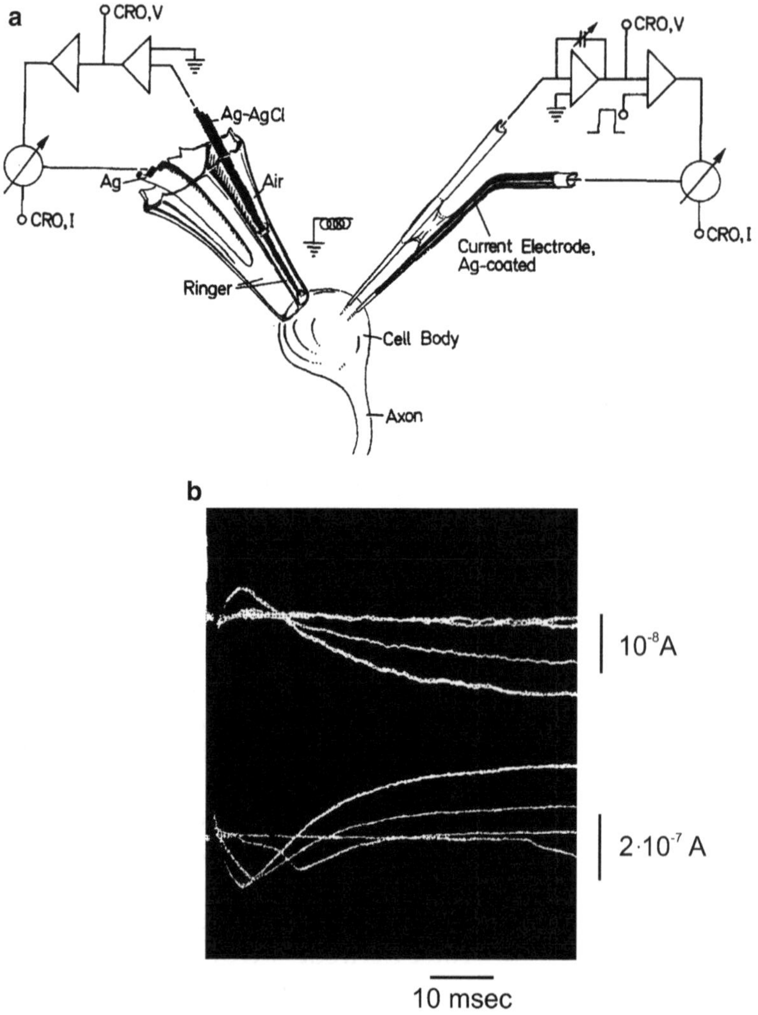

Fig. 6 Extracellular voltage-clamp recordings from snail neurones. (**a**) Schematic diagram of the electronic system, and geometry of the pipette. Current is measured with high input impedance operational amplifiers. The subesophageal ganglion of *Helix pomatia* snails was mounted in a Lucite chamber and the connective tissue sheath removed under a dissecting microscope. The dorsal surface was cleaned as carefully as possible and clusters of cells exposed. For voltage clamp the neuron was impaled with an electrode made from two microelectrodes glued together with a tip separation of 20–30 μm. One of them, the current electrode, was electrolytically coated with silver upto within 50 μm of the tip. (**b**) Simultaneous measurement of the overall intracellular clamp current (*lower traces*) and the current through the pipette (*upper traces*, inversed polarity), superimposed for 30, 40, 50, and 60 mV depolarizing pulses. Note the late inward transients in the *lowest traces*, which are not present in the corresponding *upper traces*. Active soma current is present only for depolarizing pulses of 50 mV and more. Reprinted from [47] with permission

Another experimental development was associated with various techniques of intracellular perfusion (Fig. 7), which was fundamentally important for precise separation of ion currents and controlling ion composition of the intracellular milieu, as the composition on other side of the excitable membrane, i.e., the extracellular space, could be easily controlled by bath solutions. The very first experiments replacing the cytosol of giant squid axons with an artificial solution were performed in 1961 by Peter et al. [59] (Fig. 7a). At the beginning of 1970s the intracellular perfusion technique was developed for studying individual cells; initially this method was applied to isolated mollusc neurones [60–62], and then to different types of single mammalian cells, from neurones to cardiomyocytes [63]. The very first design of intracellular perfusion was built around plastic film that separated two chambers, filled with extracellular and intracellular solutions [64] (Fig. 7b). A tiny pore of several μm in diameter was bored in the film, and the cell was placed on top of the pore; a small negative pressure applied to the "intracellular" chamber helped the cell to invade the pore. After the cell firmly occluded the pore, the membrane facing the intracellular compartment was disrupted, and thus, electrical and physical access to the cell interior was gained. This initial setup was soon modified, and the planar film was replaced by either plastic or glass pipettes [65–67], which allowed for targeting cells using micromanipulators. Further modifications of the intracellular perfusion technique included double perfusion (when the cell was fixed between two pipettes [62, 66]), which provided excellent spatial voltage clamp and fast and effective replacement of the cytosol (Fig. 7c). The plastic pipettes were also employed for extracellular recordings with the aim of measuring single channel currents [64]. The main problem of both plastic and glass versions of intracellular perfusion techniques was a relatively low seal resistance between the membrane surface and the wall of the recording pipette, which prevented low-noise recordings due to large leak currents.

Conceptually, the main problem in recording single channel currents is associated with a background electrical noise (that for microelectrodes for example could be as high as 100 pA). This noise is proportional to the clamped membrane area and, therefore, when a small area (1–10 μm^2) is isolated the extraneous noise levels can be made so low that the pico-ampere currents flowing through single ion channels can be measured directly. Using the lessons from extracellular recording techniques described above, Ervin Neher and Bert Sakmann (Fig. 1) constructed a similar extracellular electrode with relatively small tip (~1–2 μm in diameter) and by pressing this tip to the surface of an isolated skeletal muscle fiber electrically isolated a small patch of

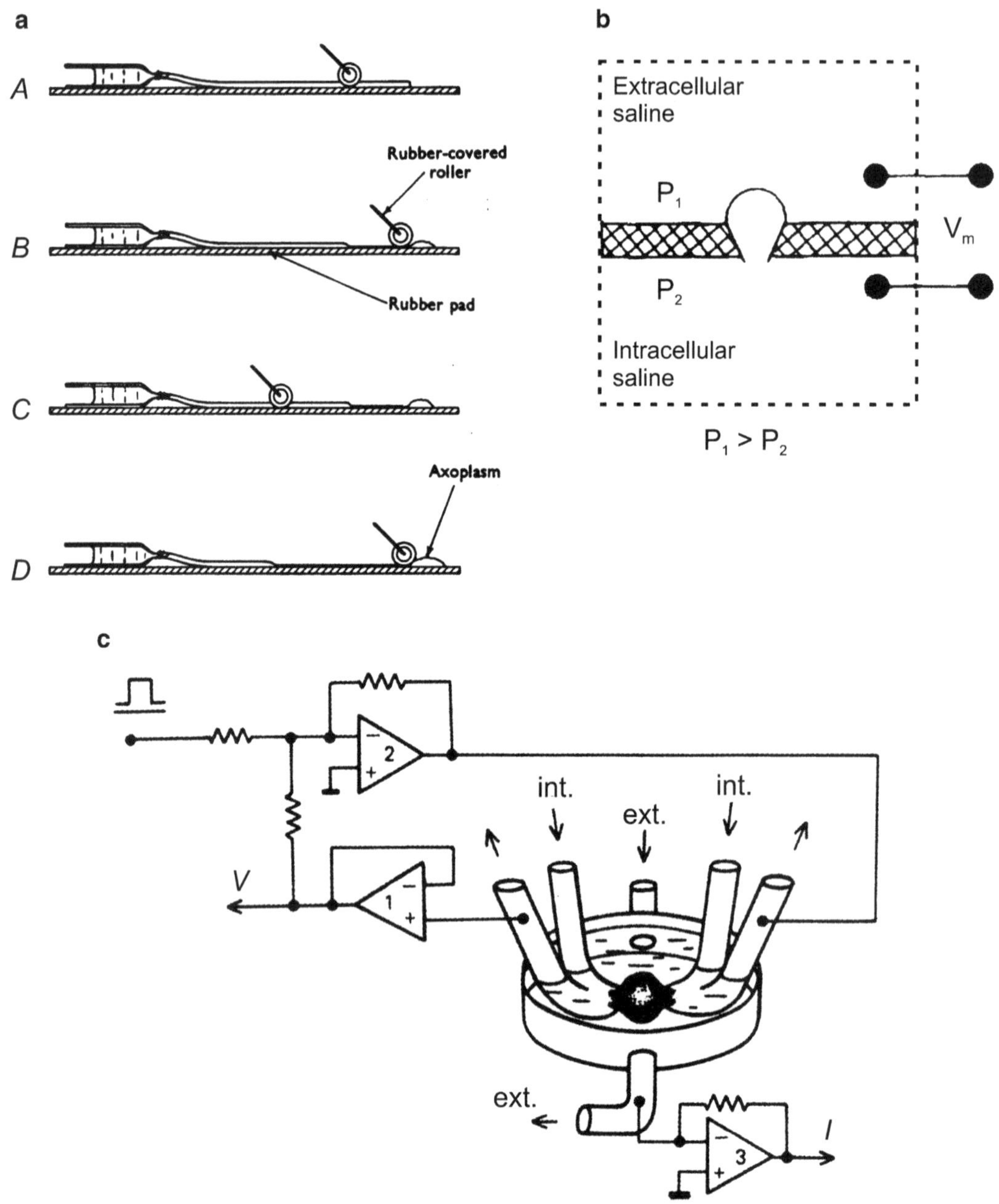

Fig. 7 Intracellular perfusion techniques. (**a**) Extrusion of the axoplasm. A cannula filled with perfusion fluid was tied into the distal end of a giant axon of 6–8 cm in length. The axon was placed on a rubber pad and axoplasm was extruded by passing a rubber-covered roller over it in a series of progressive sweeps. The first sweep started at about 1–5 cm from the cut end, the second at 3 cm and so on. At the end of this operation all the axoplasm had been removed except for a column 5–10 mm in length near the cannula. The axon was then suspended vertically in a large beaker of sea water. The cannula was connected to a mechanically driven "Agla" syringe and perfusion fluid was forced into the axon at about 6 μl/min rate. Occasionally, when the axon narrowed in the middle, the plug of axoplasm stuck and in these instances the experiment usually had to be abandoned. Modified from [59]. (**b**) Intracellular dialysis of a snail neurone. Schematic drawing of a cell fixed in the conical pore in the polyethylene film. A part of the membrane facing the compartment with artificial intracellular saline is ruptured with a pressure pulse. (**c**) Intracellular perfusion of a single cell situated between two perfusion pores. *1*, unity gain input amplifier; *2*, voltage-clamp amplifier; *3*, current–voltage transducer. The pathways of the perfusion are indicated by *arrows*. Reprinted from [47] with permission

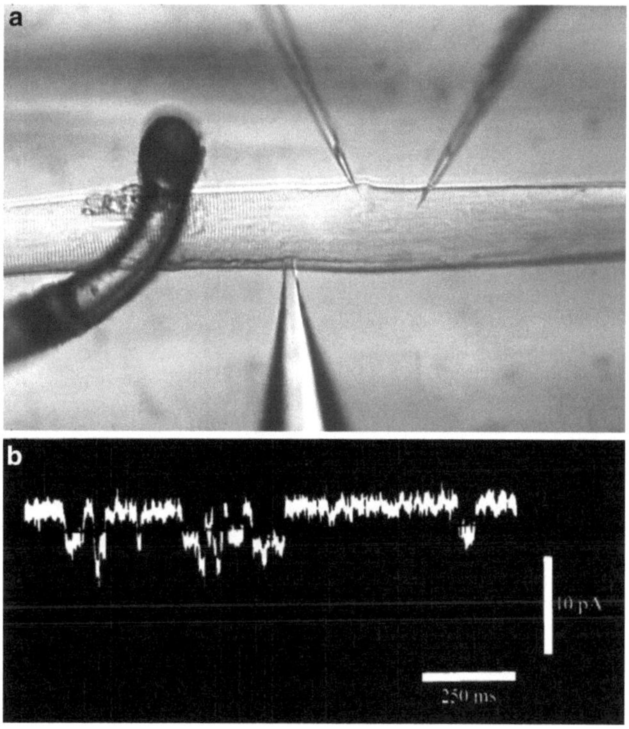

Fig. 8 First recordings of acetylcholine receptor single channel currents from denervated leopard frog (*Rana pipiens*) cutaneous pectoris muscle. (**a**) The microphotograph of the preparation. (**b**) Current recordings. The pipette contained 0.2 mM suberyldicholine, an analogue of acetylcholine which induces very long-lasting channel openings. Membrane potential, –120 mV; temperature, 8 °C. Reprinted from [47] with permission

membrane (<10 μm²) which lied beneath the opening of the electrode. With this experimental approach they were able to monitor single channel ion currents generated by the opening of nicotinic acetylcholine receptors [68] (Fig. 8). Next seminal step was made in 1980–1981 when Ervin Neher, Bert Sakmann, Frederick J. Sigworth, Alan Marty, and Owen P. Hamill [69, 70] developed the high resistance (giga-ohm; GΩ) seal between highly cleaned and very smooth micropipette tips and smooth surface cell membranes, which received the name of the "gigaseal." This involved fire-polishing the pipette tips and a gentle negative suction—that is experimental maneuvers developed in previous years. Thus, the patch-clamp technique, that revolutionized modern cellular electrophysiology, was born and rapidly became the method of choice not only to study single channels but also to monitor electrical activity of virtually any cell type both in isolation and in the in situ/in vivo settings (Fig. 9).

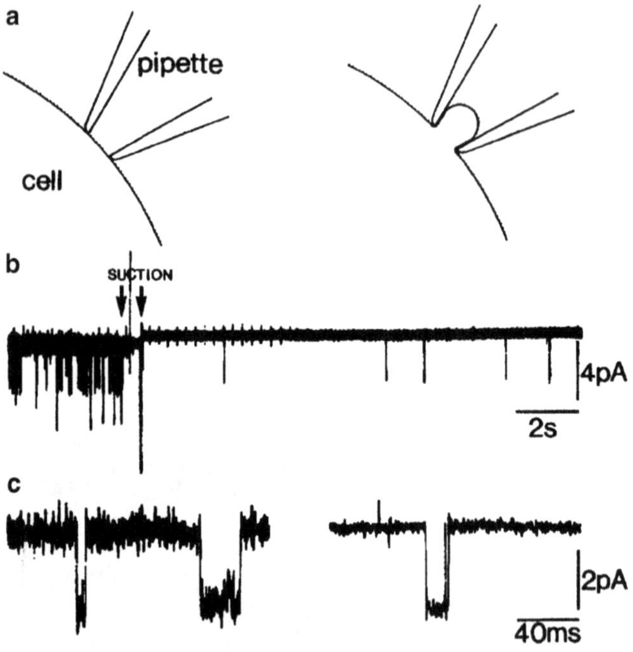

Fig. 9 Gigaseal formation between pipette tip and sarcolemma of frog muscle. Reproduction of the Fig 6 from 1981 *Pflügers Archiv* paper [69]. (**a**) Schematic diagrams showing a pipette pressed against the cell membrane when the pipette–membrane seal resistance is of the order of 50–100 MΩ (*left*), and after formation of a gigaseal when a small patch of membrane is drawn into the pipette tip (*right*). (**b**) The *upper trace* is a continuous current record before, during, and after application of suction. In this experiment a pipette–membrane seal resistance of ~50 MΩ was achieved by pressing the pipette against the membrane. Single suberyldicholine-induced channel currents are apparent. During the time indicated by the *two arrows* slight suction was applied to the pipette interior resulting in the formation of a gigaseal of 60 GΩ resistance. Note reduction in the background noise level. The decrease in the channel opening frequency presumably resulted from depletion of the agonist in the pipette tip during suction. It increased again during the minute following gigaseal formation. The two large current deflections represent artifacts. The *lower traces* show single channel currents at higher resolution before (*left*) and after (*right*) formation of a gigaseal. The single channel current pulse on the right is preceded by capacitive artifacts from a calibration pulse. All records were made at the cell's resting potential of −92 mV. They were low pass filtered at 1 kHz (*upper trace*) or 3 kHz (*lower traces*)

Acknowledgements

The authors' research was supported by Alzheimer's Research Trust (UK) to A.V. and by National Institutes of Health (The Eunice Kennedy Shriver National Institute of Child Health and Human Development award HD078678) to V.P.

References

1. Newton I (1713) Principia Mathematica, 2nd edn. University of California Press, Berkeley, CA, English translation by Andrew Motte, Sir Isaac Newton's Mathematical Principles of Natural Philosophy and his System of the World (1729, reprinted 1934 by the University of California Press)
2. Clark-Kennedy AE (1929) Stephen Hales, D.D., F.R.S. An eighteenth century biography. Cambridge University Press, London
3. Hales S (1733) Statical essays: containing haemastaticks; or, an account of some hydraulick and hydrostatical experiments made on the blood and blood-vessels of animals. W. Innys, R. Manby & T. Woodward, London
4. Haller A (1756–1760) Mémoires sur la nature sensible et irritable des parties du corps animal. Lausanne: M.-M. Bousquet
5. Piccolino M (2008) Visual images in Luigi Galvani's path to animal electricity. J Hist Neurosci 17:335–348
6. Piccolino M, Bresadola M (2002) Drawing a spark from darkness: John Walsh and electric fish. Trends Neurosci 25:51–57
7. Bresadola M (1998) Medicine and science in the life of Luigi Galvani (1737-1798). Brain Res Bull 46:367–380
8. Piccolino M (1998) Animal electricity and the birth of electrophysiology: the legacy of Luigi Galvani. Brain Res Bull 46:381–407
9. Swammerdam J (1758) The book of nature (Biblia naturae). Seyfert, London
10. Stillings D (1975) Did Jan Swammerdam beat Galvani by 134 years? Med Instrum 9:226
11. Cobb M (2002) Timeline: exorcizing the animal spirits: Jan Swammerdam on nerve function. Nat Rev Neurosci 3:395–400
12. Galvani L (1791) De viribus electricitatis in motu musculari commentarius. Bon Sci Art Inst Acad Comm 7:363–418
13. Galvani L (1841) Opere edite ed inedite del Professore Luigi Galvani raccolte e pubblicate dall'Accademia delle Science dell'Istituto di Bologna. Dall'Olmo, Bologna
14. Galvani L (1794) Dell'uso e dell'attività dell'arco conduttore. S. Tommaso d'Aquino, Bologna
15. Piccolino M (1997) Luigi Galvani and animal electricity: two centuries after the foundation of electrophysiology. Trends Neurosci 20: 443–448
16. Aldini G. (1803) An account of the late improvements in galvanism, with a series of curious and interesting experiments performed before the commissioners of the French National Institute, and repeated lately in the anatomical theaters of London, by John Aldini.
17. Aldini G (1804) Essai théorique et expérimental sur le galvanisme (2 vol). Fournier et Fils, Paris
18. Aldini G (1803) An account of the galvanic experiments performed by John Aldini, Professor of Experimental Physiology in the University of Bologna, &c. on the body of a malefactor executed at Newgate, January 17, 1803. Med Phys J 9:382–385
19. Nobili L (1828) Comparaison entre les deux galvanometres les plus sensibles, la grenouille et le moltiplicateur a deux aiguilles, suivie de quelques resultats noveaux. Ann Chim Phys 38:225–245
20. Matteucci C (1844) Traité des phenomenes electro-physiologiques des animaux suivi d'etudes anatomiques sur le systheme nerveux et sur l'organe electrique de la torpille par Paul Savi. Masson et C.ie, Paris
21. du Bois-Reymond E (1884) Untersuchungen über thierische elektricität, 1848-1884 (2 bande). Reimer, Berlin
22. Helmholtz H (1850) Note sur la vitesse de propagation de l'agent nerveux dans les nerfs rachidiens. C R Acad Sci (Paris) 30:204–206
23. Helmholtz H (1852) Messungen über fortpflanzungsgeschwindigkeit der reizung in den nerven–zweite reihe. Arch Anat Physiol Wiss Med 199–216
24. Piccolino M (2003) A "Lost time" betwen science and literature: the "Temps Perdu" from Hermann von Helmholtz to Marcel Proust. Audiologic Med 1:1–10
25. Nilius B (2003) Pflugers Archiv and the advent of modern electrophysiology. From the first action potential to patch clamp. Pflugers Arch 447:267–271
26. Bernstein J (1868) Ueber den zeitlichen Verlauf der negativen Schwankung des Nervenstroms. Pflugers Arch 1:173–207
27. Bernstein J (1871) Untersuchungen uber den Erregungsvorgang im Nerven- und Muskelsystem. Winter's Unisersitatsbuchhandlung, Heidelberg
28. Bernstein J (1912) Elektrobiologie - Die Lehre von den electrischen Vorgangen im Organismus auf moderner Grundlage dargestellt. Vieweg und Sohn, Braunschweig
29. Overton CE (1902) Betrage zur allgemaine Muskel- und Nervenphysiologie. II Uber die Unentbehrlichkeit von Natrium- (oder Litium-) Ionen fur den Contractionsact des Muskels. Pflugers Arch 92:346–380

30. Overton CE (1899) Über die Allgemeinen Osmotischen Eigenschaften der Zelle, ihre vermutlichen Ursachen und ihre Bedeutung für die Physiologie. Vierteljahrsschrift der Naturforschenden Gesellschaft in Zürich 44: 88–135
31. Gorter E, Grendel F (1925) On bimolecular layers of lipids on the chromocytes of the blood. J Exp Med 41:439–443
32. Danielli JF, Davson H (1935) A Contribution to the theory of permeability of thin films. J Cell Comp Physiol 5:495–508
33. Young JZ (1936) Structure of nerve fibres and synapses in some invertebrates. Cold Spring Harbor Symp Quant Biol 4:1–6
34. Curtis HJ, Cole KS (1940) Membrane action potentials from the squid giant axon. J Cell Comp Physiol 15:147–157
35. Hodgkin AL, Huxley AF (1939) Action potentials recorded from inside a nerve fibre. Nature 144:710–711
36. Cole KS (1949) Dynamic electrical characteristics of the squid axon membrane. Arch Sci Physiol 3:253–258
37. Marmont G (1949) Studies on the axon membrane. I. A new method. J Cell Comp Physiol 34:351–382
38. Hodgkin AL, Huxley AF (1952) The components of membrane conductance in the giant axon of Loligo. J Physiol 116:473–496
39. Hodgkin AL, Huxley AF (1952) Currents carried by sodium and potassium ions through the membrane of the giant axon of Loligo. J Physiol 116:449–472
40. Hodgkin AL, Huxley AF (1952) The dual effect of membrane potential on sodium conductance in the giant axon of Loligo. J Physiol 116:497–506
41. Hodgkin AL, Huxley AF (1952) Movement of sodium and potassium ions during nervous activity. Cold Spring Harb Symp Quant Biol 17:43–52
42. Hodgkin AL, Huxley AF (1952) Propagation of electrical signals along giant nerve fibers. Proc R Soc Lond B Biol Sci 140:177–183
43. Hodgkin AL, Huxley AF (1952) A quantitative description of membrane current and its application to conduction and excitation in nerve. J Physiol 117:500–544
44. Hodgkin AL, Huxley AF, Katz B (1952) Measurement of current-voltage relations in the membrane of the giant axon of Loligo. J Physiol 116:424–448
45. Nachmansohn D (1946) Chemical mechanism of nerve activity. Ann N Y Acad Sci 47: 395–428
46. Nachmansohn D, Neumann E (1975) Chemical and molecular basis of nerve activity. Academic, New York, NY
47. Verkhratsky A, Krishtal OA, Petersen OH (2006) From Galvani to patch clamp: the development of electrophysiology. Pflugers Arch 453:233–247
48. Mueller P, Rudin DO, Tien HT et al (1962) Reconstitution of cell membrane structure in vitro and its transformation into an excitable system. Nature 194:979–980
49. Mueller P, Rudin DO (1963) Induced excitability in reconstituted cell membrane structure. J Theor Biol 4:268–280
50. Bean RC, Shepherd WC, Chan H et al (1969) Discrete conductance fluctuations in lipid bilayer protein membranes. J Gen Physiol 53:741–757
51. Hladky SB, Haydon DA (1970) Discreteness of conductance change in bimolecular lipid membranes in the presence of certain antibiotics. Nature 225:451–453
52. Ling GN, Gerard RW (1949) The normal membrane potential of frog sartorius fibers. J Cell Comp Physiol 34:383–396
53. Purves RD (1981) Microelectrode methods for intracellular recording and ionophoresis. Academic, London
54. Pratt FH, Eisenberger JP (1919) The quantal phenomena in muscle: methods, with further evidence of the all-or-none principle for the skeletal fiber. Am J Physiol 49:1–54
55. Strickholm A (1961) Impedance of a small electrically isolated area of the muscle cell surface. J Gen Physiol 44:1073–1088
56. Strickholm A (1962) Excitation currents and impedence of a small electrically isolated area of the muscle cell surface. J Cell Comp Physiol 60:149–167
57. Frank K, Tauc L (1963) Voltage clamp studies on molluscan neuron membrane properties. In: Hoffman J (ed) The cellular function of membrane transport. Englewood Cliffs, NJ, Prentice Hall
58. Neher E, Lux HD (1969) Voltage clamp on Helix pomatia neuronal membrane; current measurement over a limited area of the soma surface. Pflugers Arch 311:272–277
59. Baker PF, Hodgkin AL, Shaw TI (1962) Replacement of the axoplasm of giant nerve fibres with artificial solutions. J Physiol 164: 330–354
60. Kostyuk PG, Krishtal OA, Pidoplichko VI (1975) Effect of internal fluoride and phosphate on membrane currents during intracellular dialysis of nerve cells. Nature 257:691–693

61. Kryshtal OA, Pidoplichko VI (1975) Intracellular perfusion of the giant neurons of snails (in Russian). Neirofiziologiia 7: 327–329
62. Kostyuk PG, Krishtal OA, Pidoplichko VI (1977) Asymmetrical displacement currents in nerve cell membrane and effect of internal fluoride. Nature 267:70–72
63. Pidoplichko VI, Verkhratsky AN (1987) Sodium currents in the membrane of isolated cardiomyocytes. Biomed Biochim Acta 46: S668–S672
64. Kryshtal OA, Pidoplichko VI (1977) Analysis of current fluctuations shunted from small portions of the membrane of a nerve cell soma (in Russian). Neirofiziologiia 9:644–646
65. Lee KS, Akaike N, Brown AM (1978) Properties of internally perfused, voltage-clamped, isolated nerve cell bodies. J Gen Physiol 71:489–507
66. Lee KS, Akaike N, Brown AM (1980) The suction pipette method for internal perfusion and voltage clamp of small excitable cells. J Neurosci Methods 2:51–78
67. Kostyuk PG, Krishtal OA, Pidoplichko VI (1981) Intracellular perfusion. J Neurosci Methods 4:201–210
68. Neher E, Sakmann B (1976) Single-channel currents recorded from membrane of denervated frog muscle fibres. Nature 260: 799–802
69. Hamill OP, Marty A, Neher E et al (1981) Improved patch-clamp techniques for high-resolution current recording from cells and cell-free membrane patches. Pflugers Arch 391:85–100
70. Sakmann B, Neher E (1984) Patch clamp techniques for studying ionic channels in excitable membranes. Annu Rev Physiol 46: 455–472

Part I

Pharmacology

Chapter 2

Whole-Cell Patch-Clamp Analysis of Recombinant NMDA Receptor Pharmacology Using Brief Glutamate Applications

Nathan G. Glasgow and Jon W. Johnson

Abstract

N-methyl-D-aspartate receptors (NMDARs) are ionotropic glutamate receptors that are essential for synaptic plasticity, learning and memory. Dysfunction of NMDARs has been implicated in many nervous system disorders; therefore, pharmacological modulation of NMDAR activity has great therapeutic potential. However, given the broad physiological importance of NMDARs, modulating their activity often has detrimental side effects precluding pharmaceutical use of many NMDAR modulators. One approach to possibly improve the therapeutic potential of NMDAR modulators is to identify compounds that modulate subsets of NMDARs. An obvious target for modulating NMDAR subsets is the many NMDAR subtypes produced through different combinations of NMDAR subunits. With seven identified genes that encode NMDAR subunits, there are many neuronal NMDAR subtypes with distinct properties and potentially differential pharmacological sensitivities. Study of NMDAR subtype-specific pharmacology is complicated in neurons, however, because most neurons express at least three NMDAR subtypes. Thus, use of an approach that permits study in isolation of a single receptor subtype is preferred. Additionally, the effects of drugs on agonist-activated responses typically depend on duration of agonist exposure. To evaluate drug effects on synaptic transmission, an approach should be used that allows for activation of receptor responses as brief as those observed during synaptic transmission, both in the absence and presence of drug. To address these issues, we designed a fast perfusion system capable of (1) delivering brief (~5 ms) and consistent applications of glutamate to recombinant NMDARs of known subunit composition, and (2) easily and quickly (~5 s) changing between glutamate applications in the absence and presence of drug.

Key words NMDA, NMDA receptor subtype, Memantine, Ketamine, Open channel blocker, Brief agonist application

1 Introduction

There is great interest in pharmacologically modulating ligand-gated ion channels to augment nervous system function or alleviate aberrant activity potentially underlying nervous system disorders. The whole-cell patch-clamp technique is essential in understanding how drugs affect ligand-gated ion channel function, cell physiology, and the nervous system under normal and

Marzia Martina and Stefano Taverna (eds.), *Patch-Clamp Methods and Protocols*, Methods in Molecular Biology, vol. 1183,
DOI 10.1007/978-1-4939-1096-0_2,

pathological conditions. Due to the great diversity of subtypes within each ligand-gated ion channel family, pharmacological analysis of a particular ligand-gated ion channel using native cells is complicated. Furthermore, the mechanisms underlying drug actions on ligand-gated ion channels may depend upon the concentration and duration of agonist exposure to receptors. Therefore, expression of recombinant ligand-gated ion channels in mammalian cell lines in conjunction with a fast perfusion system designed to deliver brief agonist applications is very useful in understanding how drugs affect ligand-gated ion channel function. Here, we describe a method for whole-cell patch-clamp analysis of ligand-gated ion channel pharmacology that allows for precise control of (1) the receptor subunit composition, (2) the agonist concentration, and (3) the duration of agonist exposure to receptors. Our method also allows for brief application of agonist in the absence and presence of drug to the same cell. Here, we demonstrate use of the system to investigate inhibition of recombinant *N*-methyl-D-aspartate receptors (NMDARs) during brief glutamate applications.

NMDARs are ionotropic glutamate receptors that exhibit voltage-dependent Mg^{2+} block, are highly Ca^{2+} permeable, and deactivate slowly. These properties contribute to the importance of NMDARs to cell survival, synaptic plasticity, and many forms of learning and memory [1]. Aberrant activation of NMDARs is implicated in neurodegenerative diseases, ischemia, depression, and neuropathic pain [2–7]. Pharmacological inhibition of NMDARs is considered to have great therapeutic potential in treating these disorders [1], although broad inhibition of NMDARs often results in undesirable side effects [8, 9]. Thus, identification of NMDAR antagonists selective for NMDARs that may be involved in a pathological state while preserving the function of NMDARs underlying normal function may be vital for successful pharmacological therapy [9–11].

NMDARs are heterotetramers composed of two GluN1 subunits either with two GluN2 subunits or with one GluN2 and one GluN3 subunit [1]. There is a single gene that encodes eight splice variants of the GluN1 subunit, four genes that encode four GluN2 subunits (GluN2A, GluN2B, GluN2C, and GluN2D), and two genes that encode two GluN3 subunits (GluN3A and GluN3B). Different combinations of GluN1, GluN2, and GluN3 subunits give rise to NMDAR subtypes with distinct properties. Combinations that include two identical GluN2 subunits form diheteromeric NMDARs (e.g., GluN1/2A) and combinations that include either two different GluN2 subunits, or mixtures of GluN2 and GluN3 subunits, form triheteromeric NMDARs (e.g., GluN1/2A/2B) [1, 10]. In principal cells in the cortex, at least three NMDAR subtypes, including GluN1/2A, GluN1/2B, and GluN1/2A/2B, are expressed and can be found postsynaptically [10, 12]. Consequently, it is difficult to study synaptic NMDAR subtype-specific pharmacology in neurons. Given this difficulty, we emulate synaptic release of glutamate using brief glutamate applications to tsA201 cells expressing recombinant GluN1/2A or GluN1/2B receptors. This approach allows for pharmacological assessment of NMDARs with known, uniform subunit compositions. The methods described in this chapter provide a powerful approach to studying ligand-gated ion channel currents in response to brief agonist applications in the absence and presence of many types of drugs.

2 Materials

2.1 Cell Culture and Transfection

1. tsA201 cell culture medium: DMEM supplemented with 10 % fetal bovine serum (FBS) and 1 % GlutaMAX.
2. Serum-free tsA201 cell culture medium: DMEM supplemented with 1 % GlutaMAX.
3. tsA201 cells (The European Collection of Cell Cultures, ECACC) are plated on 15 mm glass coverslips in 35 mm petri dishes.
4. cDNAs encoding the rat GluN1-1a (GenBank X63255 in pCDM8 vector), GluN2A (GenBank M91561 in PCDM8 vector), and GluN2B (GenBank M91562 in pCDNA1 vector) subunits are cotransfected with cDNA for enhanced green fluorescent protein (eGFP) to identify successfully transfected cells.
5. FuGene 6 Transfection Reagent.
6. D,L-2-amino-5-phosphonopentanoate (AP5) and 7-chlorokynurenic acid (7-CKA), competitive NMDAR antagonists.

2.2 Fast Perfusion System (See Fig. 1)

1. Solution reservoirs are 30 ml syringes attached to an in-house fabricated height-adjustable bracket.
2. Solution flow from reservoirs is controlled by clamping silicone tubing in solenoid pinch valves.
3. Polyethylene tubing is used to connect silicone pinch valve tubing to 2–1 Y connectors (*see* **Note 1**).
4. Polyethylene tubing (PE 50) connects Y connectors (*see* **Note 1**) to silicone tubing (outside diameter 1.2 mm and inside diameter 0.64 mm) that is attached to the back ends of individual square capillary glass (barrels) with outside diameter 0.84 mm and inside diameter 0.6 mm.
5. Four barrels are aligned and glued (Krazy Glue) into an in-house fabricated barrel holder made from a single piece of aluminum, precisely shaped to cup four barrels (*see* **Note 2**).

6. The barrel holder is attached through an in-house fabricated barrel holder arm to the shaft of a stepper motor. The barrel holder arm should give the barrels a ~1″ radius from the center of the stepper motor shaft so that stepper motor rotation translates to a nearly linear barrel movement.
7. Stepper motor rotation is controlled by a microstepping power supply set to 50,000 microsteps/revolution (*see* **Note 3**). Barrel movements are accomplished by smoothly accelerating and decelerating the frequency of brief voltage pulses sent out from a computer parallel port using software (barrel movement software) written in Basic and running in FreeDOS (www.freedos.org) (*see* **Note 4**).
8. Although the fast perfusion system is depicted and described with only two separate solutions flowing through barrel 1, 2, and 3, it is possible to have as many solutions as is experimentally necessary by using an appropriate manifold.

2.3 Whole-Cell Recordings

1. The external, control solution contains: 140 mM NaCl, 2.8 mM KCl, 1 mM $CaCl_2$, 10 mM HEPES, 10 μM EDTA, and 100 μM glycine (*see* **Note 5**). Adjust pH to 7.2 ± 0.05 with NaOH, and adjust osmolality to 290 ± 10 mOsmol/kg with sucrose.
2. The pipette solution contains: 130 mM CsCl, 10 mM BAPTA, 10 mM HEPES. Adjust pH to 7.2 ± 0.05 with CsOH. Osmolality should be 275 ± 10 mOsmol/kg.
3. Recording pipettes are fabricated using borosilicate glass (with filament) with an outer diameter of 1.5 mm and an inner diameter of 0.86 mm and lightly fire-polished.
4. Cells are imaged with an inverted fluorescence microscope with an eGFP filter set. Patch-clamp recordings are made while imaging cells and the recording pipette using a digital camera.
5. Voltage-clamp current recordings are made with an amplifier with a headstage attached to a micromanipulator and digitized with an A/D converter.

3 Methods

Brief synaptic-like agonist applications to recombinant ligand-gated ion channels expressed in tsA201 cells during whole-cell recording can be achieved using the fast perfusion system described in Subheading 2 (*see* Fig. 1). To emulate synaptic neurotransmitter release, the fast perfusion system must achieve brief agonist applications. Brief agonist applications to the entire cell under study are facilitated by "lifting" cells from the coverslip on which they are cultured. The fast perfusion system must also allow for easy changes

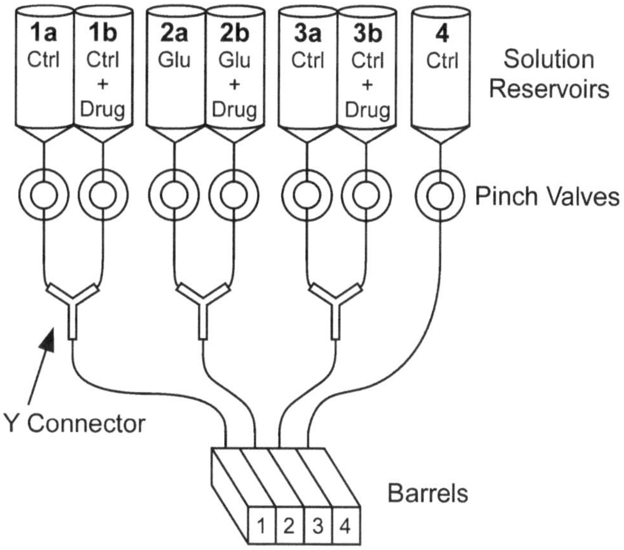

Fig. 1 Schematic of fast perfusion system designed to allow for brief applications of 1 mM glutamate (Glu) in control solution (Ctrl) in the absence and presence of a single drug concentration (Drug)

of the solutions flowing through barrels to allow for responses to brief agonist applications in the absence and presence of drug. As an example of fast perfusion system operation we focus on how NMDAR open channel blockers inhibit recombinant NMDAR responses to brief synaptic-like glutamate applications.

NMDAR open channel blockers are a class of use-dependent NMDAR antagonists. One NMDAR open channel blocker, memantine, is currently being used to treat Alzheimer's disease [13]. Memantine and ketamine, another NMDAR open channel blocker, have shown promise in the treatment of several other debilitating nervous system disorders [4–7, 14, 15]. Memantine and ketamine share the same basic mechanism of action and have similar IC_{50} values and kinetics of inhibition at NMDARs [16, 17]. However, there are subtle kinetic differences in inhibition of NMDARs by memantine and ketamine. These differences demonstrate important considerations when designing experiments to evaluate how drugs affect ligand-gated ion channel currents in response to brief agonist applications. The methods described below explain the steps used to record recombinant NMDAR currents in response to brief glutamate applications in the absence and presence of open channel blockers.

3.1 Fast Perfusion System Design

3.1.1 Brief Application Strategy

1. Rapid and continuous barrel movement from barrel 1 to barrel 3 (*see* Fig. 2a), sweeping quickly by barrel 2, delivers brief synaptic-like glutamate applications (~5 ms) to lifted transfected cells.

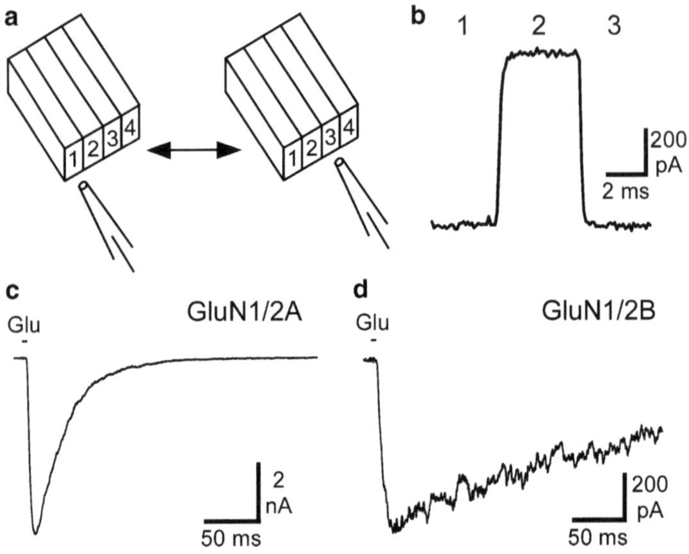

Fig. 2 Measuring the duration of glutamate application. (**a**) Schematic of barrel movement in relation to an open recording pipette. Barrel movements are from barrel position 1–3 (and from barrel position 3–1), briefly sweeping by barrel 2. (**b**) Example of a current recording from an open recording pipette in response to moving from barrel position 1–3, sweeping by barrel 2, which has solution of different osmolality than barrels 1 or 3 (application half-width, 3.7 ms; solution exchange 10–90 % current rise times, 1–2: 0.26 ms; 2–3: 0.22 ms). (**c**, **d**) Examples of whole-cell voltage-clamp recordings of lifted tsA201 cells expressing GluN1/2A receptors (**c**; 10–90 % rise time, 4.0 ms; τ_w of decay, 29.6 ms) or GluN1/2B receptors (**d**; 10–90 % rise time, 9.0 ms; τ_w of decay, 421 ms) in response to brief applications of 1 mM glutamate (Glu, *black bar*). The cells were held at −65 mV

2. Similar barrel movement from barrel 3 to barrel 1 (*see* Fig 2a) delivers another brief synaptic-like glutamate application to lifted transfected cells.
3. With careful calibration, the fast perfusion system can consistently deliver brief, repeated synaptic-like agonist applications to lifted transfected cells.
4. Lifting cells is crucial to ensure complete and rapid exchange of solution during brief agonist applications. Although the duration of agonist application is identical for recordings from attached cells and from lifted cells, the diffusionally restricted space between the bottom of an attached cell and the coverslip slows solution exchange.

3.1.2 Changing Solutions Flowing Through Barrels

1. The Y-connectors described in Subheading 2 (*see* Fig. 1) allow for one of two solutions to flow through barrels 1, 2, and 3. Importantly, pinch valves 1a, 2a, or 3a are never open concurrently with pinch vales 1b, 2b, or 3b, respectively.

2. Change the solutions flowing through each barrel by closing pinch valves 1a, 2a, and 3a and immediately opening pinch valves 1b, 2b, and 3b.
3. During changes of solution flowing through barrels 1, 2, and 3, it is advisable to move to barrel position 4 to perfuse the cell with control solution (make sure pinch valve 4 is always open). Perfusing the cell with control solution during changes of solutions flowing through barrels helps to avoid (1) releasing gas bubbles onto the cell as a result of opening and closing pinch valves, and (2) contact of the cell with glutamate + drug-containing solution due to temporary disruptions in solution flow during pinch valve opening and closing.
4. One benefit of this method is that the number of solutions that can be applied to the same cell is limited only by the number of inlets on a manifold that can replace the Y connector.

3.2 Transient Transfection of tsA201 Cells

1. tsA201 cells are maintained in culture and plated prior to transfection using standard cell culture procedures [18].
2. 12–24 h before transfection, 1×10^5 tsA201 cells are plated in 1.5 ml of tsA201 cell culture medium on uncoated 15 mm glass coverslips in 35 mm petri dishes (3 coverslips/dish).
3. Warm serum-free tsA201 cell culture medium and FuGene 6 Transfection Reagent to room temperature.
4. The following steps refer to transfection of a single dish of plated cells. If transfecting multiple dishes of plated cells, increase the volume of solutions accordingly.
5. Transfer 95 μl of serum-free tsA201 cell culture medium into a sterilized microcentrifuge tube.
6. Add 3 μl of FuGene 6 Transfection Reagent to the tube, avoiding contact with the tube wall.
7. Vortex the tube for 1 s and incubate at room temperature for 5 min.
8. Add 1 μg of cDNA total (2 μl of cDNA at a density of 0.5 μg/μl) to the tube in a ratio of 1:1:2 (eGFP–GluN1–GluN2x) (*see* **Note 6**).
9. Vortex the tube for 1 s and incubate at room temperature for 15 min.
10. Transfer 100 μl of medium–FuGene 6 Transfection Reagent–cDNA mixture from the microcentrifuge tube to a petri dish of plated cells (*see* **Note 7**).
11. Add D,L-AP5 (GluN2A or GluN2B, 200 μM; GluN2C or GluN2D, 400 μM) and 7-CKA (200 μM) to the petri dish (*see* **Note 8**).
12. Wait at least 18 h before recording from transfected cells (*see* **Note 9**).

3.3 Performing Brief Glutamate Applications in Control Solution

3.3.1 Estimating Duration of Brief Applications

1. Fill solution reservoirs 1a, 1b, 3a, and 3b with control solution and fill solution reservoirs 2a, and 2b with control solution diluted by 10 % with deionized H_2O (diluted control solution).
2. Fill a recording pipette with pipette solution, attach to the pipette holder and then apply a small amount of positive pressure (~0.5 PSI) to the side port of the pipette holder (*see* **Note 10**). Lower the pipette into the recording chamber filled with control solution.
3. Position the barrels vertically so they do not touch the bottom of the dish during movement (*see* **Note 11**). Move the pipette into the optimal vertical position for solution application (*see* **Note 12**). Position the pipette in the horizontal plane so that the tip of the pipette is about 50 μm in front of the front edge of the barrels and the tip of the pipette is aligned with the center of barrel 1. Use the barrel movement software to define that location as barrel position 1.
4. Sequentially for each of the remaining three barrels, use the barrel movement software to align the barrel with the tip of the open pipette. Use the barrel movement software to define barrel positions 2, 3, and 4.
5. Make brief solution applications to the open pipette by rapid continuous movements from barrel position 1 to 3 or barrel position 3 to 1, sweeping by the solution in barrel 2 (*see* Subheading 3.1.1). With pinch valves 1a, 2a, and 3a open, perform movements from barrel position 1 to 3 and back from barrel position 3 to 1. Measure the duration of solution application with the open pipette by measuring the current in response to the diluted control solution in barrel 2 (*see* Fig. 2b). Current changes reflect the differing solution osmolality flowing onto the open pipette tip and are used to measure the duration of barrel 2 solution application. We measured the half-width duration of solution application as 4.5 ± 0.6 ms and the solution exchange 10–90 % current rise time as <0.5 ms (*see* Fig. 2b).
6. Change the solutions flowing through the barrels by closing pinch valves 1a, 2a, 3a and opening pinch valves 1b, 2b, and 3b (*see* Subheading 3.1.2). Repeat and evaluate current measurements described in the previous point with pinch valves 1b, 2b, and 3b open.

3.3.2 Whole-Cell Recording from Lifted Cells

Patch-clamp recording from lifted cells is similar to patch-clamp recording from attached cells. For more detailed information on standard application of the patch-clamp technique *see* Hamill et al. [19].

1. Transfer a coverslip with transfected tsA201 cells to the recording chamber containing room temperature bath solution. Place the recording chamber onto the microscope stage, and then

place an efflux tube and reference electrode into the chamber (*see* **Note 13**).

2. Using the fluorescence microscope, identify an isolated eGFP-positive cell (*see* **Note 14**).
3. Position the barrels vertically to ensure that they do not make contact with the coverslip through the full range of barrel movement (*see* **Note 11**). Position the barrels in the horizontal plane so that the front edge of barrel 1 is near the cell, and the center of barrel 1 is aligned with the cell (*see* Fig. 2a). Then move the barrels axially away from the cell, without changing the alignment of barrel 1 with the cell, to avoid crashing the recording pipette into the barrels (*see* **Note 15**).
4. Position the recording pipette just above the cell. Before forming a gigaohm seal, move the barrels axially towards the cell until about 50 μm from the cell.
5. Lower the recording pipette and form a gigaohm seal. Adjust electrode capacitance, and then achieve a whole-cell configuration.
6. Set whole-cell parameters (cell capacitance and series resistance) and adjust series resistance compensation to ~80 %.
7. To lift the cell, apply a constant negative pressure of 1–1.5 PSI to the side port of the pipette holder. Slowly begin to move the pipette straight up from the coverslip. You should see the cell lift from the coverslip. Continue lifting the cell slowly until it is completely free from the coverslip. Move the pipette with the lifted cell into the optimal position for solution application (*see* **Note 12**).
8. Once the lifted cell is positioned, reduce the constant negative pressure to the side port of the pipette holder to 0.3–0.6 PSI. Readjust the whole-cell parameters, as capacitance should have decreased from lifting the cell. Also, the membrane capacitance of lifted cells often decreases throughout experiments, which may require further adjustments to whole-cell parameters.
9. Making an initial glutamate application of about 30 s is recommended to reduce response variability during the rest of the experiment.

3.3.3 Quantification of Receptor Response Time Course

1. Fill solution reservoirs 1a, 1b, 3a, and 3b with control solution, and fill solution reservoirs 2a and 2b with control solution containing 1 mM glutamate.
2. Make brief glutamate applications to the lifted cell by rapid continuous movements from barrel position 1 to 3 or barrel position 3 to 1, sweeping by the solution in barrel 2 (*see* Subheading 3.1.1).

3. Gauge the similarity to NMDAR-EPSCs of recombinant NMDAR responses by measuring the kinetics of recombinant receptor currents in response to brief glutamate applications.
4. Quantify the activation time course of recombinant NMDAR currents as the 10–90 % rise time. We measured a mean 10–90 % current rise time in response to brief glutamate applications to GluN1/2A receptors of 4.8 ± 0.6 ms (*see* Fig. 2c) and in response to brief glutamate applications to GluN1/2B receptors of 12.7 ± 5.6 ms (*see* Fig. 2d).
5. Quantify the decay time course of recombinant NMDAR currents by fitting the current decay with a double exponential function and determining the weighted time constant of decay ($\tau_w = (\tau_{fast})(\text{fraction}_{fast}) + (\tau_{slow})(\text{fraction}_{slow})$). We measured a mean τ_w in response to brief glutamate applications to GluN1/2A receptors of 27.5 ± 4.1 ms (*see* Fig. 2c) and in response to brief glutamate applications to GluN1/2B receptors of 420 ± 34 ms (*see* Fig. 2d).
6. Compare results to expected EPSC kinetics. The recombinant NMDAR response kinetics we measured are similar to previous measurements of NMDAR-EPSC kinetics and also to results of previous studies using brief glutamate applications to recombinant NMDARs in transfected cells [20–23].
7. Change the solutions flowing through the barrels by closing pinch valves 1a, 2a, 3a and opening pinch valves 1b, 2b, and 3b (*see* Subheading 3.1.2). Repeat and evaluate the kinetic measurements of recombinant NMDAR currents in response to brief glutamate applications (*see* **Note 16**).
8. After finishing an experiment, measure the duration of glutamate application to that specific cell to control for variations in solution flow rate and other potential sources of error, which may lead to exclusion of that experiment from analysis. Turn off series resistance compensation and whole-cell parameters. Return holding potential to 0 mV. Deliver >2 PSI of positive pressure to the side port of the pipette holder to remove the cell and membrane debris from the tip of the pipette. Dilute the glutamate-containing solutions in reservoirs 2a and 2b (*see* Fig. 1) with deionized H_2O by at least 10 %. Measure changes in pipette current in response to barrel movements with the open pipette (*see* Subheading 3.3.1). Make sure to measure solution applications with pinch valves 1a, 2a, and 3a open and also with pinch valves 1b, 2b, and 3b open.

3.3.4 Fast Perfusion System Optimization

1. Stepper motor controller power output. Depending on the stepper motor controller, the output power may be adjustable. If so, modifying the output power can change stepper motor operation, either introducing or eliminating oscillations that

may result from rapid acceleration and deceleration of the stepper motor. With some power settings, we observed oscillations when monitoring system performance using an open pipette that could have an undesirable impact on brief agonist applications to transfected cells.

2. Weight of barrel holder arm and barrel holder. Due to rapid acceleration and deceleration of the stepper motor, the stepper motor can overshoot desired positions or oscillate. The rotational inertia imposed by the weight of the barrel holder arm and barrel holder can strongly impact stepper motor overshoot and oscillations. The weight of the barrel holder arm and barrel holder should be minimized to reduce overshoot and oscillations if present.

3. Acceleration of stepper motor. The acceleration and deceleration of the stepper motor should be optimized for system stability and to minimize the duration of agonist application. At more rapid accelerations and decelerations, the stepper motor may overshoot desired positions or oscillate. At slower accelerations and decelerations, the duration of agonist application may be too long.

4. Rate of solution flow. Careful adjustment of the solution flow rate is essential to achieving consistent and brief agonist applications. It is important to maintain similar solution flow rates so that inconsistencies in application duration do not arise (*see* **Note 17**). Also, lifted cells are attached only to the tip of the recording pipette, making them vulnerable to being blown away if the solution flow rate is too fast.

5. Degassing solutions prior to use. Removing gas from solutions prior to starting experiments can help to (1) keep bubbles from destroying cells and (2) keep bubbles from blocking barrels, slowing or stopping solution flow. Gas bubbles can form unpredictably in tubing during experiments, and it can be difficult to determine if solution has stopped flowing from a particular barrel during an experiment. To degas solutions, pour solutions into a vacuum flask and apply negative pressure. Stop negative pressure when few gas bubbles form in solution.

6. Mixing of barrel solutions. It is important to ensure that a cell is exposed almost exclusively to the desired solution at each barrel position. Solution mixing could occur, for example, within the Y connectors, or after solutions leave the barrels if the cell is not properly positioned relative to the barrels. One way to test for mixing is to fill solution reservoirs 1a, 2a, 3a, and 4 with control solution and solution reservoirs 1b, 2b, and 3b with control solution containing agonist at a concentration orders of magnitude above its EC_{50} for the receptors under study; we use 10 mM glutamate. While whole-cell recording

from a lifted cell expressing recombinant receptors, start recording at barrel position 4 with all other pinch valves closed and determine control (in the absence of agonist) holding current. Open pinch valves 1b, 2b, and 3b and be sure that holding current does not change while the cell is in front of barrel 4. Move to barrel position 3 to observe the response to glutamate, and after current has reached steady state, be sure that there is no further change in current when moving to barrel positions 2 and 1. Move to barrel position 3, close pinch valve 3b, and open pinch valve 3a, and ensure that control holding current is observed. Repeat this procedure for the other barrels, and also change the solution flowing through adjacent barrels to be sure that the cell is exposed only to the solution flowing from the appropriate barrel. If evidence of mixing is observed, identify and correct the source of the problem (e.g., malfunctioning pinch valves or incorrect positioning of the cell relative to the barrels).

3.4 Performing Brief Glutamate Applications in the Presence of Channel Blockers

1. Use whole-cell patch-clamp recordings from lifted cells expressing GluN1/2A or GluN1/2B receptors to record responses to brief glutamate applications as described in Subheading 3.3, with modifications described below.
2. Fill solution reservoirs 1a, 3a, and 4 with control solution and reservoir 2a with control solution containing 1 mM glutamate. Fill solution reservoirs 1b and 3b with control solution + drug and reservoir 2b with control solution containing 1 mM glutamate + drug (*see* Fig. 1).
3. Choose an appropriate frequency of brief glutamate applications to lifted cells expressing a particular NMDAR subtype. The frequency must be low enough to ensure complete current decay following glutamate application and allow for recovery from desensitization before the subsequent glutamate application, yet fast enough to allow for experiments that may require many brief glutamate applications (potentially >100 applications). We used a glutamate application frequency of 0.2 Hz for both GluN1/2A (*see* Fig. 3a) and GluN1/2B (*see* Fig. 3b) receptors.
4. Measure the baseline peak current value in response to brief glutamate applications in the absence of drug (baseline current). We required ten consecutive, steady glutamate responses to establish that a stable baseline current had been reached, which were then averaged to give the baseline current mean value (*see* Fig. 4a, b).
5. Add drug to the solutions flowing through the barrels by closing pinch valves 1a, 2a, and 3a and opening pinch valves 1b, 2b, and 3b (*see* Subheading 3.1.2). Make sure to allow enough

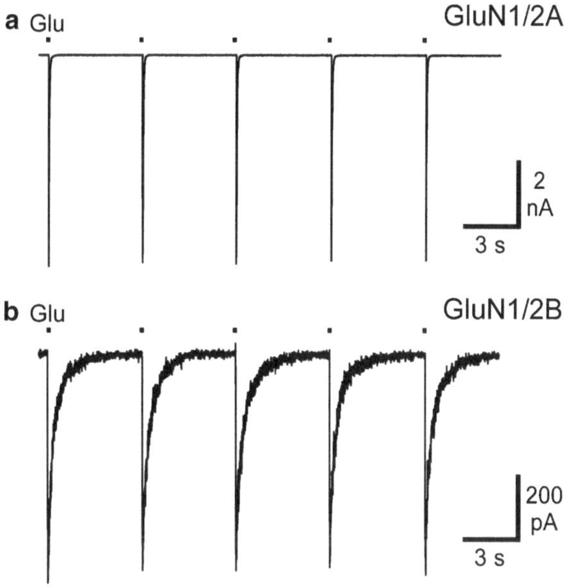

Fig. 3 Brief applications of glutamate to lifted cells expressing NMDARs. (**a**, **b**) Example whole-cell voltage-clamp recordings of lifted tsA201 cells expressing GluN1/2A receptors (**a**) or GluN1/2B receptors (**b**) in response to five brief applications of 1 mM glutamate (Glu, *black bars*) at a frequency of 0.2 Hz. The cells were held at −65 mV

time for complete changes of solutions flowing through the barrels (*see* **Note 18**).

6. Open channel blockers require that the channel be activated to bind and inhibit the channel. The number of brief glutamate applications in the presence of drug needed to reach a steady level of NMDAR inhibition depends on the drug's kinetics and must be determined for each drug and NMDAR subtype. For each successive application of glutamate in the presence of drug, the peak current should be smaller than the previous peak current until reaching a steady level (inhibited current). We required five consecutive, steady glutamate responses to establish that a stable inhibited current had been reached, which were then averaged to give the inhibited current mean value (*see* Fig. 4a, b). We used memantine and ketamine, two NMDAR open channel blockers with slightly different kinetics, to illustrate differences in the number of glutamate applications in the presence of drug needed to reach steady NMDAR inhibition. We used 20 applications of glutamate in the presence of memantine and 40 applications of glutamate in the presence of ketamine to reach steady levels of NMDAR inhibition with GluN1/2A (data not shown) and GluN1/2B receptors (*see* Fig. 4a, b).

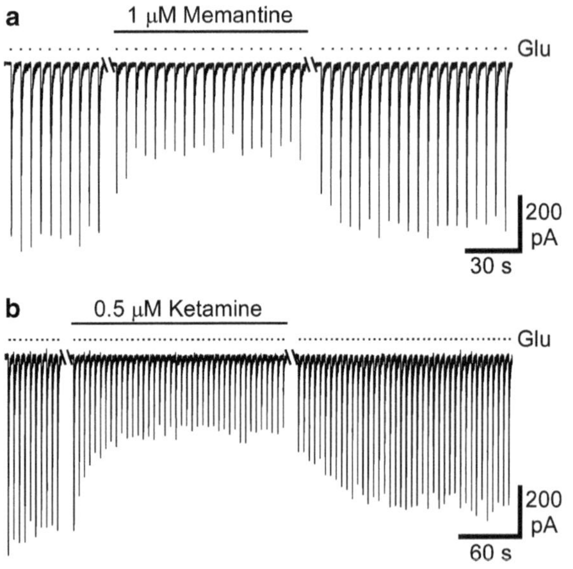

Fig. 4 Antagonist kinetics affect the number of brief glutamate applications needed to reach a steady level of current inhibition and a steady level of currents after recovery from inhibition. (**a**, **b**) Examples of whole-cell voltage-clamp recordings from lifted tsA201 cells expressing GluN1/2B receptors in response to brief applications of 1 mM glutamate (Glu, *short black bars*) at 0.2 Hz in control solution or in the presence of 1 μM memantine (**a**) or 0.5 μM ketamine (**b**) (*long black bars*). The average of peak currents from the first ten glutamate responses shown gives the baseline current mean value, the average the peak currents from the last five glutamate responses in the presence of memantine or ketamine gives the inhibited current mean value, and the average of the peak currents from the last ten glutamate responses gives the current after recovery mean value. The cells were held at −65 mV

7. Remove drug from the solutions flowing through the barrels by closing pinch valves 1b, 2b, and 3b, and opening pinch valves 1a, 2a, and 3a (*see* Subheading 3.1.2). Make sure to allow enough time for complete changes of solutions flowing through the barrels (*see* **Note 18**).
8. Open channel blockers like memantine and ketamine require channel activation to unbind and allow for recovery from inhibition (*see* **Note 19**). The number of brief glutamate applications in the absence of drug following NMDAR inhibition must be determined for each drug and NMDAR subtype. For each successive application of glutamate in the absence of drug following NMDAR inhibition, the peak current should be larger than the previous peak current until reaching a steady level after recovery from inhibition is complete (current after recovery). We required ten consecutive, steady glutamate responses to establish that a stable current after recovery had been reached, which were then averaged to give the current

after recovery mean value (*see* Fig. 4a, b). We used 20 applications of glutamate in the absence of memantine and 40 applications of glutamate in the absence of ketamine following NMDAR inhibition to reach steady levels of current after recovery with GluN1/2A (data not shown) and GluN1/2B receptors (*see* Fig 4a, b).

9. Measure peak currents in response to brief glutamate applications in the absence and presence of drug as the mean current over a 3 ms window centered at the time of peak current.

10. Calculate the percent inhibition by open channel blockers using the equation: % inhibition = 100 × (1 − (inhibited current)/(0.5 × (baseline current + current after recovery))). We averaged the values for baseline current and current after recovery to account for changes in cell properties during experiments. For Fig. 4, we used concentrations of memantine and ketamine near their IC_{50} values at NMDARs. We measured percent inhibition of responses to brief glutamate applications to GluN1/2B receptors in the presence of 1 μM memantine as 49 % (*see* Fig. 4a), and in the presence of 0.5 μM ketamine as 56 % (*see* Fig. 4b).

4 Notes

1. Silicone tubing of appropriate size is used to connect PE tubing to Y connectors.

2. Barrels were first cut to length (5 mm) by scoring with a diamond tipped scribe and then both ends were lightly fire-polished. To allow silicone tubing connections to go over the back ends of adjacent barrels, carefully heat individual barrels over a Bunsen burner and bend to a 30–60° angle. Only bend two barrels and align them in an alternating pattern of bent then straight barrels to ensure that silicone tubing will attach to the back ends of all the barrels. Glass should be cleaned in 95 % ethanol and dried before gluing to the barrel holder. Make sure to align the open edge of the barrels with each other, and ensure that there is no space between barrels.

3. With barrels at a ~1″ (~25 mm) radius from the center of the stepper motor shaft, each microstep of stepper motor rotation is translated to ~3 μm of barrel movement. Because the total range of barrel movement is about 2,500 μm, less than 1,000 microsteps (less than 1/50th revolution) are needed for total barrel movement. This translates to nearly linear barrel movement.

4. A compiled version of the stepper motor program is available from the authors by email request.

5. 10 μM EDTA is used to chelate contaminating free Zn^{2+}, which inhibits GluN1/2A receptors in the nM range. The NMDAR coagonist glycine is present in all solutions to saturate the glycine coagonist-sites on NMDARs.
6. The cDNA transfection ratio of 1:1:2 for eGFP–GluN1–GluN2x may vary depending on transfection efficiency with given vectors and subunits.
7. The volume of serum-free tsA201 medium used for transfections depends upon the cDNA solution density. The medium–FuGene 6 Transfection Reagent–cDNA mixture should be at a final volume of 100 μl for transfection of a single dish of plated cells. If the cDNA solution density differs from 0.5 μg/μl, a different volume of cDNA solution should be added to the mixture to reach 1 μg of cDNA; the amount of medium added should be adjusted to reflect this change.
8. NMDARs tonically activated by ambient glutamate present in the culture medium are excitotoxic. Therefore, we add competitive antagonists to the culture medium after transfection of tsA201 cells. Other antagonists, including elevation of the Mg^{2+} concentration of the tsA201 cell culture medium to >10 mM also may be used.
9. We find that 24–48 h after transfection offers optimal current amplitudes, cell health, and cell confluency. Depending on current amplitudes, successful recordings from transfected cells can be made up to at least 72 h after transfection. Vary the time between transfection and recording to optimize protein expression and cell health.
10. We use a 1 ml syringe connected with PE tubing to the side port of the pipette holder and connected in parallel to a pressure gauge. Pressure and suction can be applied by using the plunger of the syringe or by mouth. A stopcock on the end of the syringe can be closed to hold positive and negative pressure in the pipette.
11. The distance between the upper surface of the coverslip and the lowest point on any of the barrels changes slightly when the stepper motor rotates to cause barrel movement (*see* **Note 3**). After positioning the barrels axially so they are near the pipette tip or cell, position the barrels vertically so they are close to the coverslip, but do not touch the coverslip during movements to each barrel position. The barrels could break if they contact the coverslip during fast movements.
12. Choose the vertical position of the pipette relative to the barrel openings to optimize speed of solution changes. It is best to position the pipette so that it is near the vertical center of the barrel openings. However, note that the barrels should be angled so that they point ~30° below the horizontal plane.

The pipette should be positioned vertically so that it sits near the middle of the solution streams flowing from the barrels.

13. To maintain fluid levels in the recording chamber, we siphon solution through a glass efflux tube. The height of solution in the recording chamber is determined by the height of the waste end of the efflux tube.
14. It is important to record only from isolated eGFP-positive tsA201 cells. When recording from lifted cells, it is often difficult to tell if there are thin attachments to other cells, which could drastically alter the recordings.
15. When using lifted cells, it is possible to move the lifted cell to the barrels, even if they are placed relatively far from the starting location of the cell, to simplify barrel positioning. However, aligning the barrels as described minimizes the need to move the cell after lifting, increasing success rate.
16. Make sure that peak current amplitudes and response kinetics in response to brief glutamate applications are similar when pinch valves 1a, 2a, and 3a are open and when pinch valves 1b, 2b, and 3b are open. If significant differences are observed, further optimize the system as described in Subheading 3.3.4.
17. Differences in the rate of solution flow from barrel 1 and 3 can increase the agonist application duration while moving from barrel position 1 to 3 relative to the agonist application duration while moving from barrel position 3 to 1. Also, differences in solution flow rate from reservoirs 1a and 1b, etc. can have a significant impact on the duration of agonist application in the presence or absence of drug. Such differences could lead to complications in interpreting the effect of a drug.
18. The time required for complete changes of solutions flowing through barrels can be estimated with the following experiment. Fill solution reservoirs 1a, 2a, and 3a with control solution and fill solution reservoirs 1b, 2b, and 3b with diluted control solution. With an open pipette positioned at barrel position 1, measure the time course of current change in response to closing pinch valve 1a and opening pinch valve 1b. The current should change approximately exponentially until reaching a steady level in the presence of the diluted control solution in reservoir 1b. Measure the 10–90 % current rise time to estimate the time required for changing the solution flowing through barrel 1. We waited for 5× the 10–90 % current rise time after closing pinch valve 1a and opening pinch valve 1b to consider the change of solution flowing through barrel 1 complete. Repeat measurements of current change in response to closing pinch valve a and opening pinch valve b for barrels 2 and 3.

19. Measure recovery in all experiments to ensure that decreases in peak currents in response to brief glutamate applications in the presence of drug are due to the drug itself and not due to other changes in the cell that may decrease peak currents.

Acknowledgements

The authors would like to thank Christen Shiber for excellent technical assistance and critically reading the manuscript. The authors would also like to thank Jim Buhrman for excellent technical assistance and helpful discussions regarding fast perfusion system design.

References

1. Traynelis SF, Wollmuth LP, McBain CJ et al (2010) Glutamate receptor ion channels: structure, regulation, and function. Pharmacol Rev 62:405–496
2. Barnham KJ, Masters CL, Bush AI (2004) Neurodegenerative diseases and oxidative stress. Nat Rev Drug Discov 3:205–214
3. Lau A, Tymianski M (2010) Glutamate receptors, neurotoxicity and neurodegeneration. Europ J Physiol 460:525–542
4. Duman RS, Aghajanian GK (2012) Synaptic dysfunction in depression: potential therapeutic targets. Science 338:68–72
5. Autry AE, Adachi M, Nosyreva E et al (2011) NMDA receptor blockade at rest triggers rapid behavioural antidepressant responses. Nature 475:91–95
6. Pittenger C, Sanacora G, Krystal JH (2007) The NMDA receptor as a therapeutic target in major depressive disorder. CNS Neurol Disord Drug Targets 6:101–115
7. Collins S, Sigtermans MJ, Dahan A et al (2010) NMDA receptor antagonists for the treatment of neuropathic pain. Pain Med 11: 1726–1742
8. Palmer GC (2001) Neuroprotection by NMDA receptor antagonists in a variety of neuropathologies. Curr Drug Targets 2:241–271
9. Lipton SA (2004) Failures and successes of NMDA receptor antagonists: molecular basis for the use of open-channel blockers like memantine in the treatment of acute and chronic neurologic insults. NeuroRx 1:101–110
10. Paoletti P, Bellone C, Zhou Q (2013) NMDA receptor subunit diversity: impact on receptor properties, synaptic plasticity and disease. Nat Rev Neurosci 14:383–400
11. Lipton SA (2006) Paradigm shift in neuroprotection by NMDA receptor blockade: memantine and beyond. Nat Rev Drug Discov 5:160–170
12. Gladding CM, Raymond LA (2011) Mechanisms underlying NMDA receptor synaptic/extrasynaptic distribution and function. Mol Cell Neurosci 48:308–320
13. Witt A, Macdonald N, Kirkpatrick P (2004) Memantine hydrochloride. Nat Rev Drug Discov 3:109–110
14. Prommer EE (2012) Ketamine for pain: an update of uses in palliative care. J Palliat Med 15:474–483
15. Anitha M, Nandhu MS, Anju TR et al (2011) Targeting glutamate mediated excitotoxicity in Huntington's disease: neural progenitors and partial glutamate antagonist–memantine. Med Hypotheses 76:138–140
16. Kotermanski SE, Wood JT, Johnson JW (2009) Memantine binding to a superficial site on NMDA receptors contributes to partial trapping. J Physiol 587:4589–4604
17. Kotermanski SE, Johnson JW (2009) Mg^{2+} imparts NMDA receptor subtype selectivity to the Alzheimer's drug memantine. J Neurosci 29:2774–2779
18. Phelan MC (2006) Techniques for mammalian cell tissue culture. Curr Protoc Mol Biol Appendix 3: Appendix 3F
19. Hamill OP, Marty A, Neher E et al (1981) Improved patch-clamp techniques for high-resolution current recording from cells and cell-free membrane patches. Pflugers Arch 391:85–100
20. Vicini S, Wang JF, Li JH et al (1998) Functional and pharmacological differences

between recombinant N-methyl-D-aspartate receptors. J Neurophysiol 79:555–566
21. Erreger K, Dravid SM, Banke TG et al (2005) Subunit-specific gating controls rat NR1/NR2A and NR1/NR2B NMDA channel kinetics and synaptic signalling profiles. J Physiol 563:345–358
22. Cull-Candy SG, Leszkiewicz DN (2004) Role of distinct NMDA receptor subtypes at central synapses. Sci STKE 2004(255):16
23. Tovar KR, McGinley MJ, Westbrook GL (2013) Triheteromeric NMDA receptors at hippocampal synapses. J Neurosci 33:9150–9160

Chapter 3

Patch Clamp Combined with Voltage/Concentration Clamp to Determine the Kinetics and Voltage Dependency of *N*-Methyl-D-aspartate (NMDA) Receptor Open Channel Blockers

Chris G. Parsons and Kate E. Gilling

Abstract

Electrophysiological techniques can be used to great effect to help determine the mechanism of action of a compound. However, many factors can compromise the resulting data and their analysis, such as the speed of solution exchange, expression of additional ion channel populations including other ligand-gated receptors and voltage-gated channels, compounds having multiple binding sites, and current desensitization and rundown. In this chapter, such problems and their solutions are discussed and illustrated using data from experiments involving the uncompetitive NMDA receptor antagonist memantine. Memantine differs from many other NMDA receptor channel blockers in that it is well tolerated and does not cause psychotomimetic effects at therapeutic doses. Various electrophysiological parameters of NMDA-induced current blockade by memantine have been proposed to be important in determining therapeutic tolerability; potency, onset and offset kinetics, and voltage dependency. These were all measured using whole cell patch clamp techniques using hippocampal neurons. Full results are shown here for memantine, and these are summarized and compared to those from similar experiments with other NMDA channel blockers. The interpretation of these results is discussed, as are theories concerning the tolerability of NMDA channel blockers, with the aim of illustrating how electrophysiological data can be used to form and support a physiological hypothesis.

Key words NMDA, Uncompetitive, Concentration dependence, Concentration clamp, Voltage dependence, Kinetics

1 Introduction

Memantine is an uncompetitive *N*-methyl-D-aspartate (NMDA) receptor antagonist that is registered in Europe and the USA for the treatment of moderate to severe Alzheimer's disease (AD). It has clear symptomatic effects in both AD patients [1, 2] and animal models of AD [3] and, on the basis of its mechanism of action, is also likely to show neuroprotective activity in AD [3, 4]. This compound blocks the channel in a use-dependent manner,

Marzia Martina and Stefano Taverna (eds.), *Patch-Clamp Methods and Protocols*, Methods in Molecular Biology, vol. 1183, DOI 10.1007/978-1-4939-1096-0_3, © Springer Science+Business Media New York 2014

meaning that it can only gain access to the channel in the presence of agonist and remains trapped in the channel following removal of agonist [3, 5, 6]. Both the clinical tolerability and symptomatic effects of memantine have been attributed to its fast blocking kinetics and strong voltage dependency [3, 5–7]. These properties have been characterized by numerous groups using whole cell patch clamp recordings from primary cultures of hippocampal and cortical neurones [5, 8–15]. However, there are several factors that must be taken into account when performing such experiments to ensure the quality of the recordings and their analysis.

(a) Fast blocking kinetics can only be measured accurately when fast concentration-clamp techniques are used to apply antagonists. This is particularly problematic with primary cultures of hippocampal/cortical neurones due to their large dendritic arborization, the inability to lift such cells from the bottom of the dish and resulting problems of buffered diffusion. Fluid-in-fluid fast concentration-clamp systems with relatively large application diameters, e.g., theta glass stepping motor systems, are preferred over systems such as U-type application tubes with which offset kinetics cannot be addressed because there is no real "wash off" of compounds with the latter technique.

(b) The native cells used probably have mixed receptor populations (e.g., N1a/2A and NR1a/2B) and differences in the potency of the antagonist at each of the receptor subtypes may exist. Uncompetitive antagonists like memantine may also bind to multiple sites within the NMDA receptor channel [14, 15]. Both of these aspects can lead to double exponential blocking/unblocking kinetics which first become apparent when fast concentration-clamp techniques are used. When present, these double kinetics must be measured accurately and subsequently analyzed. In addition, the unblocking kinetics are voltage-dependent, and this aspect cannot be addressed by using ramping protocols in voltage-dependency experiments (*see* point c).

(c) Voltage-dependency experiments are often hampered by the presence of additional voltage-gated ion channels and their contribution to the currents recorded must be minimized and/or accounted for in the analysis. One way around this problem is to use tetrodotoxin (TTX) to block voltage-activated sodium channels (VASCs) and replace K^+ with Cs^+ in the recording solutions to reduce the effects of voltage-activated potassium channels (VAKCs). The contribution of voltage-activated calcium channels (VACCs) upon the currents measured can be reduced by lowering Ca^{2+} in the extracellular solution (*see* point d). The contribution of all voltage-activated channels (VACs) can also be reduced by avoiding the use of relatively fast ramping protocols and rather recording individual NMDA-induced currents after holding at different holding potentials and allowing the VACs to desensitize/inactivate for several tens of seconds following each incremental depolarizing step. For example, shifting the holding potential from −90 mV to +70 mV in steps of 10 mV, allowing at least 30 s at each new holding potential before applying agonist. Residual VAC currents can then be subtracted from the agonist-induced currents mediated via ligand-gated channels such as the NMDA receptor channel. However, such protocols are long and are associated with other potential problems—*see* points (d) and (e).

(d) NMDA receptors show various forms of Ca^{2+}-dependent desensitization (with time constants of 500 ms to several seconds) [16–18], the presence of which can interfere with experiments assessing the blocking kinetics of open channel blockers like memantine. This problem can be minimized by decreasing Ca^{2+} concentrations in the solutions used (also *see* point c), but Ca^{2+} is an important cation for membrane stability. One way around this additional problem is to reduce Ca^{2+} concentrations in the presence of agonist (0.2 mM) but maintain normal Ca^{2+} concentrations (1.5 mM) between agonist applications to allow the cell membrane to "recover" between agonist applications.

(e) Glycine is a co-agonist for NMDA receptors (at the glycine$_B$ site) with an EC_{50} of around 1 μM, and its presence is a prerequisite for receptor channel activation by glutamate or NMDA [19]. At non-saturating glycine concentrations, NMDA receptors show strong glycine-dependent desensitization, with quite fast kinetics (τ = 100–400 ms), which can impede the analysis of antagonist blocking kinetics [20]. Moreover, glycine concentrations in the solutions can be dynamically altered by the presence of microbial organisms in the perfusion setup as glycine is involved in metabolism. As such, the magnitude and rate of glycine-sensitive desensitization can change during long recordings and it is essential to keep the whole perfusion system very clean. D-serine is also a co-agonist for the glycine site [21], but it is not metabolized so easily and is the preferred co-agonist for such experiments, used at saturating concentrations. Clean perfusion systems are nonetheless very important for such experiments, e.g., contamination with previously used "sticky" (most often lipophilic) compounds should be avoided.

(f) Aside from the various forms of receptor desensitization detailed in points (d) and (f), NMDA receptors also show moderate rundown which should be minimized by the choice of appropriate intracellular (adenosine 5′-triphosphate (ATP)-regenerating) and extracellular (low Ca^{2+}) solutions. However, some form of rundown compensation is essential in the analysis to account for such dynamic changes, especially when assessing the potency of antagonists with several concentrations being tested sequentially over time.

Most scientific papers on patch clamp experiments tend to minimize the description of the methods used and minor details that could be very important for the final outcome are often not apparent from such descriptions. The aim of the present chapter is to describe, in detail, the methods used to address such aspects for memantine as an example of a fast, voltage-dependent, NMDA receptor channel blocker.

2 Materials

2.1 Cell Culture

1. Mg^{2+} free Hank's buffered salt solution stored at 2–5 °C and warmed to approximately 35 °C before use.
2. Solution of 0.05 % DNAase and 0.3 % ovomucoid in PBS, stored at −20 °C and warmed to approximately 35 °C before use.
3. Solution of 0.66 % trypsin and 0.1 % DNAase in PBS, stored at −20 °C and warmed to approximately 35 °C before use.
4. Minimum essential medium stored at 2–5 °C and warmed to approximately 35 °C before use.
5. Poly-DL-ornithine (500 μg/mL) dissolved in 0.5 M boric acid, stored at −20 °C and warmed to approximately 35 °C before use.
6. Laminin dissolved in PBS to a concentration of 10 μg/mL, stored at −20 °C and warmed to approximately 35 °C before use.
7. $NaHCO_3$/HEPES-buffered minimum essential medium supplemented with 5 % fetal calf serum and 5 % horse serum stored at 2–5 °C and warmed to approximately 35 °C before use.
8. Cytosine-β-D-arabinofuranoside stored at 2–5 °C.
9. Plasticware including flasks, petri dishes, and pipettes.

2.2 Patch Clamp

1. Borosilicate glass for recording pipettes with an outer diameter of 1.5 mm and an inner diameter of 1.275 mm.
2. Pipette puller.
3. Square-walled application pipette glass with a wall to wall measurement of 700 μm.
4. The SF-77B Perfusion Fast-Step: a stepper motor-driven, double-barrelled theta glass application pipette delivery system.
5. Silicon-glass tubing with an external diameter of 0.43 mm, and an internal diameter of 0.32 mm.

6. Polyethylene tubing of the PE-10 size.
7. Low volume manifolds.
8. Manifold valves.
9. Digital output trigger interface.
10. Amplifier.
11. Inverted microscope.
12. A valve driver circuit.
13. Software for data acquisition and analysis, Excel, fitting program, and suitable computer hardware.
14. Intracellular solution used for recording NMDA receptor-mediated currents from hippocampal neurons: 120 mM CsCl, 10 mM ethylene glycol-bis(2-aminoethylether)-*N*,*N*,*N′*,*N′*-tetraacetic acid (EGTA), 1 mM $MgCl_2$, 0.2 mM $CaCl_2$, 10 mM glucose, 20 mM tetraethyl ammonium chloride (TEACl), 2 ATP mM, and 0.2 mM adenosine 3′,5′-cyclic monophosphate (cAMP).
15. Extracellular bath solution used for recording NMDA receptor-mediated currents from hippocampal neurons: 140 mM NaCl, 3 mM CsCl, 10 mM glucose, 10 mM HEPES, 0.2 mM $CaCl_2$, and 4.5 mM sucrose, and supplemented with 0.35 μM TTX.
16. NMDA was stored as a 100 mM stock solution in distilled water at −20 °C. The stock solution of the co-agonist D-serine was prepared and stored under the same conditions, but at a concentration of 10 mM.
17. The NMDA receptor antagonists memantine, neramexane, ketamine, phencyclidine (PCP), dextromethorphan, and dextrorphan were all stored in distilled water at 2–5 °C as stock solutions of 10 mM.

3 Methods

3.1 Cell Culture

1. Hippocampal tissue was obtained from rat embryos (E20–E21) and was then transferred to Ca^{2+} and Mg^{2+} free Hank's buffered salt solution on ice.
2. Cells were mechanically dissociated in 0.05 % DNAase/0.3 % ovomucoid solution following an 8 min pre-incubation with 0.66 % trypsin/0.1 % DNAase solution.
3. The dissociated cells were then centrifuged at 18 × *g* for 10 min, resuspended in minimum essential medium and plated at a density of 150,000 cells/cm² onto poly-DL-ornithine/laminin-precoated plastic petri dishes, *see* **Note 1**.
4. The cells were nourished with $NaHCO_3$/HEPES-buffered minimum essential medium supplemented with 5 % fetal calf

serum and 5 % horse serum and incubated at 37 °C with 5 % CO_2 at 95 % humidity. The medium was exchanged completely following inhibition of further glial mitosis with cytosine-β-D-ARABINOFURANOSIDE after about 5 days in vitro (DIV).

5. Patch-clamp recordings were made after 12–15 DIV.

3.2 Patch Clamp

3.2.1 Recording

1. Voltage-clamp recordings were made in the whole-cell configuration of the patch-clamp technique at a holding potential of −70 mV, unless otherwise stated. All recordings were made at room temperature (20–23 °C; *see* **Note 2**).
2. Patch-clamp pipettes were pulled from borosilicate glass using a horizontal puller and, when filled with intracellular solution, had resistances of 1–3 MΩ.
3. Pyramidal cells were visualized using an inverted microscope under phase contrast and selected for patching based upon their position and morphology.
4. The cells were opened by suction after formation of a giga-seal between the pipette and cell membrane, and were allowed to stabilize for 1–2 min before recordings were made.
5. Currents were recorded using an amplifier, and software was used for the collection and storage of data [22, 23], *see* **Note 3**.
6. The liquid junction potential was measured for each set of solutions using an agar bridge in place of the usual silver chloride pellet as the ground electrode (*see* **Note 4**).
7. Series resistance was measured (mean value of 4.22 ± 0.14 MΩ) and accordingly compensated for in conjunction with capacitance. Fast capacitive currents were corrected for upon formation of the giga-seal, and whole-cell capacitance correction was performed after the cell was opened. These procedures were performed semi-automatically using the amplifier and software.
8. The current signal was filtered by the amplifier using the 3-pole pre-filter with Bessel 10 kHz bandwidth and the 4-pole filter set to 2.9 kHz with Bessel characteristic.
9. Current measurements were acquired at a rate of 10 kHz to avoid potential problems of aliasing.

3.2.2 Perfusion System

1. Test substances were applied by switching channels of a modified stepper motor-driven, double-barrelled theta glass application pipette. The openings of the square walled application pipette glass was reduced to 200–250 μM by pulling these glass capillaries, by hand, over a Bunsen burner and then cleanly separating "two" new perfusion pipettes by cutting them with a diamond cutter.
2. The internal dead volume of such application pipettes was reduced to a minimum by the following procedure. Silicon-glass

tubing was inserted as far as possible towards the tips of the theta glass "pipettes" (around 2–3 mm from the tip). The non-tip, open ends of the silicon-glass tubing was blocked with acrylic glue and the pipettes were then reverse filled with molten wax almost to the open tips of the silicon-glass tubing.

3. After this procedure, wax was cleaned from the outside of the perfusion pipettes and the glued ends of the silicon-glass tubing were cut free and attached via conventional polyethylene tubing to the low volume 6-port manifolds. These manifolds were then connected to an automatic perfusion array gated via manifold valves to gravity-fed syringes containing the solutions of interest.
4. Perfusion was controlled using a digital output trigger interface in conjunction with the amplifier and the data acquisition system. Valves were driven by a custom-made valve driver to provide the necessary power (490 mW per valve) and spiked voltage jumps. 20 V for 10 ms then held at 10 V.
5. Optimal positioning of the pipette was practiced using solutions of different osmolarity to visualize the interface between solutions under phase contrast microscopy, *see* **Note 5**.
6. The solution exchange time of this perfusion system as measured using small, lifted cells was approximately 20 ms. Complete exchange of the perfused solutions to be applied via the application pipette was of the order of 1–2 s. The level of bath solution was kept constant using a vacuum driven glass suction pipette. This is important to avoid changes in recording pipette capacitive characteristics.

3.2.3 Solutions

1. The composition of the intracellular solution used when recording NMDA receptor-mediated currents from hippocampal neurons is given in Subheading 2.2, **item 14**. The absence of intracellular K^+ and the presence of intracellular TEA should block VAKCs. ATP and cAMP were included to decrease rundown although more elaborate ATP regenerating systems can be used for more problematic receptors such as neuronal nicotinic receptors [24].
2. The corresponding extracellular bath solution composition is given in Subheading 2.2, **item 15**. TTX was included at 0.35 μM in order to block VASCs, and D-serine was present at 10 μM in all extracellular solutions—this concentration being sufficient to saturate the glycine$_B$ site which should remain stable during the course of the experiments.

3.2.4 Cumulative Protocols

Cumulative protocols can be used for faster determination of concentration dependency of blockade and were shown to produce very similar IC_{50} determinations as kinetic protocols (*see* Table 1).

Table 1

Summary of the results from experiments measuring the antagonistic properties of various known NMDA receptor antagonists against NMDA-induced currents recorded from *Xenopus* oocytes and cultured hippocampal neurons

Substance	Cumulative IC_{50} (μM)	Cumulative hill	Kinetic IC_{50} (μM)	Kinetic hill	K_{on} (10^4 M^{-1} s^{-1})	K_{off} (s^{-1})	K_d (μM)	IC_{50} (0mV) (μM)	δ	β
Memantine	1.87 ± 0.17	0.80 ± 0.04	1.27 ± 0.08	0.93 ± 0.06	5.94 ± 0.35	0.125 ± 0.019	2.100 ± 0.450	17.35 ± 1.78	0.83 ± 0.04	0.08 ± 0.02
Neramexane	1.21 ± 0.04	0.88 ± 0.02	0.85 ± 0.05	0.81 ± 0.03	9.30 ± 0.90	0.120 ± 0.008	1.330 ± 0.200	16.99 ± 2.09	0.96 ± 0.04	0.10 ± 0.03
Ketamine	1.47 ± 0.22	0.82 ± 0.08	0.98 ± 0.09	1.01 ± 0.08	3.95 ± 0.50	0.070 ± 0.005	1.670 ± 0.320	13.16 ± 0.90	0.85 ± 0.05	0.08 ± 0.02
Phencyclidine	0.13 ± 0.01	0.78 ± 0.08	0.11 ± 0.03	0.70 ± 0.05	7.00 ± 0.19	0.010 ± 0.003	0.143 ± 0.004	1.26 ± 0.11	0.65 ± 0.04	0.10 ± 0.17
Dextromethorphan	1.95 ± 0.17	1.37 ± 0.15	2.11 ± 0.17	1.16 ± 0.10	1.74 ± 0.16	0.127 ± 0.027	7.470 ± 1.820	15.34 ± 1.51	0.84 ± 0.05	0.14 ± 0.02
Dextrorphan	0.34 ± 0.05	1.02 ± 0.10	0.36 ± 0.08	1.00 ± 0.17	19.90 ± 0.78	0.063 ± 0.004	0.317 ± 0.023	3.39 ± 0.70	0.61 ± 0.06	−0.01 ± 0.14
Mg^{2+}	N.C.	N.C.	37.62 ± 1.41	0.82 ± 0.02	11.28 ± 0.75	21.61 ± 0.60	191.6 ± 13.8	5,521 ± 261	0.95 ± 0.02	−0.32 ± 0.01

Only compounds produced by Merz Pharmaceuticals GmbH were tested using *Xenopus* oocytes. The hippocampal IC_{50} and Hill coefficient values and onset and offset kinetics are calculated from recordings in which the NMDA currents were allowed to recover between applications of various concentrations on antagonist. All data are shown in the form of: value ± S.E., N.C. = not calculated

In these protocols, 5–6 sequentially increasing concentrations (in a log 3 progression, e.g., 0.3, 1, 3, 10, and 30 μM) of memantine or other standard uncompetitive NMDA receptor antagonists are applied in a cumulative regime, each for 10–30 s, in the continuous presence of NMDA (200 μM) for 100–200 s and recovery is only recorded after the last and highest concentration. With such protocols, cells do not have to remain stable for such long recording durations. However, they give little useful information on the kinetics of block, and desensitization/rundown is more of a problem.

3.2.5 Kinetic Experiments

Kinetic experiments were performed by applying various single concentrations of memantine or standard uncompetitive NMDA receptor antagonists for 10–30 s in the continuous presence of NMDA (200 μM) for 30–120 s. When using this protocol, currents are allowed to recover after each application of antagonist, and onset and offset kinetics can be measured.

3.2.6 Voltage Dependency

Fractional block of currents by memantine (10 μM) at various holding potentials was used to determine the voltage dependency of this effect. The holding potential was changed every 120 s from −80 to +60 mV in 10 mV increments and NMDA (200 μM) was applied for 41 s at each holding potential. Memantine (10 μM) was applied for 11 s during each NMDA application period. During the recovery period, 15 s following the removal of memantine, neurones were clamped to +70 mV for 5 s in the continuing presence NMDA to facilitate complete recovery from antagonism. Similar experiments were performed with PCP, dextrorphan, and (+)MK-801 except that the application and recovery times had to be increased for these very slow channel blockers.

In order to subtract any residual VAC currents, mirror voltage-clamp (P5) protocols with smaller (20 %) voltage steps in the opposite direction were run between agonist/antagonist applications at each holding potential, e.g., the equivalent for a step from −90 to +70 mV (difference of +160 mV) was −90 to −122 mV (difference of −32 mV).

3.2.7 Analysis of Data

1. An analysis program was used for the quantification of individual current amplitudes and kinetics. A spreadsheet program was used to pool these data and a fitting program used to fit pharmacological equations and pooled kinetic K_{on} and K_{off} values. For all data points, the value given is the mean of results from 4 to 8 cells per concentration.
2. Rundown was usually not extreme for these NMDA receptor currents recorded from cultured hippocampal neurons (normally less than 10 % over a 1 min period in the continuous presence of agonist). However, in order to produce the most accurate assessment of potency, the analysis nevertheless corrected for extrapolated linear current rundown—this was

increasingly important for high affinity compounds where the duration of the agonist exposure was prolonged to assure that low concentrations of antagonist reached steady-state blockade. Antagonism of NMDA receptor-mediated currents was measured as the magnitude of the steady state blocked current as a percentage of the control current. For non-kinetic antagonistic protocols with five cumulatively increasing concentrations of antagonist, the control current for each antagonist concentration was calculated by a linear projection from the steady-state current before and after antagonist application, i.e., for the first antagonist concentration control current = 0.85 × current before antagonist + 0.15 × recovery current. For the second antagonist concentration these same currents were multiplied by 0.7 and 0.3 respectively, and so on for all concentrations. For kinetic protocols, single concentrations of antagonist were applied in the continuing presence of agonist, and the control current was taken as the mean of the steady-state current before and after antagonist application.

3. Potency of compounds was assessed by plotting the mean percentage current magnitude, calculated with standard error (S.E.) against antagonist concentration and a curve was fit using the logistic equation for which the variable parameters IC_{50} and Hill coefficient (n) were free and the range and background values were normally set to 100 and 0 respectively;

$$\%\ \text{current} = \frac{\text{range}}{1+\left(\frac{[\text{antagonist}]}{IC_{50}}\right)^{n}} + \text{background}. \quad (1)$$

4. For kinetic protocols, exponential fits to both onset and offset kinetics were made using an analysis program and most onset and offset responses were often well fit by a single exponential function, where a and b are current amplitudes, and c represents the time constant, τ;

$$\text{amplitude} = a + b \cdot e^{(\text{time}/c)}. \quad (2)$$

When double exponential fits described the data better, these were fit with the following equation;

$$\text{amplitude} = a + b \cdot e^{(\text{time}/c)} + d \cdot e^{(\text{time}/f)}. \quad (3)$$

where a, b, and d are current amplitudes and c and f are time constants (c represents τ_{fast} and f is τ_{slow}). These were then integrated to produce a single time constant (t_{combined}) according to Eq. 4.

$$t_{\text{combined}} = \left(c \cdot \frac{b}{b+d}\right) + \left(f \cdot \frac{d}{b+d}\right). \quad (4)$$

The rational for this weighting procedure was to allow for simple comparison between the rate kinetics produced by the present investigation, and also with previous data for memantine and other channel blockers for which similar techniques or simple single exponentials were used to fit the data. Another reason for this weighting was to allow for a simple calculation of calculated K_d (*see* Eq. 6).

5. Mean values of the time constant (τ) were calculated for each concentration of antagonist and $1/\tau$ was plot against concentration. These data were fit using linear equations, where m is the gradient of the line, which corresponds to K_{on} for the onset fit, and g is the intercept at the y axis, which corresponds to K_{off} for the offset rate fit;

$$y = m \cdot x + g. \tag{5}$$

All compounds tested showed concentration-dependent open channel blocking kinetics, whereas the offset rate was essentially concentration-independent. As expected, the fit of τ_{on} against concentration intercepted the y axis at similar values to the fit for τ_{off} against concentration. Calculation of the ratio K_{off}/K_{on} was used to reveal an apparent K_d which was then compared to the IC_{50} calculated at equilibrium.

$$K_d = K_{off} / K_{on}. \tag{6}$$

6. To assess voltage dependency, single concentrations of antagonist were applied to currents in the plateau phase and then the cell was allowed to recover before stimulation was repeated at a more positive holding potential (holding potential was increased by increments of +10 mV each time). Blockade was expressed as a percentage of the mean control current recorded before and after application of antagonist, as in the kinetic experiments as described earlier in this section. The pooled data were then fit by the following equation, where $IC_{50}(0)$ is the IC_{50} at 0 mV, β is the fraction of voltage-independent sites, δ is the fraction of the electric field sensed by the voltage-dependent site, and all other parameters have their normal meaning (z, valency; F, Faraday's constant; R, universal gas constant; and T, absolute temperature):

$$\text{fractional current} = \frac{1-\beta}{1+\dfrac{[\text{antagonist}]}{IC_{50}(0)\cdot e^{(-z\delta FV/RT)}}}. \tag{7}$$

3.2.8 Results

The application of NMDA (200 μM) in the continuous presence of D-serine (10 μM) to hippocampal neurons evoked currents that were attenuated by all of the uncompetitive NMDA receptor antagonists used in this study. The data are summarized in Table 1.

3.2.9 Potency

Initial experiments involved application of five sequentially increasing concentrations of antagonist. However, kinetic measures of both onset and offset rates for each concentration of antagonist cannot be determined using this application protocol, so the kinetic protocol was employed which involved application of antagonist in the continuous presence of NMDA and D-serine with a period of current recovery between each antagonist concentration. This protocol could also be considered to give a more accurate measure of potency, since the magnitude of blockade by each concentration is assessed compared to the current before and after each application of antagonist, but the difference between values determined from each set of results is clearly only slight (*see* Table 1).

Memantine antagonized NMDA currents concentration-dependently with an IC_{50} value of 1.87 ± 0.17 μM and a Hill coefficient of 0.80 ± 0.04 when the cumulative protocol was used (*see* Fig. 1). From the data produced using the kinetic protocol, as shown in Fig. 1b, a new concentration–response curve was produced, and the IC_{50} value calculated from this curve was only marginally lower (1.27 ± 0.08 μM) although the difference does reach a level of significance ($p = 0.019$, $t = 3.19$; Students t test using raw data). The Hill coefficient is slightly higher (0.93 ± 0.06) than that derived from the earlier data, but values derived using these two protocols are not significantly different ($p > 0.05$; Students t test using raw data).

3.2.10 Kinetics

Simple analysis of the kinetic data show memantine to have concentration-dependent blocking and concentration-independent unblocking kinetics ($K_{on} = 5.94 \pm 0.35 \times 10^4$ M^{-1} s^{-1} and $K_{off} = 0.125 \pm 0.019$ s^{-1}). From these values, K_d can be calculated as 2.10 ± 0.45 μM according to the equation $K_d = K_{off}/K_{on}$, which correlates well with the IC_{50} values calculated by both of the concentration–response curves.

However, memantine did indeed sometimes show double exponential kinetics (*see* Fig. 2). The onset and offset kinetics of blockade following concentration jumps with memantine

Fig. 1 (continued) and the curve fit using Eq. 1 (*filled circles*). (**b**) Attenuation of currents by memantine (0.3–30 μM) with current recovery between antagonist applications. Mean control-corrected percentage blockade (±S.E.) was plotted against antagonist concentration on a logarithmic scale and the curve fit using Eq. 1 (*open circles*). The onset and offset kinetics followed a single or double exponential time course. Mean $1/\tau$ values (±S.E.) were plotted against antagonist concentration and fit according to Eq. 5 from which K_{on} and K_{off} were read. These two values were then used to calculate K_d. (**c**) Voltage dependence of antagonism of the NMDA currents by a single concentration of memantine (10 μM). The initial holding potential of −80 mV was increased for each stimulation in increments of 10 mV until it reached +60 mV. In the example recording shown, it can be seen that a voltage step to +70 mV was added during current recovery but this was not deemed necessary for further experiments. The mean percentage blockade of control currents (±S.E.) was plotted against holding potential and fit using Eq. 7

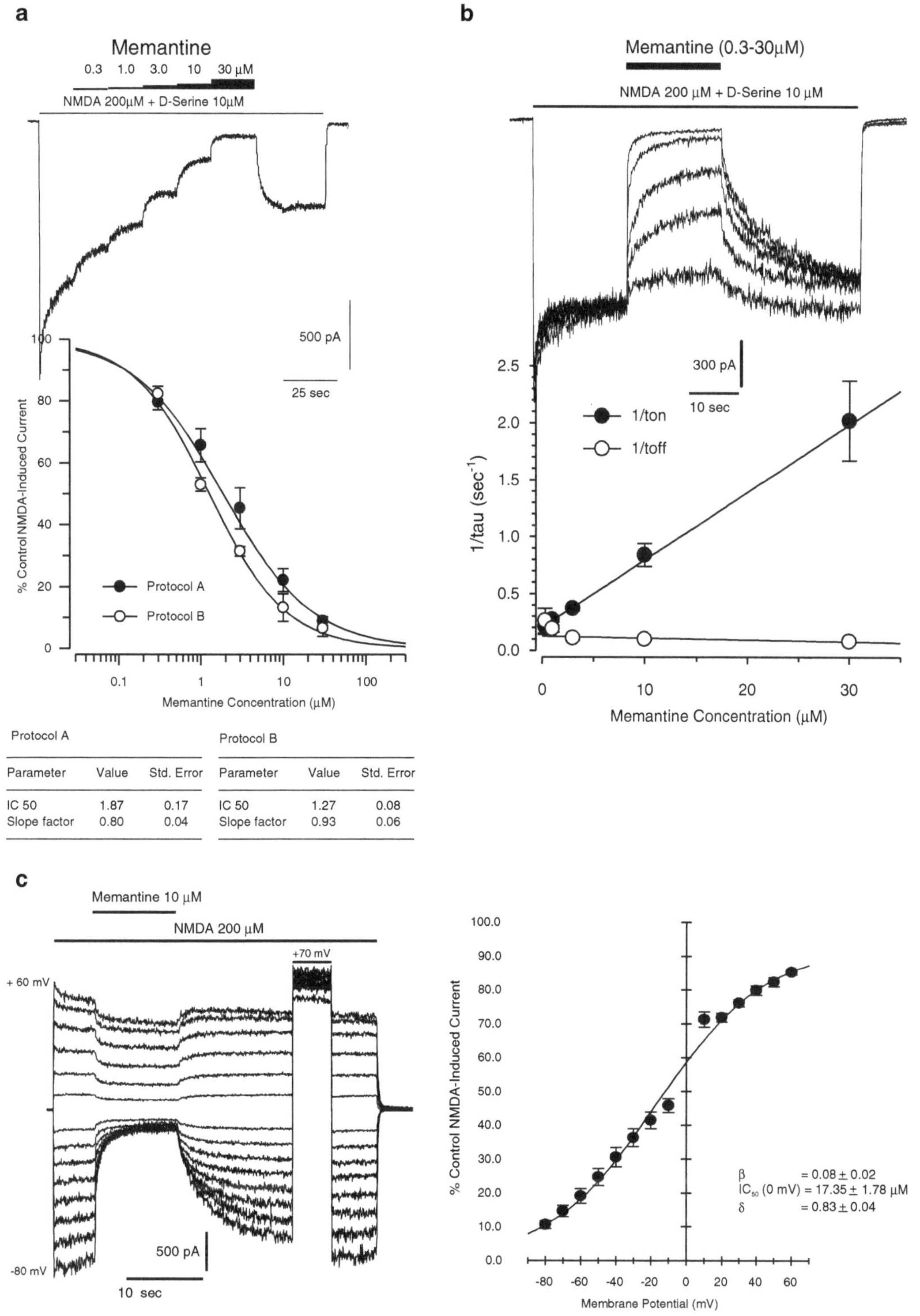

Protocol A

Parameter	Value	Std. Error
IC 50	1.87	0.17
Slope factor	0.80	0.04

Protocol B

Parameter	Value	Std. Error
IC 50	1.27	0.08
Slope factor	0.93	0.06

Fig. 1 Antagonism of NMDA currents recorded from cultured hippocampal neurons by memantine. (**a**) Stepwise application of increasing concentrations of antagonist during constant application of agonists. Mean rundown-corrected percentage blockade (±S.E.) was plotted against antagonist concentration on a logarithmic scale

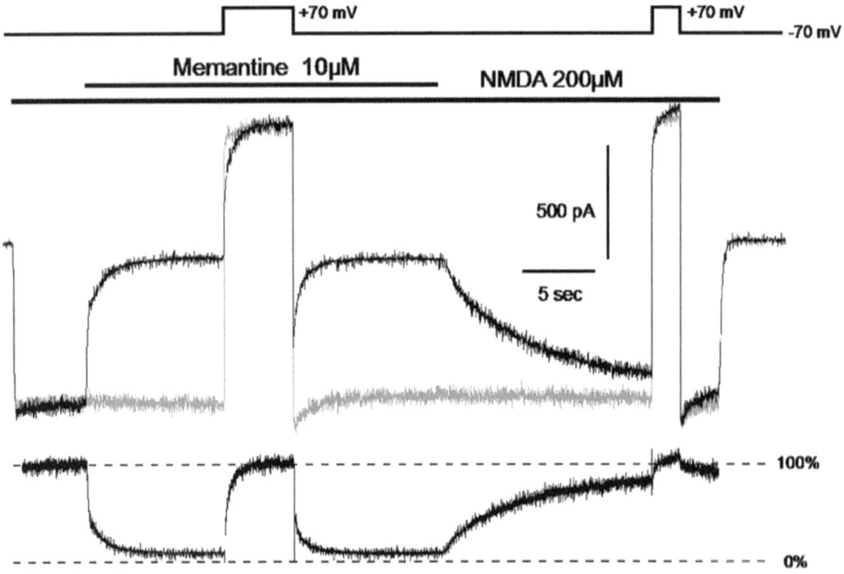

Fig. 2 Double exponential kinetics of blockade by memantine in cultured hippocampal neurones. Traces are averages of ten recordings and residual capacitive artifacts were subtracted. The bottom trace was constructed by basing recordings in the presence of memantine (*black*) as a percentage of those in the absence of memantine (*grey*). This was then used to fit the kinetics of the data. NB: neurones were shortly clamped to +70 mV during the memantine washout phase to facilitate recovery

(10 μM at −70 mV) showed double exponential kinetics: τ_{on} fast 86.9 ± 6.3 ms (64.7 %); τ_{on} slow 1,383 ± 122 ms; τ_{off} fast 834 ± 321 ms (22.9 %); τ_{off} slow 4,795 ± 921 ms. It should be noted that the kinetics of blockade by memantine in these experiments with D-SERINE (10 μM at −70 mV) were somewhat faster than those using glycine at a sub saturating concentration (1 μM at −70 mV cf. [8]). This is not surprising as channel gating kinetics also influence the kinetics of open channel blockade. The relief of blockade following voltage jumps to +70 mV in the continuous presence of memantine (10 μM) also showed rapid, double exponential kinetics: τ_{off} fast 98.7 ± 38.1 ms (43.9 %) a; τ_{off} slow 725 ± 122 ms. Reblock by memantine following jumps back to −70 mV was extremely fast: τ_{on} fast 7.7 ± 2.2 ms (41.6 %); τ_{on} slow 285 ± 41 ms. The extremely rapid reblock by memantine and additional minor effects of such voltage-steps on the kinetics of NMDA currents in the absence of antagonist necessitated fitting responses with memantine as the ratio between responses in the presence/absence of memantine. The difference in onset kinetics between the concentration clamp and voltage-step protocols may be explained by a second, extracellular holding site for memantine on the NMDA receptor: *see* [3]. Alternatively, it may just indicate that the concentration clamp was not fast enough.

3.2.11 Voltage Dependency

Blockade of NMDA receptor-mediated currents by memantine was voltage-dependent, as illustrated by the δ value of 0.83 ± 0.04, and an approximately tenfold higher IC_{50} value at 0 mV compared to −70 mV ($IC_{50(0mV)} = 17.35 \pm 1.78$ μM) (*see* Fig. 1c). The proportion of the voltage-independent sites, β, was very small (0.08 ± 0.02). To ensure that memantine had completely left the channel and the current was fully recovered, a 5 s voltage step to +70 mV was added during the current recovery, as can been seen in the trace shown, but this was not considered necessary for subsequent experiments with moderate affinity blockers which show similar fast kinetics. However, such procedures can be very useful when determining the voltage dependency of more potent, slower blockers such as (+)MK-801 (*see* [3]).

Interpretation/Application of Electrophysiological Data

The reason for the better therapeutic safety of memantine compared to other channel blockers such as (+)MK-801 and phencyclidine is still a matter of debate, and data such as those presented in this chapter (*see* Table 1) have been utilized to support several theories. The interpretation of these electrophysiological data and how they have been used to form hypotheses concerning actions of compounds in vivo are described in the following:

1. Memantine and other well tolerated open channel blockers such as amantadine, dextromethorphan, (*S*)-*alpha*-phenyl-2-pyridine-ethanamine dihydrochloride (ARL 15896AR) and (+-)-5-aminocarbonyl-10,11-dihydro-5H-dibenzo[a,d] [a,d] cyclohepten-5,10-imine (ADCI) show much faster open channel blocking/unblocking kinetics than compounds burdened with negative psychotropic effects such as (+)MK-801 or phencyclidine [3, 5, 7, 25–27]. The kinetics of (+)MK-801 and phencyclidine are too slow to allow them to leave the channel upon depolarization, which is reflected in apparently weaker voltage dependency. These two parameters are directly related to affinity, with lower affinity compounds such as memantine showing faster kinetics and apparently stronger voltage dependency, as reflected in estimated δ value [8]. The δ value describes the percentage of the transmembrane field the drug experiences when blocking the NMDA receptor channel [3]. The unblocking rate of memantine in the continuous presence of this antagonist following depolarizing voltage steps is very rapid and well within the time course of a NMDA receptor-mediated excitatory postsynaptic potential (EPSP).
2. Memantine blocks and unblocks open NMDA receptor channels with double exponential kinetics. The amplitude and speed of the fast component of block increases with memantine concentration. In contrast, the speed of fast unblock remains constant but the amplitude decreases with memantine concentration [11, 12, 14, 15, 28]. Moreover, the

predominant effect of depolarization is to increase dramatically the weight of the faster recovery time-constant [10, 11, 28]. These data indicate that memantine binds to at least two sites within the channel [14, 15].

3. Both Lipton's and Rogawski's groups have proposed that the ability of low affinity open channel blockers to gain rapid access to the NMDA receptor channel is important in determining their therapeutic safety in ischæmia and epilepsy [7, 25, 26]. However, this hypothesis alone cannot explain the better therapeutic profile of memantine as, even if receptors are only blocked following pathological activation, they would then remain blocked in the continuous presence of memantine, and therefore be unavailable for subsequent physiological activation. Physiologically, NMDA receptors are transiently activated by mM concentrations of glutamate [29] following strong depolarization of the postsynaptic membrane which rapidly relieves their voltage-dependent blockade by Mg^{2+} [30], whereas during pathological activation, NMDA receptors are activated by lower concentrations of glutamate but for much longer periods of time [31–36]. Unfortunately, the voltage dependency of the divalent cation Mg^{2+} is so pronounced that it also leaves the NMDA channel upon moderate depolarization under pathological conditions. Although uncompetitive antagonists also block the NMDA receptor channel, high affinity compounds such as (+)MK-801 have much slower unblocking kinetics than Mg^{2+} and less pronounced functional voltage dependency and are therefore unable to leave the channel within the time course of a normal NMDA receptor-mediated excitatory post synaptic potential. As a result, (+)MK-801 blocks both the pathological and physiological activation of NMDA receptors [3].

4. We were the first to suggest that the combination of fast offset kinetics and strong voltage dependency allow memantine to rapidly leave the NMDA channel upon transient physiological activation by mM concentrations of synaptic glutamate but block the sustained activation by μM concentrations of glutamate under moderate pathological conditions [5, 8, 9]. This hypothesis is further supported by the fact that although the predominant component of offset kinetic at near resting membrane potentials is still too slow to allow for synaptic activation—i.e., around 5 s—the relief of blockade in the continuous presence of memantine upon depolarization is much faster due to an increase in the weight of the faster recovery time-constant [11, 14, 15, 28]. These kinetics are likely to be even faster in vivo due to higher temperatures [37]. Furthermore, the rate of recovery from memantine blockade is dependent on the open probability of NMDA channels [13]

and therefore would be faster in the presence of higher, synaptic concentrations of glutamate [29]. Given the crucial role of NMDA receptors in neuronal plasticity, the fact that memantine improves cognition and neuronal plasticity seems paradoxical at the first glance. It should be realized, however, that Mg^{2+} is an endogenous NMDA channel blocker and its removal from the channel leads both to an impairment in neuronal plasticity [38, 39] as well as neuronal death [40]. Any dysfunction of postsynaptic neurons leading to weakened blockade by Mg^{2+}, e.g., due to partial depolarization as a consequence of an energy deficit, may trigger such functional (plasticity) and structural (neuronal loss) deficits [4, 41]. Because memantine is more potent and slightly less voltage-dependent than Mg^{2+} it may thus serve as a more effective surrogate for Mg^{2+} [5]. As a result of its somewhat less pronounced voltage dependency, memantine is more effective than Mg^{2+} in blocking tonic pathological activation of NMDA receptors at moderately depolarized membrane potentials. However, following strong synaptic activation, memantine like Mg^{2+}, can leave the NMDA receptor channel with voltage-dependent, fast unblocking kinetics. In turn, memantine suppresses synaptic noise but allows the relevant physiological synaptic signal to be detected. This provides both neuroprotection and symptomatic restoration of synaptic plasticity by one and the same mechanism [3, 4]. Antagonists that have "too high" affinity for the channel or "too little" voltage dependence, such as dizocilpine ((+)MK-801), thus produce numerous side effects since they essentially act as an irreversible plug of the NMDA receptor channel and block both pathological and physiological function.

5. A moderate potentiation of NMDA-induced outward currents by memantine at positive potentials in hippocampal neurones has also been reported [10] (data not shown here). This could be related to the finding that Mg^{2+} and ketamine increased NMDA receptor-mediated currents in cultured mouse hippocampal neurones and HEK-293 cells expressing NMDA $\xi1/\varepsilon2$ receptors by increasing the affinity of the glycine$_B$ site for the co agonist [42]. Such facilitation would be predicted to be more pronounced with lower concentrations of glycine. This could have important functional implications as the differentiation between block of NMDA receptors at near resting membrane potentials and the lesser block following strong synaptic depolarization to around −20 mV would be enhanced by such a mechanism and would facilitate the ability of memantine to differentiate between pathological and physiological activation of NMDA receptors. Such a potentiation was not seen in the present study, most likely due to the use of saturating concentrations of D-serine.

6. It should also be noted that a third theory was proposed in an excellent paper by [12] *see* [6] for review and supported by data from [15] (not shown here). The data indicate that memantine and amantadine appear to have a lesser tendency to be trapped in NMDA receptor channels than do phencyclidine or (+)MK-801. This difference was attributed to the ability of channel blockers to increase the affinity of NMDA receptors for agonist and the faster kinetics of the aminoadamantanes. Receptors blocked by memantine retain agonist and thereby open and release memantine following removal of both agonist and memantine from the extracellular solution—*see* also [13]. This partial trapping is less pronounced for higher affinity compounds as their slower unblocking kinetics do not allow them to leave the channel quickly enough following agonist removal. The relief of block in the absence of agonist was greater in the experiments of Sobolevsky et al. [15]. This however, may have been due to the use of higher concentrations of aspartate which would have increased the proportion of liganded receptors at the time of agonist/antagonist removal. Blanpied et al. [12] proposed that partial trapping may underlie the better therapeutic profile of memantine as a proportion of channels—around 15–20 %—would always unblock in the absence of agonist and thereby be available for subsequent physiological activation. In other words, the antagonism by memantine is like that of a low intrinsic activity partial agonist in that it doesn't cause 100 % blockade of NMDA receptors. Although this theory is attractive, it is only relevant for the therapeutic situation if partial trapping also occurs in the continuous presence of memantine. This point had not been addressed previously. This prompted us to perform experiments on partial trapping in the continuing presence of memantine and the results of these studies were very similar to those reported by Blanpied et al. [12], i.e., around 15 % of channels released memantine following agonist removal [3]. However, although this theory can be used to explain the therapeutic tolerability of memantine, it provides no mechanism of action for the symptomatic effects observed in AD patients.

4 Notes

1. These dishes were precoated by treating dishes overnight at 37 °C with poly-DL-ornithine, washing twice with PBS and then incubating with laminin solution overnight at 37 °C. Excess solution was aspirated and dishes washed with PBS followed by the cell medium before cell plating.
2. The kinetics of drug–receptor interactions are highly dependent on temperature.

3. Offset compensation was performed for each open pipette in order to ensure that the command potential is equal to the membrane potential.
4. Liquid junction potential was measured by filling one perfusion chamber with intracellular solution as the reference, and the other chamber with extracellular solution. For the solutions used for recording from the hippocampal neurons, the liquid junction potential was measured to be 3.4 mV.
5. The best angle was found to be 45 % and care was taken to keep the lower edge of the application pipette tip parallel to, and as close as possible to, the bottom of the dish without scratching the plastic during the stepping motor movement. The lower edge of the application pipette was positioned with the start channel centered some 150–250 μM from the neuron of interest.

References

1. Reisberg B, Doody R, Stoffler A et al (2003) Memantine in moderate-to-severe Alzheimer's disease. N Engl J Med 348:1333–1341
2. Tariot PN, Farlow MR, Grossberg GT et al (2004) Memantine treatment in patients with moderate to severe Alzheimer disease already receiving donepezil: a randomized controlled trial. JAMA 291:317–324
3. Parsons CG, Danysz W, Quack G (1999) Memantine is a clinically well tolerated N-methyl-D-aspartate (NMDA) receptor antagonist - a review of preclinical data. Neuropharmacology 38:735–767
4. Danysz W, Parsons CG (2003) The NMDA receptor antagonist memantine as a symptomatological and neuroprotective treatment for Alzheimer's disease preclinical evidence. Int J Geriatr Psychiatry 18:S23–S32
5. Parsons CG, Gruner R, Rozental J et al (1993) Patch clamp studies on the kinetics and selectivity of N-methyl-D-aspartate receptor antagonism by memantine (1-amino-3,5-dimethyladamantan). Neuropharmacology 32:1337–1350
6. Johnson JW, Kotermanski SE (2006) Mechanism of action of memantine. Curr Opin Pharmacol 6:61–67
7. Rogawski MA (1993) Therapeutic potential of excitatory amino acid antagonists - channel blockers and 2,3-benzodiazepines. Trends Pharmacol Sci 14:325–331
8. Parsons CG, Quack G, Bresink I et al (1995) Comparison of the potency, kinetics and voltage-dependency of a series of uncompetitive NMDA receptor antagonists in vitro with anticonvulsive and motor impairment activity in vivo. Neuropharmacology 34:1239–1258
9. Parsons CG, Panchenko VA, Pinchenko VO et al (1996) Comparative patch-clamp studies with freshly dissociated rat hippocampal and striatal neurons on the NMDA receptor antagonistic effects of amantadine and memantine. Eur J Neurosci 8:446–454
10. Parsons CG, Hartmann S, Spielmanns P (1998) Budipine is a low affinity, N-methyl-D-aspartate receptor antagonist: patch clamp studies in cultured striatal, hippocampal, cortical and superior colliculus neurones. Neuropharmacology 37:719–727
11. Bresink I, Benke TA, Collett VJ et al (1996) Effects of memantine on recombinant rat NMDA receptors expressed in HEK 293 cells. Br J Pharmacol 119:195–204
12. Blanpied TA, Boeckman FA, Aizenman E et al (1997) Trapping channel block of NMDA-activated responses by amantadine and memantine. J Neurophysiol 77:309–323
13. Chen HS, Lipton SA (1997) Mechanism of memantine block of NMDA-activated channels in rat retinal ganglion cells: uncompetitive antagonism. J Physiol 499:27–46
14. Sobolevsky AI, Koshelev SG, Khodorov BI (1998) Interaction of memantine and amantadine with agonist-unbound NMDA-receptor channels in acutely isolated rat hippocampal neurons. J Physiol 512:47–60
15. Sobolevsky A, Koshelev S (1998) Two blocking sites of amino-adamantane derivatives in open N-methyl-D- aspartate channels. Biophys J 74:1305–1319

16. Grantyn R, Lux HD (1988) Similarity and mutual exclusion of NMDA- and proton-activated transient Na<+>-currents in rat tectal neurons. Neurosci Lett 89:198–203
17. Clark GD, Clifford DB, Zorumski CF (1990) The effect of agonist concentration, membrane voltage and calcium on N-methyl-D-aspartate receptor desensitization. Neuroscience 39: 787–797
18. Zilberter Y, Uteshev V, Sokolova S et al (1991) Desensitization of N-methyl-D-aspartate receptors in neurons dissociated from adult rat hippocampus. Mol Pharmacol 40:337–341
19. Johnson JW, Ascher P (1987) Glycine potentiates the NMDA response in cultured mouse brain neurons. Nature 325:529–531
20. Parsons CG, Zong XG, Lux HD (1993) Whole cell and single channel analysis of the kinetics of glycine-sensitive N-methyl-D-aspartate receptor desensitization. Br J Pharmacol 109:213–221
21. Hashimoto A, Oka T (1997) Free D-aspartate and D-serine in the mammalian brain and periphery. Prog Neurobiol 52:325–353
22. Hamill OP, Marty A, Neher E et al (1981) Improved patch-clamp techniques for high-resolution current recording from cells and cell-free membrane patches. Pflugers Arch 391:85–100
23. Sigworth FJ, Affolter H, Neher E (1995) Design of the EPC-9, a computer-controlled patch-clamp amplifier. 2. Software. J Neurosci Methods 56:203–215
24. Albuquerque EX, Pereira EFR, Castro NG et al (1995) Nicotinic receptor function in the mammalian central nervous system. Ann N Y Acad Sci 757:48–72
25. Rogawski MA, Yamaguchi SI, Jones SM et al (1991) Anticonvulsant Activity of the Low-Affinity Uncompetitive N-Methyl-D-Aspartate Antagonist (+/-)-5-Aminocarbonyl-10,11-dihydro-5H-dibenzo<a, d>cyclohepten-5, 10-imine (ADCI) - comparison with the Structural Analogs Dizocilpine (mK-801) and Carbamazepine. J Pharmacol Exp Ther 259: 30–37
26. Chen HSV, Pellegrini JW, Aggarwal SK et al (1992) Open-channel block of N-methyl-D-aspartate (NMDA) responses by memantine - therapeutic advantage against NMDA receptor-mediated neurotoxicity. J Neurosci 12:4427–4436
27. Black M, Lanthorn T, Small D et al (1996) Study of potency, kinetics of block and toxicity of NMDA receptor antagonists using fura-2. Eur J Pharmacol 317:377–381
28. Frankiewicz T, Potier B, Bashir ZI et al (1996) Effects of memantine and MK-801 on NMDA-induced currents in cultured neurones and on synaptic transmission and LTP in area CA1 of rat hippocampal slices. Br J Pharmacol 117:689–697
29. Clements JD, Lester RAJ, Tong G et al (1992) The time course of glutamate in the synaptic cleft. Science 258:1498–1501
30. Nowak L, Bregestovski P, Ascher P et al (1984) Magnesium gates glutamate-activated channels in mouse central neurons. Nature 307: 462–465
31. Benveniste H, Drejer J, Schusboe A et al (1984) Elevation of the extracellular concentrations of glutamate and aspartate in rat hippocampus during transient cerebral ischemia monitored by intracerebral microdialysis. J Neurochem 43:1369–1374
32. Andine P, Sandberg M, Bagenholm R et al (1991) Intra- and extracellular changes of amino acids in the cerebral cortex of the neonatal rat during hypoxic-ischemia. Dev Brain Res 64:115–120
33. Globus MYT, Busto R, Martinez E et al (1991) Comparative effect of transient global ischemia on extracellular levels of glutamate, glycine, and gamma-aminobutyric acid in vulnerable and nonvulnerable brain regions in the rat. J Neurochem 57:470–478
34. Globus MYT, Ginsberg MD, Busto R (1991) Excitotoxic index - a biochemical marker of selective vulnerability. Neurosci Lett 127:39–42
35. Buisson A, Callebert J, Mathieu E et al (1992) Striatal protection induced by lesioning the substantia-nigra of rats subjected to focal ischemia. J Neurochem 59:1153–1157
36. Mitani A, Andou Y, Kataoka K (1992) Selective vulnerability of hippocampal CA1 neurons cannot be explained in terms of an increase in glutamate concentration during ischemia in the gerbil: brain microdialysis study. Neuroscience 48:307–313
37. Davies SN, Martin D, Millar JD et al (1988) Differences in results from in vivo and in vitro studies on the use-dependency of N-methyl-aspartate antagonism by MK-801 and other phencyclidine receptor ligands. Eur J Pharmacol 145:141–152
38. Coan EJ, Irving AJ, Collingridge GL (1989) Low-frequency activation of the NMDA receptor system can prevent the induction of LTP. Neurosci Lett 105:205–210
39. Frankiewicz T, Parsons CG (1999) Memantine restores long term potentiation impaired by tonic N-methyl-D-aspartate (NMDA)

receptor activation following reduction of Mg^{2+} in hippocampal slices. Neuropharmacology 38:1253–1259
40. Furukawa Y, Okada M, Akaike N et al (2000) Reduction of voltage-dependent magnesium block of N-methyl-D-aspartate receptor-mediated current by in vivo axonal injury. Neuroscience 96:385–392
41. Rogawski MA, Wenk GL (2003) The neuropharmacological basis for the use of memantine in the treatment of Alzheimer's disease. CNS Drug Rev 9:275–308
42. Wang LY, Macdonald JF (1995) Modulation by magnesium of the affinity of NMDA receptors for glycine in murine hippocampal neurones. J Physiol 486:83–95

Chapter 4

State-of-the-Art Automated Patch Clamp: Heat Activation, Action Potentials, and High Throughput in Ion Channel Screening

Sonja Stoelzle-Feix

Abstract

A successful robotic approach of the patch clamp technique is based on planar patch clamp chips where a glass pipette, as used in conventional patch clamping, is replaced by a thin planar glass sheet with a small hole in the middle. Automated patch clamp (APC) systems utilizing this chip design offer higher throughput capabilities and ease of use and thus have become common in basic research, drug development, and safety screening. Further development of existing devices and introduction of new systems widen the range of possible experiments and increase throughput. Here, two features with different areas of applications that meet the needs of drug discovery researchers and basic researchers alike are described. The utilized system is a medium throughput APC device capable of recording up to eight cells simultaneously. The temperature control capability and the possibility to perform recordings not only in the voltage clamp but also in the current clamp mode are described in detail. Since eight recordings can be generated in parallel without compromising data quality, reliable and cost-effective and time-effective screening of compounds against ion channels using voltage clamp and current clamp electrophysiology can be performed.

Key words Automated patch clamp, Action potential, Current clamp, Stem cell-derived cardiomyocytes, High throughput, Ion channel screening

1 Introduction

The traditional manual patch clamp technique using glass pipettes was developed by Erwin Neher and Bert Sakmann [1]. However, conventional patch clamp is notoriously low throughput and requires a highly specialized operator because it is technically demanding. Automation of the patch clamp technique increases throughput and ease of use [2–4]. Providers of automated patch clamp platforms are pushing the boundaries by continually improving existing platforms and implementing new and innovative features; some of the available devices have even revolutionized the field [5].

Automated electrophysiology is now commonplace in many laboratories and is not limited to the pharmaceutical industry.

Marzia Martina and Stefano Taverna (eds.), *Patch-Clamp Methods and Protocols*, Methods in Molecular Biology, vol. 1183, DOI 10.1007/978-1-4939-1096-0_4,

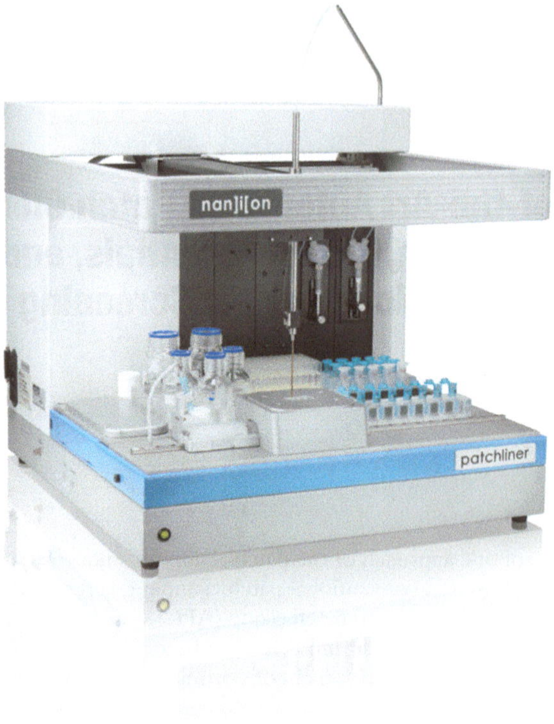

Fig. 1 The automated planar patch-clamp platform Patchliner is a benchtop device and consists of the pipetting robot, amplifiers, and a computer. Three NPC®-16 patch-clamp chips can be loaded on the chip wagon for one session

Universities and research institutions also implemented the use of APC devices [6, 7], combining increased throughput, sophistication, and high quality. Available platforms are NPC®-16 Patchliner® and SyncroPatch® 96 (Nanion Technologies GmbH, Munich), CytoPatch™ (Cytocentrics AG, Rostock); PatchXpress® 7000A, IonWorks® Quattro, and IonWorks Barracuda™, (Molecular Devices, LLC); Dynaflow® HT (Cellectricon AB, Mölndal); QPatch HT (Sophion A/S, Copenhagen); IonFlux HT (Fluxion Bioscience Inc, USA), which have in common the capability to generate patch clamp recordings with increased throughput. In this chapter, procedures and experimental protocols of Nanion's NPC®-16 Patchliner® are described.

The NPC®-16 Patchliner® (Fig. 1) is a successful higher-throughput APC device because it is particularly flexible with regard to experimental protocols. The system contains a well controlled and fast solution exchange, and its temperature control allows for measurements in a more physiologically relevant environment. Furthermore, the availability of the current clamp mode enables the investigation of not only standard cell lines in the

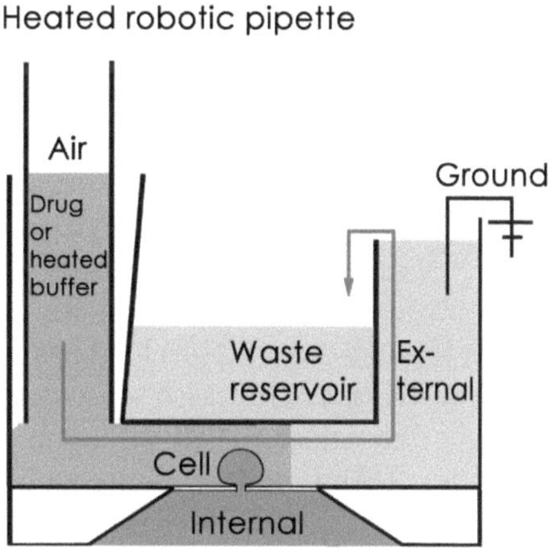

Fig. 2 Image of one chamber of a NPC®-16 patch-clamp chip with a heatable pipette, extracellular and intracellular microfluidic chambers, waste reservoir as well as the position of the planar glass chip with the aperture for the cell

voltage clamp mode but also action potential recordings of excitable cells such as stem cell-derived cardiomyocytes [8]. The planar patch clamp approach using the Patchliner requires vital and homogeneous cells in suspension. Since cell catch on the patch clamp chip is a random process, the manual cell preparation has to be optimal in order to receive a cell preparation with healthy cells and thus achieving the possibility of high success rates. Cell clusters and pieces of slices and tissue are not suitable. The following steps after cell preparation and allocation in the *cell hotel* (*see* **Note 1**) involves automatic liquid handling including delivery of cells, solutions and compounds onto the planar patch clamp chip which is a borosilicate sheet of glass with a micron-sized aperture in the center. The glass is embedded in microfluidic chambers within a chip cartridge (*see* Fig. 2). A single cartridge contains 16 individual microfluidic chambers, 8 of these are used simultaneously for each experiment. Once the patch clamp solutions and cells have been robotically pipetted into the external and internal microfluidic channels, negative pressure is applied from underneath each chamber to attract a single cell onto each chamber's aperture. Pressure is applied to each chamber individually to obtain maximum success rates when capturing cells on the aperture, forming seals and establishing the required patch configuration (cell attached, whole cell, or perforated whole cell).

Standard cell lines such as HEK or CHO cells [6], neuroblastoma cell lines [9–11], primary cells [6], and excitable cells such as stem cell-derived cardiomyocytes [8] have been successfully used in the voltage-clamp and current-clamp modes with the view to finding new drug candidates.

This chapter describes in more detail the application of automated planar patch clamp protocols involving the temperature responsive TRPV3 channel, using fast solution application protocols combined with temperature control.

The ligand-gated transient receptor potential cation channel, subfamily V, member 3 (TRPV3), exhibiting 43 % sequence identity to TRPV1, is a non-selective cation channel [12–14]. TRPV3 is found in keratinocytes in skin but has been also detected immunologically in the CNS where it is suggested to be co-localized with TRPV1. Given its threshold for temperature activation of >34 °C [12–14] it has been speculated that TRPV3 acts in cooperation with sensory neurons to perceive warmth. In addition to heat, TRPV3 can also be activated by the ligand 2-Aminoethoxydiphenyl borate (2-APB), and furthermore, can be inhibited by the nonspecific blocker ruthenium red [15]. TRPV channels play an important role in chronic pain and therefore are receiving more and more attention as potential therapeutic candidates for chronic pain therapies.

To overcome the hurdles in working with acutely dissociated cardiomyocytes from human donors or animals, the use of stem cell-derived cardiomyocytes on an APC is an attainable option [8] for safety screening of larger-compound libraries. This approach offers the key advantage that drugs and their inhibitory effect on relevant cardiac ion channels can be analyzed not only in the voltage clamp mode, but their impact on the AP shape can be seen in the same experiment. Overexpressing ion channels in, for example, HEK cells can be useful for determining ion channel specificity of a particular compound, whereas for example human induced pluripotent stem cell-derived cardiomyocytes (hiPS-derived cardiomyocytes) give an overview of all cardiac relevant ion channels. Thus, stem cell-derived cardiomyocytes could have an advantage over heterologous expression systems because of their greater physiological relevance as multiple ion channels are expressed simultaneously in an environment more representative of the native situation. Furthermore, with stem cell-derived cardiomyocytes being readily accessible compared to freshly isolated cardiomyocytes, and at the same time the automated patch clamp assay being scalable, this opens the way for the pharmaceutical industry to perform high-throughput screening of new drug candidates with regard to cardiotoxicity [8]. Up to 48 cells can be analyzed without user intervention making the Patchliner® also an ideal tool for safety pharmacology.

2 Materials

2.1 Equipment and Reagents

1. NPC®-16 Patchliner Quattro/Octo with PatchControlHT software, PatchMaster Software for multichannel data acquisition and Igor Pro for data analysis.
2. Patch clamp amplifiers (EPC-10 Quadro).

3. NPC®-16 patch-clamp chips (single use, disposable). NPC®-16 internal and external electrodes (*see* **Note 2**).
4. Dimethyl sulfoxide (DMSO).
5. BayK 8644.
6. Tetrodotoxin (TTX).

2.2 Cell Culture

1. hTRPV3-HEK, hTRPV3 stably expressed in HEK cells.
2. Cor.4U, hiPS-derived cardiomyocytes (*see* **Note 3**).
3. Cell culture flasks T-12.5, T-75.
4. Cor.4U Complete Culture Medium (*see* **Note 3**).
5. Dulbecco's modified Eagle's medium/Ham's F12 with L-GLUTAMINE, FCS, geneticin (G418).
6. Dulbecco's phosphate buffer saline Ca^{2+} and Mg^{2+} free (1× PBS) with and without 2 mM EDTA.
7. Dulbecco's phosphate buffer saline with Ca^{2+} and Mg^{2+} (1× PBS).
8. Fibronectin.
9. TrypLE Express, 1×.
10. Trypsin, 0.5 g/L with 0.2 g/L EDTA.
11. Standard incubator at 37 °C with humidified atmosphere (5 % CO_2).

2.3 Patch-Clamp Recording Solutions

Prepare solutions in deionized water, adjust the pH, sterile filter (*see* **Note 4**), and measure the osmolarity (*see* **Note 5**). Solutions can be stored at 4 °C for up to 1 week, for longer storage, freeze the solutions at −20 °C. Solutions should be warmed to room temperature (20–24 °C) before use.

1. Standard extracellular solution: 140 mM NaCl, 4 mM KCl, 1 mM $MgCl_2$, 2 mM $CaCl_2$, 5 mM D-Glucose monohydrate, 10 mM HEPES; adjust to pH 7.4 with NaOH. Osmolarity 298 mOsmol (*see* **Note 5**).
2. Seal Enhancer for enhancing the seal resistance when applied externally: 80 mM NaCl 3 mM KCl, 10 mM $MgCl_2$, 35 mM $CaCl_2$, 10 mM Hepes/NaOH pH 7.4. Osmolarity 298 mOsmol.
3. Standard intracellular solution hiPS-derived cardiomyocyte recordings: 50 mM KCl, 10 mM NaCl, 60 mM KF, 20 mM EGTA, 10 mM HEPES; adjust to pH 7.2 with KOH Osmolarity: 285 mOsmol (*see* **Note 5**).
4. Standard intracellular solution for hTRPV3 cells: 50 mM CsCl, 10 mM NaCl, 60 mM CsF, 20 mM EGTA, 10 mM HEPES, adjust to pH 7.2 with CsOH, Osmolarity 285 mOsmol (*see* **Note 5**).

3 Methods

3.1 Culturing Cells for Planar Patch Clamp

The use of vital cells is very important to reach a high success rate in planar patch clamping, and therefore, cell lines should be passaged every 2–3 days and kept to a confluence not more than 60–80 % (*see* **Note 6**).

3.1.1 hTRPV3-HEK

hTRPV3-HEK are cultured and passaged in standard T75 tissue culture flasks in DMEM supplemented with 10 % FBS and 100 μL/mL geneticin in a standard incubator at 37 °C with humidified atmosphere (5 % CO_2). The cells should be passaged every 2–3 days using trypsin and kept to a confluence of 60–80 % (*see* **Note 6**).

3.1.2 hiPS-Derived Cardiomyocytes

hiPS-derived cardiomyocytes Cor.4U are seeded at a density of 10^5 viable cells per square centimeter culture area in one T-12.5 cell culture flask with 3 mL Cor.4U Complete Culture Medium (*see* **Note 3**).

1 day after seeding, a continuous rhythmic contraction of the cells (beating) can already be observed (*see* **Note** 7). The cells should be pre-cultured for 3–12 days prior patch clamp experiments. Other than HEK cells, the hiPS-derived cardiomyocytes need a non-conventional cell treatment and is thus described in detail here (*from Axiogenesis Cor.4U preculture protocol, modified*):

1. Prepare two fibronectin coated T-12.5 flasks for one vial of one million Cor.4U® cardiomyocytes.
2. Dilute an appropriate amount of fibronectin solution 1:100 in PBS (with Ca^{2+}/Mg^{2+}). Apply 2 mL of diluted fibronectin solution to each of the T-12.5 flasks.
3. Incubate the flasks for 3 h at 37 °C in the incubator or incubate the flasks with the coating solution overnight at 4 °C.
4. Pre-warm approx. 50 mL of Cor.4U® medium to 37 °C in the water bath. Pipette 8 mL of Cor.4U® culture medium in a 50 mL tube.
5. Remove the cells from the liquid nitrogen and thaw the cells immediately in the 37 °C water bath.
6. Transfer the cell suspension from the vial into the pre-warmed culture medium (37 °C) in the 50 mL tube using a P1000 pipette. Rinse the cell vial with 1 mL Cor.4U® medium and transfer to the 50 mL tube (total volume of 10 mL).
7. Spin cells immediately at 200 × *g* for 5 min. Aspirate the supernatant carefully without disrupting the cell pellet. Loosen the cell pellet by carefully tapping on the tube and then add 2 mL of pre-warmed Cor.4U® medium. Resuspend the suspension by first tapping on the tube and then very carefully pipetting up and down with a P1000 pipette.

8. Remove the fibronectin coating solution from the cell culture flasks and apply 2 mL of Cor.4U® medium to each flask.
9. After centrifugation, resuspend the Cor.4U® cardiomyocytes in 2 mL of Cor.4U® medium and transfer 1 mL cell suspension from the one million cells into each of the two T-12.5 flasks. Make sure that the cell suspension is distributed evenly in the flask.
10. Place the flask(s) into an incubator and culture the Cor.4U Cardiomyocytes at 37 °C, 5 % CO_2, and 95 % humidity for 24 h.
11. After 24 h, check the flask(s) under the microscope to ensure that a majority of the cells has attached to the culture surface. It is normal to see non-adherent, floating cells within the culture.
12. Perform a medium change by removing the medium from the flask and adding fresh Cor.4U® medium (3 mL for a T-12.5 flask). Most of the non-adherent cells will be removed with the media exchange.
13. After 48 h, perform a medium change by removing the medium from the flask and adding fresh Cor.4U medium supplemented with 1.5 μg/mL puromycin. Leave the puromycin containing medium on the cells for exactly 24 h (*see* **Note 7** and ref. 16).
14. The next day, after 24 h, remove the puromycin containing medium, wash twice with Cor.4U medium (without puromycin), and add fresh Cor.4U medium.
15. Grow the cells for 4–7 days in total and make a medium change every second day.

3.2 Harvesting Cells for Planar Patch Clamp

Vital cells are very important to reach a high success rate in planar patch clamping. Since the capture of cells onto the planar chip is a random process, unlike conventional patch clamping where healthy cells can be visually selected, the harvesting method should result in a homogeneous cell suspension with a high percentage of vital cells. The health of the cell impacts on the quality of the seal formed between the cell membrane and the glass chip, which also influences the quality of the recording.

3.2.1 TRPV3 Cells

1. Remove the medium from the cells and gently wash twice with PBS (without Ca^{2+} and Mg^{2+}) with 2 mM EDTA.
2. Pipette 1 mL of trypsin on to the cells, incubate at 37 °C, and wait for detachment of the cells (1–3 min) (*see* **Note 8**).
3. As soon as cells are detached, add 9 mL culture media and pipette up and down gently (*see* **Note 9**).
4. Transfer the cell suspension to a 15 mL conical centrifuge tube and centrifuge at room temperature for 2 min at $100\times g$. Discard the supernatant by decanting.

5. Resuspend the cell pellet in a 1:1 mixture of standard extracellular solution and culture media (*see* **Note 10**) at a density of 1×10^6–5×10^7 cells/mL (*see* **Note 11**).
6. Transfer the cells to the teflon *cell hotel* on the Patchliner. The cell suspension is continuously pipetted up and down to maintain single cells and viability (*see* **Note 1**).

3.2.2 hiPS-Derived Cardiomyocytes

1. Remove the medium from the cells and gently wash twice with PBS (without Ca^{2+} and Mg^{2+}) with 2 mM EDTA.
2. Pipette 2 mL of PBS (without Ca^{2+} and Mg^{2+}) with 2 mM EDTA and incubate at room temperature for 5 min.
3. Remove PBS/EDTA.
4. Add 3 mL of pre-warmed (37 °C) TrypLE, rotate the flask to evenly distribute the enzyme solution, then remove most of the enzyme so that only a film of solution is left.
5. Incubate for 2–3 min at 37 °C, until the cells start to float.
6. Tap the flask to remove the remaining cells.
7. Add 2 mL of 4 °C media and 2 mL of 4 °C external patch clamp solution (cell density at the end should be approx.: 300.000 cells/mL). Pipette the cells up and down only once and very gently, to get them into suspension.
8. Gently transfer the cell suspension into 1 well of a new, non-coated 6-well plate. If the cells are not handled with care at this pipetting step, the cell membranes will be ruptured and seals will not be formed.
9. For recovery, incubate the cells for 20 min in the fridge (4 °C).
10. Pipette the cells up and down (blue tip/1 mL pipette) four times to resuspend cells and transfer into the Patchliner's *cell hotel* without centrifuging (*see* **Note 12**).

3.3 Automated Patch-Clamp Using Patchliner

In order to control the robot and conduct the electrophysiological measurements on the Patchliner there are two programs running on the computer simultaneously: PatchControlHT and PatchMaster.

PatchMaster is used to control the amplifiers as well as to design the electrophysiological pulse protocols, online analysis and conduct the recordings. PatchControlHT is used to define the complete experiment from catching and sealing the cells to going whole cell and obtaining the desired recordings as well as controlling the microfluidics utilizing the pipettor for internal and/or external solution exchanges.

Preprogrammed experimental protocols can be loaded and modified for optimization with different cell types/characteristics.

The routine for catching and sealing the cells is rather independent of the cell type, whereas the part of the *Tree* (PatchControl^{HT} *Tree*) involving the measurement routine needs to be adjusted for the particular ion channel specific experiment.

Depending on the cell size and whether or not cell-attached recordings will be performed, the aperture resistances of the NPC-16 patch-clamp chips can be chosen. The resistances used here were 1–2 MΩ (hiPS-derived cardiomyocytes) and 2–3 MΩ (HEK cells). NPC®-16 chip resistance can be manufactured within a range from 1 to 15 MΩ, making them suitable for a variety of cell types of different sizes. Chips with multiple holes are also available, with four or eight holes per chip and, because the currents are summated, these are ideal for enhancing current size when recording from cells which express very small currents or from cells which are transiently transfected with a low transfection rate. Three of the chips on can be placed on the motorized chip wagon, allowing for 48 unattended recordings. When a *Tree* is activated, recording solutions are dispensed by the pipetting arm of the robot into the microfluidic chambers of the patch clamp chip. The chip cartridge is moved into the measuring head, which contains the external and internal electrodes and the suction lines for applying negative pressure to the cells. The recording site can address eight chip sites in parallel. Initially, a slight positive pressure is applied independently to each chip chamber and the offset currents resulting from the different ionic compositions of the internal and external recordings solutions are corrected. Once the cell suspension is added, a pressure of −50 mBar is applied to attract cells to the holes, and this results in an increase of the seal resistance. The seal resistance can be increased when the external Ca^{2+} concentration is elevated by perfusing seal enhancing solution, though this step is not a must in case elevated Ca^{2+} concentrations are interfering with second messenger cascades resulting in non-wanted physiological states of the cell (*see* **Note 13**). Along with suction pulses and the application of a negative voltage, a GigaOhm seal is reached. If no cell-attached recordings are performed at this stage, whole-cell access is gained by physical rupture of the cell-attached patch or by permeabilization with a perforator such as amphotericin B or nystatin. The software adapts the applied negative pressure to each cell individually, according to parameters such as series resistance and slow capacitance. In this way, it is automatically determined if an individual cell has been sealed to the chip aperture and whether the parameters correspond to a successfully attained whole-cell configuration. Cells that do not meet selected user-defined quality control parameters like seal resistance or series resistance will be disabled at this stage. The user can adjust pressure, voltage, and quality control values according to the cell type and channel characteristics and include those in any step of the *Tree*.

3.3.1 General Experimental Setup

1. Load a preprogrammed example *Tree* as a template and select the edit mode to make modifications according to your day-specific experiment, to your specific cell type and/or ion channel characteristics (*see* **Note 14**). The full range of commands

can be edited and inserted into the *Tree* using a generic drag-and-drop function.

2. Select the appropriate NPC®-16 chips with regard to resistance and number of holes (single hole or multihole) for your cells and place three chips on the chip wagon.
3. Prepare all recording and compound solutions and place in position, according to those defined in the job list (*see* **Note 15**).
4. Place the cell suspension into the *cell hotel*, where they will be aspirated every 30 s to maintain vitality and prevent clumping.
5. Select and activate the *Tree* from the topmost position (initialize Hardware folder) to initialize hardware and software settings.
6. Type in/select parameters using the automatically appearing dialogue windows (*Job* window and *compound position* window) such as compound positions, compound names, speed and volume of solutions, and voltage protocols (*see* **Note 15**).
7. The experiment will start after having typed in all necessary information and when the *Tree* is automatically moved on to the fill chip position. At the end of each run, it will loop back to the start and continue this process until all preprogrammed jobs are done or until all chips on the chip wagon have been used.

3.3.2 Heat Activation of TRPV3 Cells

The solution inside the pipette, the measuring head and the chip wagon can be heated simultaneously to maintain physiological temperatures throughout the experiment. Alternatively, for the study of heat-activated channels, the solution in the pipette can be heated independently, thus offering the capability to expose the cells to transient temperature increases.

1. Load a preprogrammed "Temperature control" *Tree* for solution application combined with temperature control. Temperature settings can be adjusted in the *Tree*.
2. Select medium resistant single hole chips (2–3 MΩ) for TRPV3 cell recordings and load three chips onto the chip wagon.
3. Within the *Tree*, adjust the holding potential to −70 mV and select a ramp protocol, with a 0 s sweep interval, from the pulse generator file in the job list to record the heat-activated currents.
4. Follow **steps 1–6** in Subheading 3.2.1.

Prior looking at heat activation of TRPV3 channels, current responses of an individual cell to 200 ms voltage ramps (−100 mV to +100 mV) and activation by application of increasing concentrations of 2-APB are shown in Fig. 3. A concentration response curve, using th current amplitude at +90 mV for calculation, revealed an EC_{50} for 2-APB activation of 61.1 ± 7.6 μM

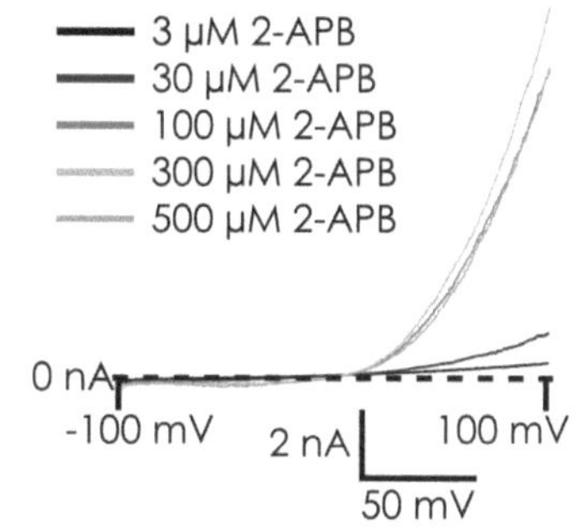

Fig. 3 Activation of TRPV3 by increasing concentrations of 2-APB

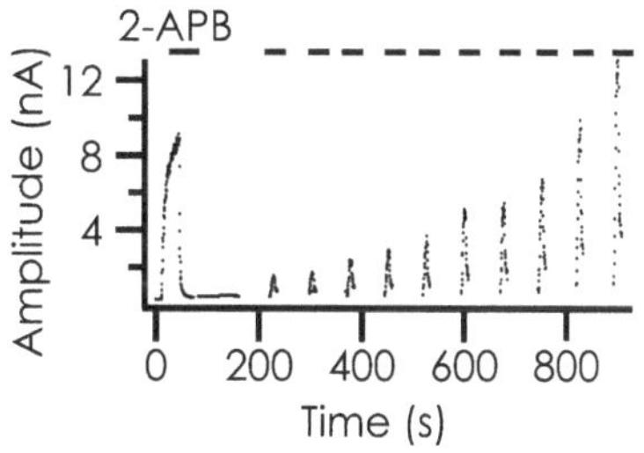

Fig. 4 Repeated exposure of TRPV3 currents to temperature caused sensitization of currents (*lines* indicate 2-APB or heat exposure)

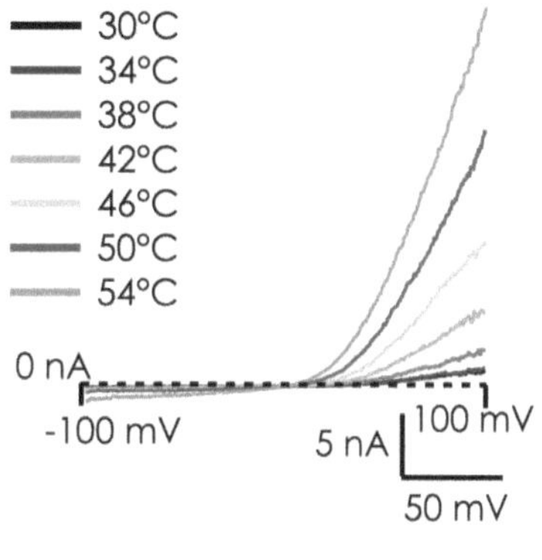

Fig. 5 Activation of TRPV3 by increasing temperature. The solution inside the pipette was heated to the temperature indicated and applied to the cell. TRPV3 was activated at temperatures of 38 °C and above

($n = 11$, data not shown), in good agreement with the literature [15, 17]. As previously reported [12, 15] TRPV3 currents displayed sensitization to repeated application of 2-APB (data not shown) or heat (Fig. 4).

Figure 5 shows the activation of TRPV3 channels by increasing temperature. Outwardly rectifying currents started to activate at 38 °C and increased in amplitude as temperatures increased, up

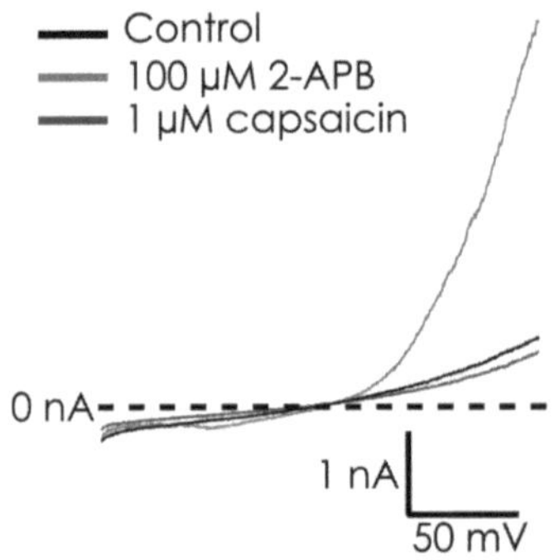

Fig. 6 TRPV3 was activated by 100 μM 2-APB but not 1 μM capsaicin

to 54 °C [12, 15, 17]. To rule out the involvement of TRPV1 a control was performed using capsaicin. Currents activated by 100 μM 2-APB were not activated by 1 μM capsaicin (Fig. 6) as previously reported [12].

3.3.3 Voltage- and Current-Clamp Recordings on hiPS-Derived Cardiomyocytes

hiPS-derived cardiomyocytes can be measured on Patchliner in the voltage-clamp mode and current-clamp mode, to assess the electrophysiological properties of individual ion channel currents and APs, which resemble the interplay of all channels, respectively.

1. Load a preprogrammed *Tree* for cardiomyocytes. Current–voltage relations of for example potassium (I_K) or sodium (I_{Na}) currents are first recorded in the voltage-clamp mode (Fig. 7). APs and pharmacological effects channel/AP modulating compounds are subsequently recorded in the current-clamp mode (Fig. 7, *see* **Note 16**).
2. Select low resistant single hole chips (1–2 MΩ, cell size ~30 pF) for hiPS-derived cardiomyocytes recordings and load three chips onto the chip wagon.
3. Follow **steps 1–10** in Subheading 3.2.2.

4 Notes

1. The *cell hotel* is a teflon made container for up to 4 mL of cell suspension, where the cells are pipetted up and down on a regular basis. Speed, frequency, and volume of the pipetted solution can be predefined individually depending on cells' characteristics, density, and available volume.
2. Place electrodes in bleach filled chloridation chambers for 20 min and then rinse with deionized water. Air-dry. Avoid scratching or using sandpaper with the electrodes.
3. Cor.4U® cardiomyocytes are human induced pluripotent stem (hiPS) cell-derived, primary-like cardiomyocytes. Cor.4U

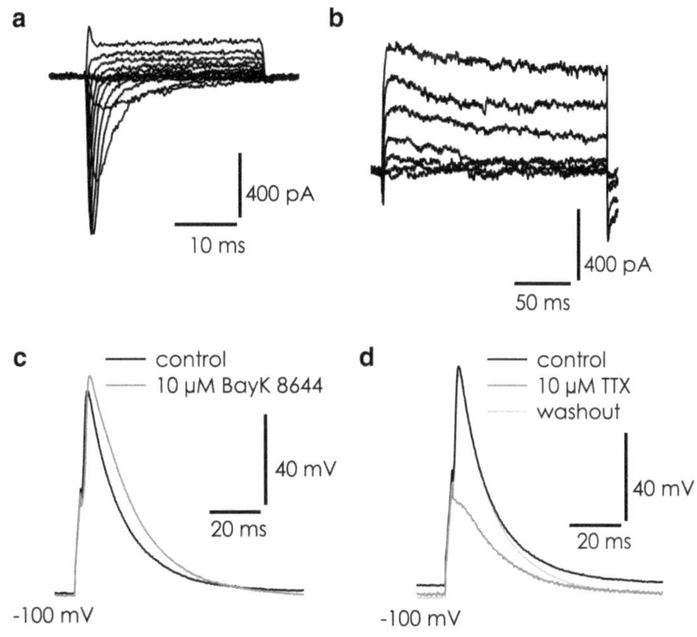

Fig. 7 (**a**–**d**) Parallel voltage- and current clamp recordings of hiPS-derived cardiomyocytes in the automated patch clamp system Patchliner. I_{Na} (**a**) and I_K (**b**) current traces. (**c**) APs in control solution and in the presence of the Ca^{2+} channel potentiator BayK8644, which increases the AP amplitude as well as the duration. APs are reversibly inhibited upon the application of the Na^+ channel inhibitor TTX (**d**)

culture medium is a complete medium that has been optimized for use with Cor.4U hiPS-derived cardiomyocytes. Axiogenesis is the supplier of Cor.4U® cardiomyocytes; note that there are other suppliers of human induced pluripotent stem cells such as CDI (iCells™) or ReproCell (ReproCardio cells) or Cellectis (hiPS-CM) and suppliers of human embryonic stem cells, GE Healthcare (Cytiva cells).

4. All recording solutions should be filtered with a 0.22 μm-pore diameter filter. Storage at 4 °C for up to 1 week or at –20 °C for a longer period of time.

5. Recording solution osmolarity is measured using a freezing-point osmometer; internal solutions should have 285 mOsmol and external solutions between 290 and 310 mOsmol.

6. Cells should be passaged every 2–3 days to prevent over-confluency, i.e., cells left to grow to greater than 80 % confluency may lead to cell clusters which could result in poor capture rates. Enzymes like TrypLE or Trypsin or also a lower temperature (4 °C, after enzymatically lifting the cells off of the culture flask) encourage cell separation and isolation while cell harvesting. Seal rates can sometimes be improved by optimizing cell culture methods, e.g., by using alternative enzymes or cooling of cells.

7. The beating of all seeded iPS-derived cardiomyocyte after 1 day is a good indicator that the cells are pure cardiomyocytes and are not interspersed with residual non-cardiac cells like proliferating fibroblasts or non-differentiated cells. In such a case, the addition of puromycin in the culture medium for 1 additional day is sufficient to obtain 100 % pure cardiomyocyte cultures.
8. Trypsin yields good results for robust cell types and channels, but can have degrading effects on cells and/or channels, which can sometimes be reversed by a recovery of cells for 20 min at 4 °C. Therefore Trypsin treatment should always be as short as possible. Several different other detachment agents can be used for harvesting the cells, e.g., accutase or TrypLE, although in our experience, the standard Trypsin solution works very well for most cell lines.
9. Check cells with microscope to see if most cells are single/non-clustered. If most of the cells are still aggregated, pipette ten more times to mechanically separate cells from each other. In case cell clusters are still predominant after approximately 15 times pipetting, a good option is to let cells recover at 4 °C for 20 min, since lower temperatures encourage cell separation as well. In general a single growth of cells on the culture plate of flask and letting cells not grow over-confluent prevents cell clusters at the outset (*see also* **Note 6**).
10. Cells resuspended in a mixture of recording solution and culture media (1:1) remain viable for up to 5 h when stored at room temperature in the *cell hotel* (HEK TRPV3 cells) or in a cell culture plate at 4 °C (e.g., stem cell-derived cardiomyocytes).
11. Cell densities should be 1×10^6–5×10^7 cells/mL for use on the Patchliner. For some cell types (especially more challenging primary cells or non-proliferating, expensive cells like stem cell-derived cells), it might however be difficult to obtain this number of cells. For these cells, an optimized cell application procedure can be used [18].
12. When resuspending the cardiomyocyte pellet, pipette up and down very slowly and gently for five times. Following the recovery period the cells need to be pipetted more frequently in the *cell hotel* to keep them single. This feature is already incorporated into the preprogrammed *Tree* for stem cell-derived cardiomyocytes.
13. Seal enhancer solution increases the chance of giga-seal formation with most of cells. The extracellular seal enhancer solution contains an elevated Ca^{2+} concentration and it can be replaced once a giga-seal is achieved and before establishing whole-cell access. In most of the cases the temporary presence of elevated

Ca does not harm the experiment and the seal enhancer is usually replaced by control recording solutions after gaining whole cell access. Alternatively, Ca could be replaces by Ba or Mg since those divalent cations also stabilize membranes and thus act as seal-enhancing reagents. Furthermore, when having an optimized cell handling resulting in most healthy cells the addition of seal enhancer solution in order to obtain a good success rate (60–80 % GigaOhm seals) is not necessary.

14. A full range of preprogrammed *Tree* templates for specific cell and channel types are installed on a Patchliner system and accessible in PatchControlHT, e.g., for ligand- or voltage-gated channels, for recording in voltage camp or current clamp mode.
15. Compound position and name as well as volume and speed of solution application and also the electrophysiological pulse protocol are selected in the job list within the *Tree*. The corresponding interface windows appear automatically once a *Tree* is executed and a user input is required prior to starting of experiment.
16. After identifying ion channel currents in the voltage clamp mode, the *Tree* switches the amplifiers to the current clamp mode. To induce AP responses in hiPS-derived cardiomyocytes, cells are held at a constant hyperpolarized membrane potential. The program then finds the stimulus threshold for each individual cell by applying 1 ms depolarizing pulses. The individual values for the thresholds are then set in the pgf in Patchmaster, and thus each individual cell is stimulated accordingly.

References

1. Neher E, Sakmann B (1976) Single channel currents recorded from membrane of denervated frog muscle fibres. Nature 260:799–802
2. Dunlop J, Bowlby M, Peri R et al (2008) High-throughput electrophysiology: an emerging paradigm for ion-channel screening and physiology. Nat Rev 7:358–368
3. Farre C, Haythornthwaite A, Haarmann C et al (2009) Port-a-patch and patchliner: high fidelity electrophysiology for secondary screening and safety pharmacology. Comb Chem High Throughput Screen 12:24–37
4. Stoelzle S, Obergrussberger A, Brüggemann A et al (2011) State-of-the art automated patch-clamp devices: heat activation, action potentials, and high throughput in ion channel screening. Front Pharmacol 2:1–11
5. Polonchuk L (2009) Toward a new gold standard for early safety: automated temperature-controlled hERG test on the Patchliner®. Front Pharmacol 3:3
6. Milligan CJ, Li J, Sukumar P, Majeed Y et al (2009) Robotic multiwell planar patch-clamp for native and primary mammalian cells. Nat Protoc 4:244–255
7. Balansa W, Islam R, Fontaine F et al (2010) Ircinialactams: subunit-selective glycine receptor modulators from Australian sponges of the family Irciniidae. Bioorg Med Chem 18:2912–2919
8. Stoelzle S, Haythornthwaite A, Kettenhofen R et al (2011) Automated patch-clamp on mESC-derived cardiomyocytes for cardiotoxicity prediction. J Biomol Screen 16:910–916
9. Farre C, Stoelzle S, Haarman C et al (2007) Automated ion channel screening: patch-clamping made easy. Expert Opin Ther Targets 11:557–565

10. Brüggemann A, Farre C, Haarmann C, Haythornthwaite A et al (2008) Planar patch-clamp: advances in electrophysiology. Methods Mol Biol 491:165–176
11. Majeed Y, Bahnasi Y, Seymour V et al (2010) Rapid and contrasting effects of rosiglitazone on transient receptor potential TRPM3 and TRPC5 channels. Mol Pharmacol 79: 1023–1030
12. Peier A, Reeve AJ, Andersson DA et al (2002) A heat-sensitive TRP channel expressed in keratinocytes. Science 296:2046–2049
13. Xu H, Ramsey IS, Kotecha SA et al (2002) TRPV3 is a calcium-permeable temperature-sensitive cation channel. Nature 418:181–186
14. Smith GD, Gunthorpe MJ, Kelsell RE et al (2002) TRPV3 is a temperature-sensitive vanilloid receptor-like protein. Nature 418:186–190
15. Chung MK, Lee H, Mizuno A et al (2004) 2-aminoethoxydiphenyl borate activates and sensitizes the heat-gated ion channel TRPV3. J Neurosci 24:5177–5182
16. Kolossov E, Bostani T, Roell W et al (2006) Engraftment of engineered ES cell-derived cardiomyocytes but not BM cells restores contractile Function to the infarcted myocardium. J Exp Med 203:2315–2327
17. Hu H, Grandl J, Bandell M et al (2009) Two amino acid residues determine 2-APB sensitivity of the ion channels TRPV3 and TRPV4. Proc Natl Acad Sci U S A 106:1626–1631
18. Becker N, Stoelzle S, Göpel S et al (2013) Minimized cell usage for stem cell-derived and primary cells on an automated patchclamp system. J Pharmacol Toxicol Meth 68(1):82–87, pii: S1056-8719(13)00232-3

Chapter 5

Industrializing Electrophysiology: HT Automated Patch Clamp on SyncroPatch® 96 Using Instant Frozen Cells

Liudmila Polonchuk

Abstract

Patch-clamping is a powerful technique for investigating the ion channel function and regulation. However, its low throughput hampered profiling of large compound series in early drug development. Fortunately, automation has revolutionized the area of experimental electrophysiology over the past decade. Whereas the first automated patch-clamp instruments using the planar patch-clamp technology demonstrated rather a moderate throughput, few second-generation automated platforms recently launched by various companies have significantly increased ability to form a high number of high-resistance seals. Among them is SyncroPatch® 96 (Nanion Technologies GmbH, Munich, Germany), a fully automated giga-seal patch-clamp system with the highest throughput on the market. By recording from up to 96 cells simultaneously, the SyncroPatch® 96 allows to substantially increase throughput without compromising data quality. This chapter describes features of the innovative automated electrophysiology system and protocols used for a successful transfer of the established hERG assay to this high-throughput automated platform.

Key words SyncroPatch® 96, Automated patch clamp, Drug discovery, High throughput screening, Electrophysiology, hERG

1 Introduction

The gold standard for assessing ion channel functionality is the patch-clamp technique. The high-resolution conventional manual method involves a glass micropipette filled with an ionic solution that electrically connects an electrode wire to a small patch of cell membrane. However, need for a skillful operator and a low throughput technology significantly restricted functional early profiling of large compound series for proper clinical candidate selection. The availability of a cost-effective electrophysiology assay based on efficient high-resolution technology is thus of major importance for the pharmaceutical industry.

The first automated patch-clamp (APC) instruments focused on increasing throughput by the parallel formation of high-resistance seals using the planar patch clamp (reviewed in [1]).

Marzia Martina and Stefano Taverna (eds.), *Patch-Clamp Methods and Protocols*, Methods in Molecular Biology, vol. 1183, DOI 10.1007/978-1-4939-1096-0_5, © Springer Science+Business Media New York 2014

Recently various companies have launched second-generation automated platforms, including SyncroPatch® 96 (Nanion Technologies GmbH, Munich, Germany), a fully automated patch-clamp system which supports true giga-seal recordings from up to 96 cells in parallel. Despite obvious advances in technologies, APC still faces some challenges including a requirement for a large number of dissociated cells, with high cell quality and robustness being a critical factor for a success.

Because automated systems select cells in a random manner, it is important that they are of extremely high quality with uniform size and homogeneity. A constant throughput and success rate can only be achieved under highly standardized cell culture procedures. Various experimental parameters such as the handling of cells and the number of passages influence protein expression, seal formation and stability of electrophysiological recordings. Developing or adapting a cell line for a new automated platform, therefore, requires time and significant effort. Having realized this, some APC system vendors already started to evaluate and certify commercially available cell lines based on cell and platform performance. Development of the assay from the frozen cell stocks in a large scale production may help to alleviate many of the shortcomings. Utility of frozen cell lines in medium-throughput electrophysiology screening have recently been exploited at several platforms such as CytoPatch [2] and PatchXpress [3]. These studies have shown that instant cells may improve the stability and reduce costs without compromising experimental quality especially important for standard electrophysiological tests run in a large-scale industrial environment.

The hERG test has become the routine in vitro approach for assessing the cardiac safety of new drugs in early development and has been established at various technology platforms with patch-clamp being the only method available to analyze the ion channel function directly. Therefore, this standard cardiac safety test has been selected as an example of upscaling an electrophysiological assay from medium (PatchXpress 7000A) to the high throughput SynchroPatch® 96 platform in operation with the frozen cell stocks in the industrial environment where the high reproducibility of the results for thousands of compounds needs to be warranted.

2 Materials

2.1 Equipment

1. SynchroPatch® 96, connected to the benchtop computer with associated data acquisition (PatchControl 96) and analysis (DataControl 96) software. Single-use disposable NPC-96 seal-chips and NPC-96 electrode set from Nanion Technologies GmbH.
2. Temperature-controlled water bath.

3. 50 mL centrifugation tubes.
4. 10, 100 and 1,000 μL pipettes.
5. Glass coated 96-well plates, for example.

2.2 Supplies

1. IMDM, Iscove's Modified Dulbecco's Medium.
2. Dimethyl sulfoxide (DMSO).
3. $FeCl_3 \cdot 6H_2O$.

2.3 Cells

The CHO crelox hERG cell line (ATCC reference Nr. PTA-6812) was generated and validated at Roche [4]. Recombinant hERG K^+ channels originally cloned from human heart are stably expressed in CHO cells grown in sterile tissue flasks (CellSTACK, Corning) in a standard DMEM/F-12 (1:1) medium (Invitrogen, USA) supplemented with 10 % (v/v) heat-inactivated fetal calf serum (FCS) (Hyclone, USA) and 500 μg/mL gentamicin solution (Gibco, UK) at 35–37 °C in 5 % CO_2. The cryopreservation of the CHO-hERG cells can be done either by a commercial service provider (CCS GmbH, Germany, Invitrogen Life technologies, GeneScript, etc.) or using self-developed methods and protocols. The instant cells should be stored in liquid nitrogen in single-use aliquots (*see* **Note 1**).

2.4 Solutions

1. The extracellular solution contains: 150 mM NaCl; 4 mM KCl; 1 mM $CaCl_2$; 1 mM $MgCl_2$; 10 mM HEPES; 5 mM glucose. Adjust pH to 7.2–7.6 with NaOH and adjust osmolarity to 290–330 mOsm.
2. The internal solution contains: 50 mM KCl; 60 mM KF; 10 mM NaCl; 10 mM HEPES; 20 mM EGTA. Adjust the pH at 7.0–7.4 with KOH and adjust osmolarity to 260–300 mOsm.
3. The extracellular solution for seal enhancing contains: 80 mM NaCl; 60 mM *N*-Methyl-d-Glucamine, 4 mM KCl; 5 mM $CaCl_2$; 1 mM $MgCl_2$; 10 mM HEPES; 5 mM glucose. Adjust the pH at 7.2–7.6 with NaOH and adjust osmolarity to 290–330 mOsm.
4. All solutions should be filtered and stored at +2–8 °C (*see* **Notes 2–5**).

3 Methods

3.1 Cell Culture

3.1.1 Thawing and Recovery Procedure for Frozen Ready-to-Use Cells

1. Immerse a vial with cryopreserved cells in a water bath at 37 °C for 2–3 min (*see* **Note 6**).
2. When only a tiny part of ice is visible, transfer the vial to the sterile hood.
3. Disinfect vial surface with ethanol and open it under sterile condition ensuring that no cells are left in the cap.

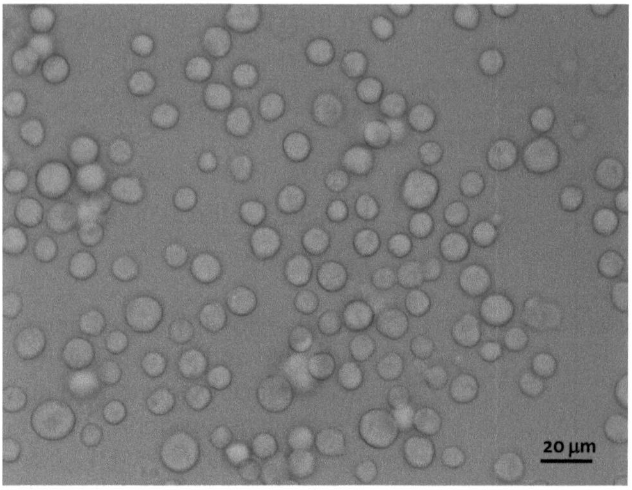

Fig. 1 Microscopic view of instant CHO-hERG cell after thawing

4. Resuspend the cells by pipetting and transfer the cell suspension into a 50 mL tube.
5. Drop-wise add 20 mL of pre-warmed culture medium while gently shaking the tube.
6. Take an aliquot of the suspension for visual inspection (Fig. 1), cell counting, and viability tests (*see* **Note 7**).
7. Centrifuge the suspension for 1 min at 180×*g*, RT.
8. Resuspend the cells pellet in a suitable amount of the assay medium or buffer (*see* **Note 8**).
9. Transfer the cells to the Cell hotel on SyncroPatch® 96. The cells can be used for up to 3 h after thawing (*see* **Note 9**).

3.2 Automated Patch-Clamp Procedures on the SyncroPatch® 96

3.2.1 Basic Concepts

The key element of every planar patch-clamp system is a seal chip. SynchroPatch® 96 shown in Fig. 2 utilizes NPC®-96 chips based on a 96-well plate format (Fig. 3) with bottom made from borosilicate glass known for its excellent dielectric and adhesive properties. The plate has 96 wells and each well has a tiny hole at the bottom (aperture) as shown in Fig. 3 insert. The planar NPC®-96 chip electrode is designed to control giga-Ohm seal formation and maintain stable whole-cell recordings with low access resistance over an experiment. Selection of the appropriate seal chip type is based on the resistance of the aperture and the size of the cells (*see* **Note 10**).

PatchControl 96 software controls both the mechanical and electrical functionalities of the system and allows data acquisition to be managed through commands in the experimental protocol (PatchControl 96 Tree). The motorized stage (chip cartridge) accommodates one seal chip plate allowing up to 96 parallel recordings.

Fig. 2 SyncroPatch® 96 in the working environment

The plate is moved into the measuring head, which contains the pneumatic and electric contacts for all 96 wells.

Cells and solutions are automatically distributed into well plates by a 16-channel pipettor arm (*see* **Note 11**). Once the cells are added to all wells, a negative pressure of −50 mBar is usually applied by a pressure controller to attract cells to the holes resulting in a small increase in the seal resistance. The subsequent addition of seal enhancing solution along with application of suction pulses and negative voltage to the cells further increase seal resistance to a giga-Ohm seal. The applied negative pressure and voltage have to be selected according to the cell type (*see* **Note 12**).

Continuous recordings during applications and fast (within 100 ms) solution exchange allows for screening of compound action on both ligand- and voltage-gated ion channels. Monitoring such parameters as seal resistance, series resistance, and slow capacitance helps to determine if a cell has been sealed to the chip aperture and whether the whole-cell access is achieved. The quality control functions can be set up to disable cells that do not meet selected quality control parameters (e.g., seal resistance, series

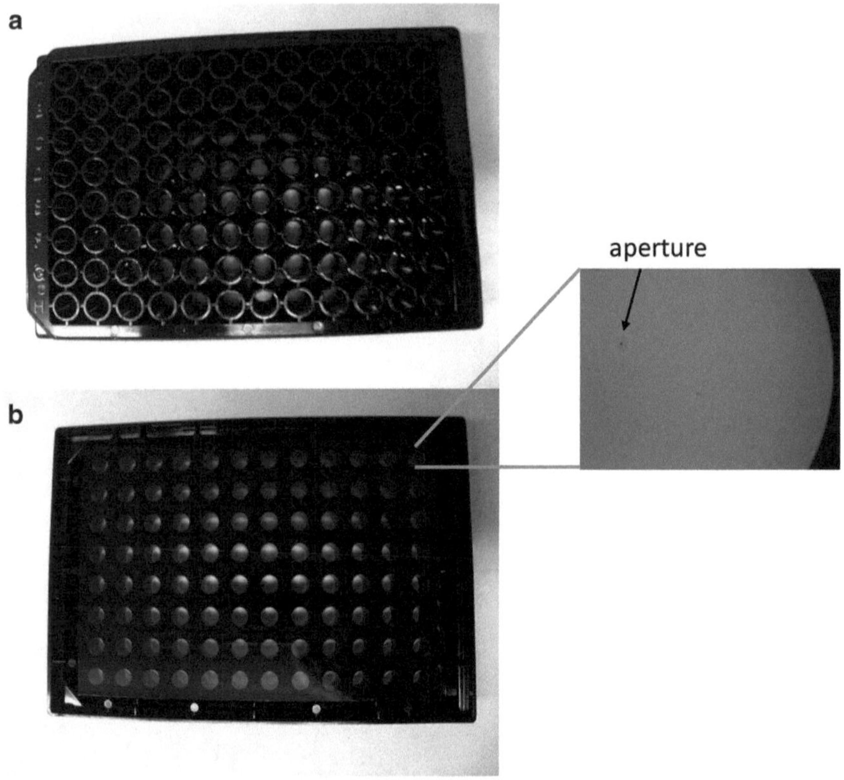

Fig. 3 Top (**a**) and bottom (**b**) view of the NPC-96 seal chip. A closer microscopic view of the well is shown in the *insert*

resistance, and current amplitude) at certain stages. This QA function greatly facilitates subsequent data overview and analysis done with DataControl 96 software.

3.2.2 Initialize the System

1. Turn on the power switch at the machine.
2. Switch on the computer.
3. Check if the system water is full and waste is NOT full.
4. Start *PatchControl96.exe* with a double-click on its shortcut on the desktop.
5. Load the appropriate tree and select the edit mode to make modifications using the full range of commands, which can be inserted into a protocol by a drag-and-drop function.
6. Initialize the hardware by clicking on the appropriate folder and press play.
7. Check if electrodes need to be chloridized:
   (a) Remove the cover of the measurement chamber and unplug the cables from the board.
   (b) Take out the electrode board and check for the AgCl status.

(c) Push back the black metal cover, pull out the ground electrodes, and check their chloridation status.

(d) Place electrodes in $FeCl_3$ solution for chloridation if needed and then rinse them with dd H_2O and gently wipe with a tissue.

(e) Put everything back in place.

8. Flush the System Flush with dd H_2O (*see* **Note 13**).

(a) Place an old chip in the chip tray.

(b) Place all tubes for internal solution in take in a bottle with dd H2O.

(c) Initiate the system flush. Use the waiting time to enter for example compound information.

(d) Open the Compound dialog and edit the compound positions.

3.2.3 Experimental Procedure and Results

1. Prepare and place solutions on the appropriate positions.
2. Place a tube for the internal solution intake in the bottle with the internal solution (roboter with a liquid detector).
3. Load an appropriate seal chip plate in the chip cartridge on the chip wagon.
4. Place an aliquot of the cell suspension in the cell hotel and activate the cell hotel.
5. Start the experiment by clicking on the appropriate folder in the selected tree and press play.

Low resistance seal-chips showed the best performance with the CHO-hERG instant cells. Series resistance was typically 5–9 MΩ and not compensated. The cells were held at a resting voltage of −80 mV and stimulated by a voltage pattern to activate hERG channels and conduct outward IK_{hERG} current (Fig. 4), at a stimulation frequency of 0.1 Hz (6 bpm). The reported current amplitudes represent the maximal peak tail current. After the cells stabilized for a few minutes and currents were steady, IK_{hERG} amplitude and kinetics were recorded under control conditions (vehicle control) for 3–5 min. Thereafter, a single test item concentration was added and incubated for at least 3 min until a new steady state current level was reached. At the end of the compound incubation, a 1 μM solution of standard IK_{hERG} blocker E-4031 was applied twice to each cell for 1 min to suppress IK_{hERG} entirely.

An example of a vehicle control hERG experiment on the Synchropatch® 96 is shown in Fig. 5. For compound testing IK_{hERG} amplitudes were recorded at four drug concentrations. The hERG current was measured as the average current from five sweeps collected at the end of vehicle or compound addition. After subtracting current levels from that of full block (i.e., positive control) they were compared to the vehicle control values

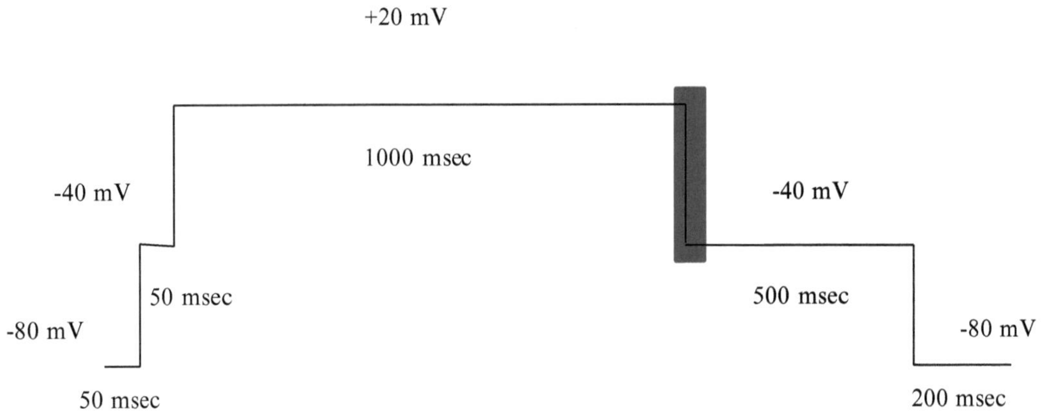

Fig. 4 A voltage pulse protocol to activate and conduct the hERG current (the area for the peak current measurement is *highlighted* in *red*) (color figure online)

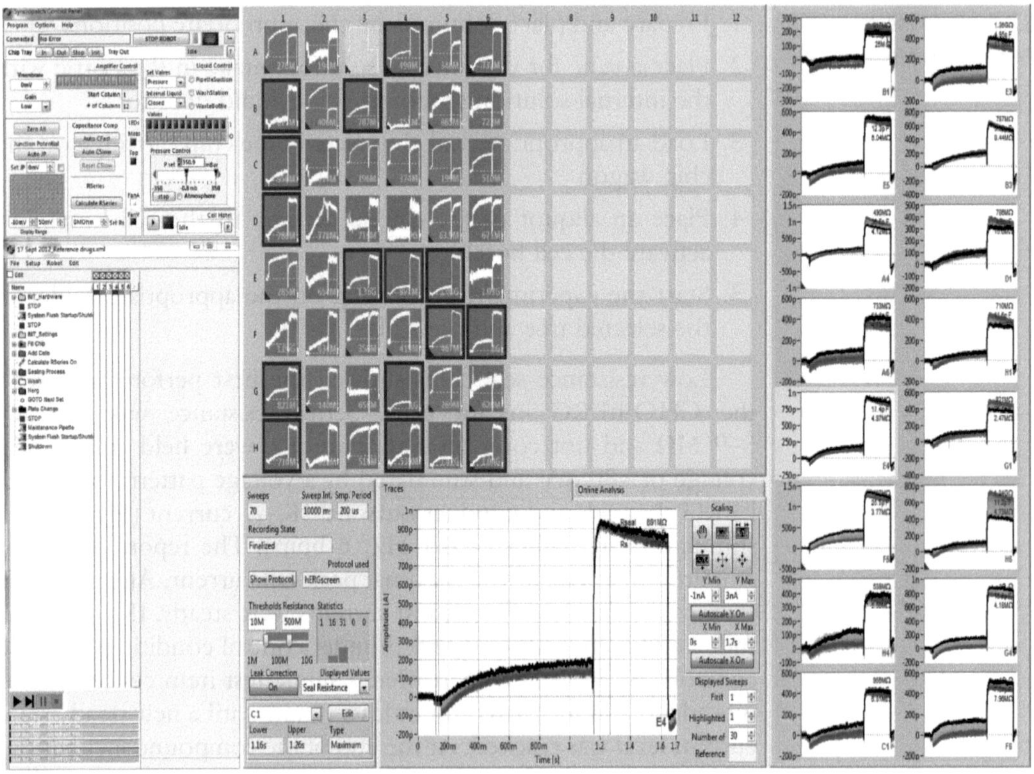

Fig. 5 Typical outward K^+ current traces obtained with the SynchroPatch® 96 in frozen CHO-hERG cells (vehicle control experiment)

(taken as 100 %) to define fractional blocks. Offline analysis of the peak tail current was performed using DataControl 96 software (Nanion Technologies GmbH, Germany). Data were expressed as mean ± SEM for each drug concentration and were used to

Table 1
Effects of reference drugs on the hERG K^+ current

Compound	Test conc. (μM)	hERG IC_{50} SP (μM) Mean ± S.E.	hERG IC_{50} PX (μM)
Alprazolam	0.3–1–3–10	3.5 ± 0.5	5.2
Amiodarone	0.01–0.03–0.1–0.3	0.19 ± 0.07	0.22
Amoxicillin	0.3–3–30–300	>300	>300
Astemizole	0.001–0.003–0.01–0.03	0.003 ± 0.0005	0.003
Bepridil	0.01–0.03–0.1–0.3	0.02 ± 0.003	0.065
Captopril	1–10–100–1000	>1,000	>1,000
Cisapride	0.001–0.01–0.1–1	0.012 ± 0.002	0.026
Clemastine	0.001–0.01–0.1–1	0.028 ± 0.003	0.027
Dofetilide	0.001–0.01–0.1–1	0.020 ± 0.005	0.015
E-4031	0.0003–0.003–0.03–0.3	0.017 ± 0.0002	0.018
Erythromycin	1–10–100–1000	>1,000	>1,000
Fluoxetine	0.1–0.3–1–3	0.40 ± 0.03	0.8
Glibenclamide	0.2–0.6–2–6	>6	>6
Haloperidol	0.001–0.01–0.1–1	0.018 ± 0.005	0.028
Ibuprofen	0.3–3–30–300	37 ± 11	74
Isradipine	0.15–0.5–1.5–5	2 ± 0.2	5
Moxifloxacin	0.3–3–30–300	25 ± 4	83
Olanzapine	0.3–1–3–10	2.1 ± 0.7	1.6
Quinidine	0.1–0.3–1–3	0.5 ± 0.1	0.47
Tamoxifen	0.1–0.3–1–3	0.35 ± 0.04	0.42
Terfenadine	0.003–0.01–0.03–0.3	0.028 ± 0.0004	0.06
Thioridazine	0.03–0.1–0.3–1–3	1.07 ± 0.04	0.67
Verapamil	0.01–0.003–0.1–0.3	0.59 ± 0.006	0.14

determine the concentration–response relationship for the test compounds. Concentration–response curves were fitted by non-linear regression analysis and the IC_{50} values were reported.

A set of reference compounds was used to validate the novel technology in combination with ready-to-use frozen CHO cells stably transfected with hERG. The hERG IC_{50} values obtained at SynchroPatch® 96 were compared with the in-house reference data (*see* Table 1).

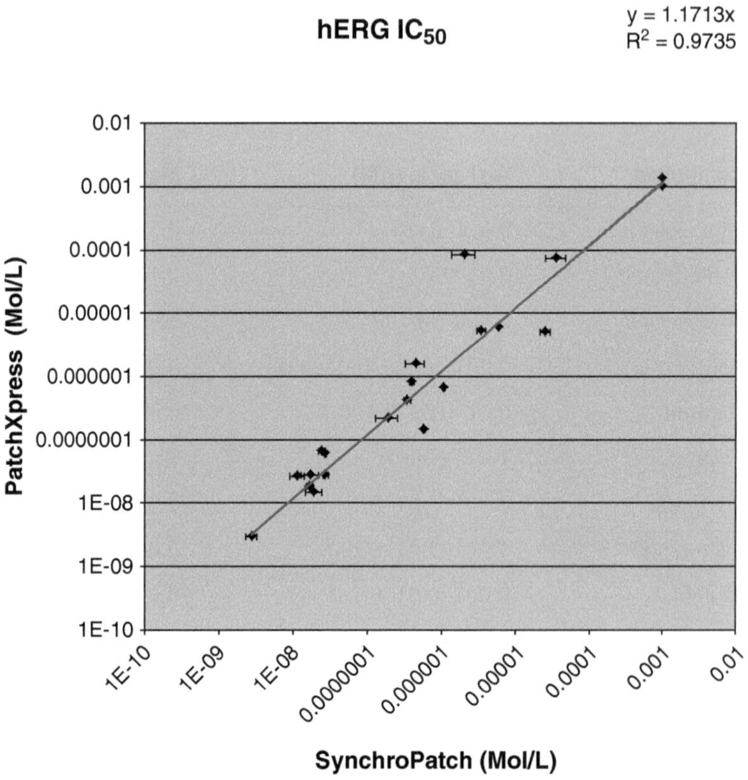

Fig. 6 Correlation of the hERG IC_{50} values obtained in the SynchroPatch® 96 (SP) with those measured using PatchXpress 7000A (PX)

A good correlation was found between IC_{50} values for the reference drugs obtained with the SynchroPatch® 96 platform and those generated in the previously used medium-throughput automated patch clamp system PatchXpress 7000A (Fig. 6). Implementation of the SynchroPatch® 96 for the hERG screening enables to perform high quality tests, increases the throughput as well as reduces the costs and minimizes the amount of compounds used that greatly facilitates compound progression in the early development.

4 Notes

1. The frozen stocks need to have about 10^7 cells/vial to assure a sufficient volume of cell suspension with the optimal cell density for the SynchroPatch® 96 ($0.2–0.4 \times 10^6$ cell/mL).
2. The tenfold-stock extracellular solution is suitable for multiple uses over a half a year when stored below 25 °C. For microbial and chemical stability reasons, bottles should be kept tightly closed after each withdrawal.

3. The internal solution should be prepared freshly and used within 1 day. It contains fluoride used to maintain stability of recordings on planar seal chips.
4. Seal enchanser solution contains high Ca^{2+} and NMDG known to improve giga-seal formation. It is exchanges with a standard extracellular solution after establishment of a whole cell configuration to avoid calcium-dependent inactivation of the hERG channels.
5. Osmolarity is an important factor in regulating the cell shape and it should be checked when the assay solutions are prepared. External osmolarity should always exceed the internal one. Addition of sorbitol (up to 50 mM) to the extracellular solution proved to be helpful in adjusting osmolarity and maintaining stability of recordings. Replacement of Na^{+} with NMDG may also have a positive effect on seal formation and stability. However, one should be careful with using non-physiological compounds in the buffers as they may have some unknown and unexpected effects on ion currents.
6. The cell hawing should not take longer than 2–3 min. However, defrosting time may vary depending on how deep a vial is immersed in the warm water. It is, therefore, recommended to use a vial holder rack and check for the water level in the water bath.
7. Cell viability is a critical experimental parameter. It is not recommended to use cell cultures with viability below 95 % for automated electrophysiological recording.
8. Increasing a volume of cell suspension used at the SynchroPatch® 96 above 15 mL has a positive influence on the cell performance.
9. While some cell types may need time to recover after defrosting and should not be used directly after thawing them, it may be essential to start experiments as soon as possible for others. Therefore, one should test different approaches to define the best one for the specific cell type.
10. Selection of the appropriate NPC-96 seal chip type for specific cell type is crucial for the HT electrophysiology. Nanion's seal chips can be customized and are generally available in the following ranges: low resistance (1–2 MΩ), medium resistance (2–4 MΩ) and high resistance (5–6 MΩ). CHO cells seem to work best with low-medium resistance chips (2–2.5 MΩ). Seal chips with multiple holes in a single well are also available for enhancing current recorded from cells with low current signals.
11. Simultaneous liquid handling at 16 channels significantly speeds up the experimental turnover and the system throughput, but implies some restrictions on the compound and solution addition to single wells.

12. Negative pressure (suction) applied to capture cells is usually in the range of 10–50 mb. And some cells may even do not need it. Adjusting the suction settings for a giga-seal formation with 96 cells simultaneously is a challenging task. It strongly depends on a cell type used and needs some additional experimental testing. This is when the cell homogeneity gets really important.
13. If the system has not been used for a while, one should run a flush with 70 % ethanol followed by several dd H_2O water flushes.

Acknowledgments

The author would like to thank the Nanion Technologies GmbH team, especially Drs. Andrea Brüggemann, Sonja Stoelzle, and Timo Stengel, for their help. I am also grateful to Drs. Claudia McGinnis and Thomas Weiser from Roche Non-Clinical Safety for supporting the acquisition and validation of the system, and to Evgenia Gissinger for technical assistance.

References

1. Möller C (2010) Impact of new technologies for cellular screening along the drug value chain. Drug Discov Today 15:384–390
2. Scheel O, Himmel H, Rascher-Eggstein G et al (2011) Introduction of a modular automated voltage-clamp platform and its correlation with manual human Ether-à-go-go related gene voltage-clamp data. Assay Drug Dev Technol 9:600–607
3. Donovan BT, Bakshi T, Galbraith SE et al (2011) Utility of frozen cell lines in medium throughput electrophysiology screening of hERG and Nav1.5 blockade. J Pharmacol Toxicol Methods 64:269–276
4. Guthrie H, Livingston FS, Gubler U et al (2005) A place for high-throughput electrophysiology in cardiac safety: screening hERG cell lines and novel compounds with the ion works HTTM system. J Biomol Screen 10:832–840

Chapter 6

Planar Patch Clamp for Neuronal Networks—Considerations and Future Perspectives

Alessandro Bosca, Marzia Martina, and Christophe Py

Abstract

The patch-clamp technique is generally accepted as the gold standard for studying ion channel activity allowing investigators to either "clamp" membrane voltage and directly measure transmembrane currents through ion channels, or to passively monitor spontaneously occurring intracellular voltage oscillations. However, this resulting high information content comes at a price. The technique is labor-intensive and requires highly trained personnel and expensive equipment. This seriously limits its application as an interrogation tool for drug development. Patch-clamp chips have been developed in the last decade to overcome the tedious manipulations associated with the use of glass pipettes in conventional patch-clamp experiments. In this chapter, we describe some of the main materials and fabrication protocols that have been developed to date for the production of patch-clamp chips. We also present the concept of a patch-clamp chip array providing high resolution patch-clamp recordings from individual cells at multiple sites in a network of communicating neurons. On this chip, the neurons are aligned with the aperture-probes using chemical patterning. In the discussion we review the potential use of this technology for pharmaceutical assays, neuronal physiology and synaptic plasticity studies.

Key words Ion channels, Synapses, Neuronal networks, Patch-clamp chips, Microfluidic networks, Chemical patterning

1 Introduction

The conventional patch-clamp has proven to be the state-of-the-art technique for high resolution interrogation of electrophysiological activity. This technique permits to monitor the activity of single ion channels (cell-attached configuration) and the variations in current and potential of an entire cell (whole-cell configuration). In addition, it represents a formidable tool to study many neuronal functions (action potential propagation, neurotransmitter release, synaptic transmission and plasticity, etc.) and to screen how these functions are modulated by diseases and drugs. However, this technique requires highly qualified personnel, is labor intensive,

Marzia Martina and Stefano Taverna (eds.), *Patch-Clamp Methods and Protocols*, Methods in Molecular Biology, vol. 1183, DOI 10.1007/978-1-4939-1096-0_6, © Springer Science+Business Media New York 2014

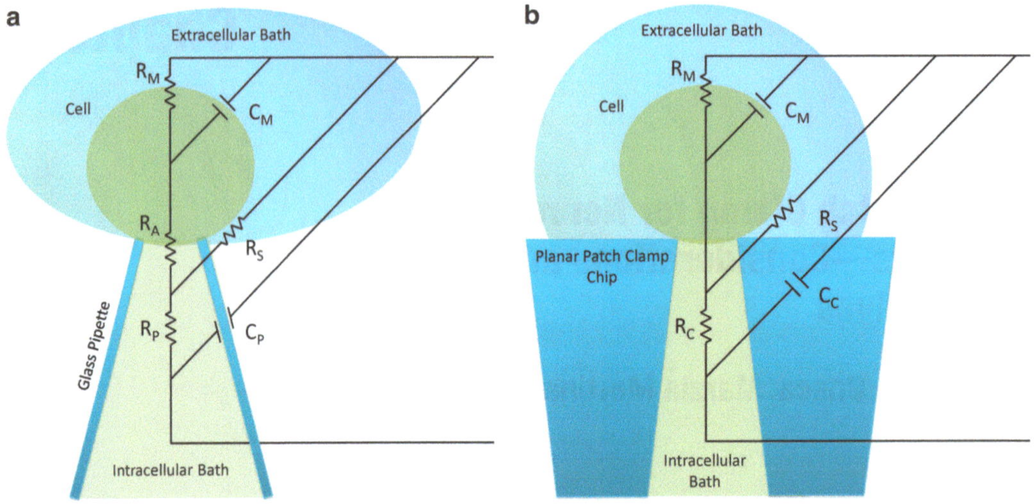

Fig. 1 Schematics and equivalent circuit of conventional and planar-patch-clamp (PPC) experiments. (**a**) Schematics and equivalent circuit of a conventional patch-clamp experiment: a glass pipette with micrometric aperture is used to access the inside of the studied cell and measure the ionic current flow between intracellular and intracellular solution by means of a current probe. The key components of such a circuit are the resistance R_P and the capacitance C_P associated with the patch pipette, the access resistance R_A (determinate mainly by the diameter of the glass pipette), the resistance R_M and the capacitance C_M of the cell membrane. (**b**) Planar analogous in which the resistance and the capacitance of the pipette are replaced by whom of the chips

slow, and expensive; these are factors which limit its use for high-throughput drug development assays.

To overcome these limitations, over the last decade, the planar-patch-clamp (PPC) technology has been developed [1–3]. In the planar version of the patch-clamp technique the pipette is replaced by an insulating self-supported thin film separating two chambers (extracellular bath and intracellular bath) filled with ionic solutions. On this partition a micron sized aperture, the topological equivalent of the tip of a glass pipette, has been micro-machined. The cell is attracted over the aperture via suction (negative pressure); the seal is successively induced by additional negative pressure. The whole-cell configuration can be obtained by rupturing the cell membrane as in a classical patch-clamp experiment using a pulse of current or negative pressure. An alternative method comprising in the addition of pore forming antibiotics in the intracellular solution can be used (perforated patch clamp). In the intracellular bath, an electrode permits the recording of the membrane potential (*see* Fig. 1).

The PPC technology has various major advantages. Firstly, it does not require micromanipulators, optical microscope, and antivibration table, greatly reducing the cost and footprint of the experimental setup. Secondly, it can be automated and extended to large arrays, allowing for high-throughput electrophysiological measurements. The planar geometry also includes the potential for more sensitive recordings. Indeed, if the design of the device is aimed to minimize the shunt capacitance and access resistance (*see* Fig. 1), the noise raised from the random voltage fluctuation of the amplifier can be knocked down [4]. In addition, compared to conventional patch-clamp experiments, the physical contact between the cell membrane and the aperture is more stable and the electrical path from the electrode to the inside of the cell shorter, producing lower access resistances [5]. A planar substrate is easily accessible and facilitates the integration of other techniques such as fluorescence [6] or atomic force microscopy [7] to the electrophysiological recordings. This also enables the integration of an extracellular microfluidic system for the perfusion of drugs and the precise control of cell culture environment. Finally, the ability to integrate a planar device in a microfluidic system allows for the automation of protocols and the execution of parallel independent measurements; improvements particularly desirable for the commercial development of automated patch-clamp systems with increased throughput [8]. The main disadvantage is that this technology is so far restricted to non-adherent isolated cells lacking synaptic communication.

Multi-electrode arrays (MEAs) permit the study of the electrical activity of large neuronal networks by detecting extracellular field potentials in cell cultures, brain slices and in vivo preparations [9, 10]. MEAs are powerful tools for investigating spontaneous and evoked activity of neuronal networks in normal and altered conditions (diseases and/or drugs) [11, 12]. Recent advances in Complementary Metal-Oxide Semiconductor (CMOS) technology have allowed for the creation of large MEAs integrating thousands of electrodes with subcellular spatial resolution [13–17]. Progress in nanotechnology has allowed for the development of lower impedance electrodes resulting in remarkable improvements in signal to noise ratios [18–20]. Despite these advances, MEAs still lack intracellular control and the specific information pertaining to ion channels is difficult to be extract from the multiplexed data.

Considering that the majority of drugs acting on ion channels targets synaptic transmission, combining the key benefits of patch-clamp and MEAs is highly desirable. In particular, developing a PPC technology providing high resolution patch-clamp recordings of individual cells at multiple sites in a network of communicating neurons has enormous potential for the development of pharmaceutical assays aimed to investigate in vitro models of diseases, neuronal physiology and synaptic plasticity [21].

In order to interrogate multiple individual adherent cells connected in a network, we need to create an array of tightly packed planar patch apertures connected to independent subterranean microfluidic channels [22]. These requirements can be fulfilled using a rigid insulator (glass or quartz) as base substrate on which

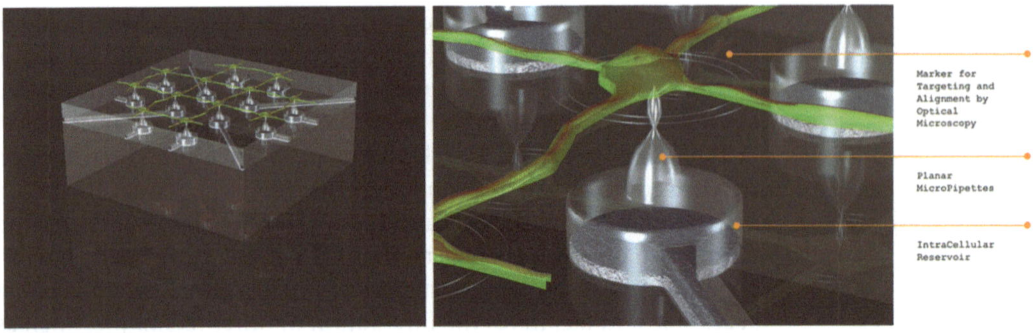

Fig. 2 PPC micropipettes array. A microfluidic system provided with an array of planar micropipettes and suction channels is assembled on a rigid support. On the support a planar microelectrode array will address the electrical signal originating from every cell (in *green*) coupled with the patch sites through a small intracellular microfluidic reservoir. The position and the connection between neurons can be controlled by patterning the polymeric surface with adhesion molecules

a metallic planar microelectrode array has been created. On this support a microfluidic system can be integrated to separate the electrodes and the solutions coming from every cell and providing independent suction channels for parallel patch-clamp interrogation. On the microfluidic system an array of apertures has to be assembled (*see* Fig. 2b). A major challenge is coupling the planar device with the component of the neuronal network. In an ideal scenario every cell of the network has to be positioned above every aperture of the array to enable the monitoring of every neuron of the network (*see* Fig. 2a). We can achieve this condition by patterning adhesion molecules (i.e., PDL) on the extracellular (plating) substrate [23, 24]. This will permit to control the position and the interconnection scheme of the cells, facilitating the alignment of the network with the device. Guiding cell growth in a predetermined pattern could be useful to investigate specific properties and developmental cues of the network [25].

2 Materials

In this section the main materials used for the fabrication of PPC chips are briefly introduced. In the following section, detailed examples of chip fabrication processes are provided for each material system.

2.1 Choice of Materials for PPC Fabrication

The first PPC devices to be fabricated were in silicon [26, 27]. Silicon has been extensively used for microelectronics and MEMS systems [28], and therefore its micro-fabrication can rely on well-established processes with high yield and critical process control. In addition, it is convenient for the integration of multiple functionalities such as on-chip amplification [29]. Exploiting recent advances in micro-fabrication technologies, silicon can be also machined to mimic the tip of a micropipette [30, 31]. In recent years silicon patch-clamp chips have been largely planar, with a strong emphasis on optimizing aperture shape to obtain high-quality seals [32–36]. The major disadvantage of using such a semi-conductive substrate is the high capacitive coupling between the intracellular and extracellular (culturing) media. To overcome this limitation, different insulator materials such as glass or quartz [2, 37], polyimide (PI) [38] or polydimethyl siloxane (PDMS) [39] can be used. Glass and quartz have excellent dielectric properties; since pipettes are made of these materials, there is a large body of knowledge on how to optimize adhesion to the cell membrane. Conventional micro-fabrication techniques are not well established for these materials and the most successful fabrication attempts initially involved serial methods like ion track milling [37] or laser machining [40]; lately, plasma etching on fused quartz wafers has proved to yield very high-quality apertures [41], suggesting that batch fabrication of quartz PPC chips is feasible. PI is desirable because of its cytocompatibility, low dielectric constant (3.5 at 1 MHz), and high chemical and mechanical stability. Since PI processing technology is standard in the microelectronic industry, PI films have been used to fabricate microfluidics devices [42, 43] such as perforated MEAs [44]. PDMS also has excellent biocompatibility, electrical and optical properties, and therefore can be used for electrophysiological recordings. In addition, PDMS is waterproof and permeable to gases. It is compatible with optical and fluorescence microscopy, and its dielectric constant (2.6 at 1 MHz) is lower than most glasses and quartz (3.8 at 1 MHz). Because of these characteristics it makes an excellent culturing substrate. It can be easily and inexpensively micro-structured by soft lithography in a standard laboratory [45], even though moving the material to a standard micro-fabrication foundry may presents problems such as contamination and dimensional control.

2.1.1 Equipment

1. Clean room.
2. Vacuum chamber.
3. Air/Oxygen plasma cleaner.
4. Spin coater.
5. Mask aligner system.
6. Vacuum furnace.
7. Low-Pressure Chemical Vapor Deposition (LPCVD) system.
8. Low-Stress Plasma-Enhanced Chemical Vapor Deposition (PECVD) system.
9. Reactive Ion Etching (RIE) system.
10. Air molding apparatus.

11. Focused Ion Beam (FIB) system.
12. e-beam/Thermal metal evaporator.
13. Flip-chip bonder system.

2.1.2 Solutions

1. PDMS (Sylgard 184).
2. SU-8 negative photoresist.
3. Silane derivative (i.e., hexamethyldisilazane, HDMS).
4. Polyimide (PI).
5. PI adhesion promoter.
6. Silicon wafer.
7. Dichlorosilane and ammonia gases.
8. Etching solution: 30 % KOH.
9. 70 % ethanol.
10. Rinsing solution: deionized (DI) water plus 10 % sodium dodecyl sulfate (SDS).
11. Nitrogen.
12. Standard Phosphate Buffered Saline (PBS) solution: 137 mM NaCl, 2.7 mM KCl, 10 mM Na_2HPO_4 and 2 mM KH_2PO_4. The pH is adjusted to 7.4.
13. Poly-D-Lysine (PDL) saline solution: 33 mg of PDL in 1 L of standard PBS.

2.2 Primary Cortical Neurons Culture

2.2.1 Instruments

1. Filter Stericups.
2. Conical Tubes (15 mL, 50 mL).
3. Water bath.
4. Guillotine.
5. Surgical scissors (one large and one small).
6. Two 145 mm petri dishes.
7. Three 100 mm and six 60 mm petri dish.
8. Three pairs of forceps (two blunt end tips and one pair with sharp tips).
9. Two pairs of curved end forceps.
10. One 20-gauge needle.
11. Centrifuge with capability to fit 50 mL tubes and rotate at 1,000 rpm.
12. Trypan blue.
13. One 1.5 mL Eppendorf vial.
14. Hemocytometer.
15. Microscope for counting of cells.

2.2.2 Solutions

1. Plating medium: 400 mL of EMEM, 50 mL of Horse Serum (10 %), 50 mL of Fetal Bovine Serum (FBS, 10 %) and 25 mM Glucose (*see* **Note 1**). The medium should be filtered using a sterilized filter with Stericup.
2. Anesthetic gas containing isoflurane 4 %.

3 Methods

The fabrication processes for PPC chips in silicon, PI and PDMS are described. PDMS being the most common material used for microfluidics, it is useful regardless of the material chosen for the membrane which will be perforated to produce PPC apertures. We will therefore first describe a protocol for the fabrication of a PDMS microfluidic system.

3.1 Microfluidic System

Starting from a single micro-structured master for the microfluidic system, replica molding can be used to obtain multiple low-cost polymeric copies with defined geometry [46]. Usually the master is a silicon wafer on which microstructures have been photo lithographically defined in a photoresist (i.e., SU-8, Microchem, widely used for Bio-MEMS applications). While the creation of the master requires a controlled environment (clean room) and specific machinery, replicas can be created using the master and standard lab equipment. In order to facilitate the removal of the polymer from the master, and hence improve its durability, a silane derivative is used to treat its surface. After this treatment the master can be stored indefinitely in a dry environment.

PDMS is usually obtained by mixing a polymer base and a cross-linker in 10:1 weight ratio. After mixing, in order to remove the absorbed air and to give to diffusion the time to finish the process, degassing in a vacuum chamber for at least 10 min is advised. Cross-linking and successive solidification of the polymer occurs at room temperature in about 24 h. This process can be accelerated by increasing the temperature (80 °C for 2 h). Error in mixing the exact ratio of polymer and cross-linker could cause the shrinking of the PDMS (~1.5 %) during curing and a deformation of the microstructures, making wafer level assembly over several centimeters difficult. To reduce shrinking variations, a multilayer soft lithography process can be used [47]. In this process, a thick PDMS slab prepared independently is used as a support for a thinner layer containing the channels. After a 5-min partial cure at 95 °C of the channels, the two PDMS layers are covalently bonded by treating the two surfaces with oxygen (or alternatively air) plasma and immediately assembling them in conformal contact. The resulting microfluidic chip is peeled off the master and fully cured. PDMS has been employed as the material for the insulating self-supported thin film in which apertures are being machined. Micromolding [39] and air molding [48] techniques have been reported to machine vertical apertures. Lateral patches through apertures patterned horizontally in the walls separating microchannels have also been developed [67]. Other groups have simply integrated a Silicon PPC fabricated in a microfluidic PDMS system [32, 33, 49].

3.2 PPC Silicon Chip

A common feature of all PPC chips is the membrane, perforated with PPC apertures, which separate the extracellular (culturing) chamber from the intracellular chamber. In the case of a silicon-based chip, this membrane should have a high dielectric rigidity, should be thin (because of the small dimension of the apertures) and mechanically strong (because it should be suspended). Silicon nitride in this respect is a better candidate than silicon dioxide.

The first fabrication process for a PPC chip in silicon was reported in 1999 [26]; we will describe here a more recent process leading to a lower shunt capacitance [35] (*see* also Fig. 3b). In this process, a 1 μm thick silicon nitride film is coated on both sides of a double-side polished (100) silicon wafer using low-pressure chemical vapor deposition (LPCVD) at 800–850 °C.

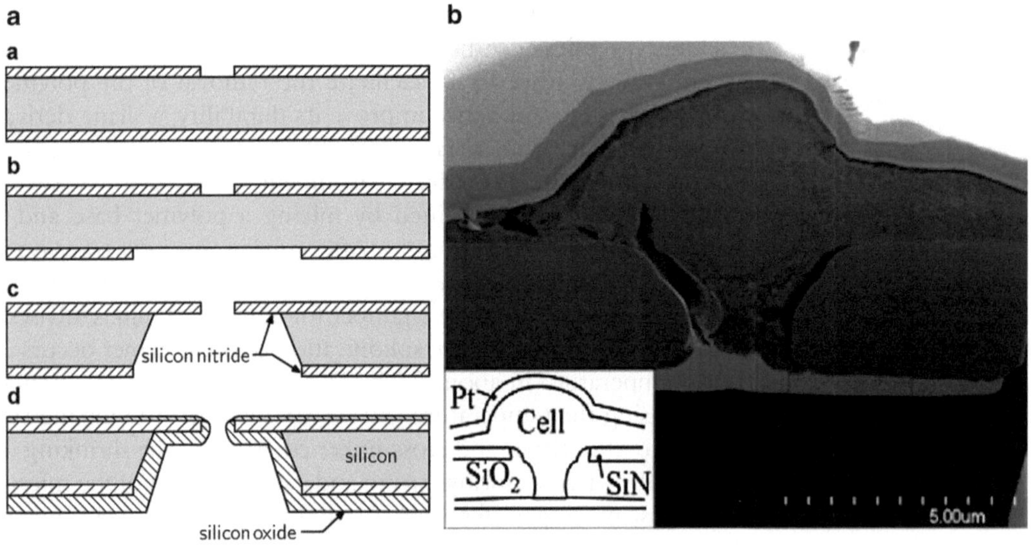

Fig. 3 Silicon patch-clamp chip fabrication process. (**a**) Fabrication process of a single-aperture silicon patch-clamp chip. (*a*) A circular 4 μm aperture was etched in the top silicon nitride film; (*b*) A square 600 μm × 600 μm was etched in the bottom silicon nitride film, aligned with the aperture; (*c*) KOH etches a pyramidal well and leaves silicon nitride self-standing; (*d*) A 5.4 μm thick silicon dioxide film passivates the well, rounding the edges of the aperture. A 0.1 μm layer (not shown) is also sputtered on the front side to facilitate cell attachment. (**b**) Scanning Electron Micrograph of a Focused Ion Beam section of a cultured P19 cell over an aperture in a SiN/SiO_2 stack membrane (see *insert*). To avoid damage during sectioning, the assembly is sandwiched between two platinum (*light gray*) layers. While some damage occurred during fixing, the cell visibly lowers into the rounded conical aperture. Adapted from [66]

LPCVD uses dichlorosilane and ammonia as source gases to grow dense films resistant to KOH etch and free of pinholes and defects. A photolithographic process is then used to create a photoresist mask and, by reactive ion etching (RIE) using CF_4 chemistry, open 3–4 μm apertures. A similar process opens large square windows in the silicon nitride film at the back of the wafer, aligned with <100> crystallographic axes and centered with the apertures in the front. The silicon bulk is then etched in 30 % KOH at 80 °C at a rate of about 1 μm/min; this anisotropic KOH etch results in an inverted truncated pyramid-shaped well with 54.74° facets on which a 100×100 μm^2 silicon nitride membrane is suspended. The chips are then "passivated" on the backside by a 5.4 μm thick low-stress plasma-enhanced chemical vapor deposition (PECVD) silicon dioxide film. The coating reduces the diameter of the aperture by 2 μm while rounding its edges. Finally, a 0.1 μm thick silicon dioxide film is deposited by the same method on the front side. The resulting surfaces are smooth, with a surface roughness of about 6–8 Å over a 1 μm^2 area, and hence suitable for obtaining a high proportion of high-quality cell to aperture seals; in Fig. 3a the first visual evidence of an intimate cell to aperture interaction is provided.

These single-aperture chips are assembled in dedicated Plexiglas packages to allow for easy perfusion and cell culture [50]. Packaged chips are sterilized prior to perfusion in an air plasma cleaner for 15 min to oxidize all surfaces and facilitate fluidic loading. The package permits to properly insulate the two membrane sides, avoiding leaks and minimizing the shunt capacitance of the chips, nominally 17 pF. This value, generally much lower than other silicon PPCs, is still higher than the typical 1–10 pF of a standard glass pipette. However, this disadvantage is compensated by the lower access resistance (~1.5 MΩ measured with a 150 nM PBS solution).

3.3 PPC Polyimide Chip

In order to minimize the capacitance and improve the recording quality of the measurements, the choice of a polymer membrane with a low dielectric constant is advantageous. Stett and collaborators [38] were the first to report the fabrication of a polyimide (PI) self-standing membrane mounted on a PDMS (Sylgard) backing. Here we describe a more recent microfluidic system that allows for the simultaneous probing of two synaptically connected cells cultured in the same bath, and that, through dedicated subterranean microfluidic channels, enables the monitoring of their synaptic connection [47]. In order to prevent the collapse of the microfluidic system, square pillars are integrated in the microfluidic channels as support (*see* Fig. 4b). Multiple considerations should be taken into account when designing microfluidics. On the one hand, the shunt capacitance is proportional to the section of the microfluidic channels and the inverse of the membrane thickness;

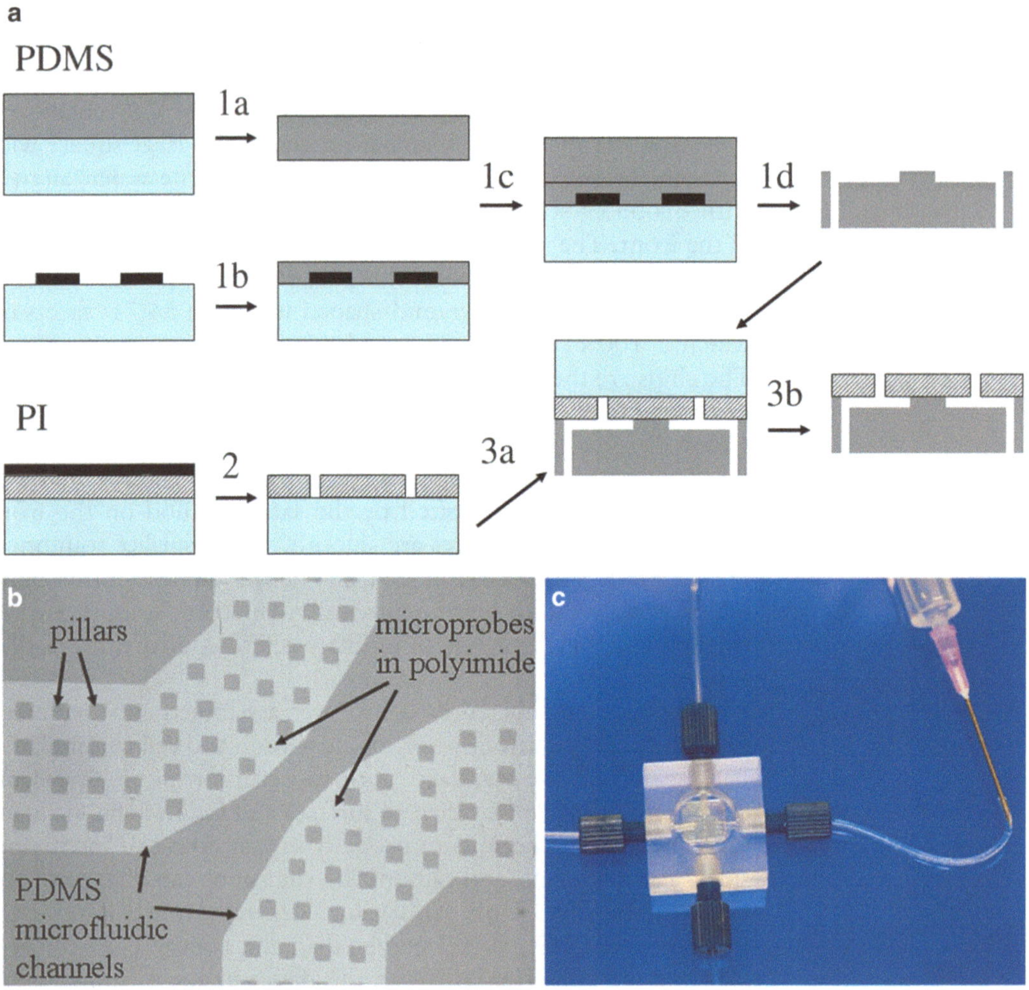

Fig. 4 PI–PDMS chip fabrication process. (**a**) PI–PDMS chip fabrication process: PDMS (*dark gray*, step 1) and PI (*hatched*, step 2) are processed independently on two Si wafers (*light green*); alignment-bonding (step 3a) and release (step 3b) form the final microchip. (**b**) PI-PDMS chip with two apertures, each with a dedicated microfluidic channel for perfusion and measurement, seen through the transparent PI film in *lighter gray*. The channels are supported by many square pillars part of the PDMS structure. (**c**) Plexiglass enclosure with a top cell culture vial and a dedicated input and output for each aperture

on the other hand, the access resistance that is the sum of the resistance of the aperture (directly proportional to the membrane thickness) and the resistance of the microfluidic channel (inversely proportional to its section). Increasing the thickness will decrease capacitance but increase access resistance. These two parameters can however be minimized independently by designing deep and narrow microfluidic channels. A 3 μm thick PI membrane with 3 μm round apertures suspended on 200 μm wide and 10 μm deep PDMS micro-channels will result in 7.7 pF capacitance, similar to that of a glass pipette, and an access resistance of 1.6 MΩ (with 150 mM PBS solution); these values are compatible with high fidelity patch-clamp recordings [1].

The chip fabrication process [47] for both microfluidics (*see* Subheading 3.1) and PI membrane is summarized in Fig. 4a. The PI membrane is patterned by photolithography and reactive ion etching (RIE). First, a 3 μm thick PI film is spin-coated on a silicon wafer and hard cured in three temperature stages up to 350 °C. Second, an aluminum mask layer is evaporated on the film, defined by lithography and transferred through the PI by oxygen RIE; the residual aluminum is subsequently removed in a metal etchant. Third, the PI film is aligned and bonded with the microfluidics. This is achieved at wafer level using a flip-chip bonder after treating PDMS with air plasma and PI with an adhesion promoter. The assembly is then heated at 65 °C for 2 h to stabilize the bond and peeled off the Si substrate. Finally, the wafers are diced in single chips, bonded to a lower glass support, and glued to a machined Plexiglas package fitted with connectors for easy micro-channel access (*see* Fig. 4c). After 5 min sterilization in air plasma, the microfluidic channels can be filled with electrophysiological recording solution.

3.4 PPC PDMS Chip

Using PDMS as a base material for the membrane has also several advantages. PDMS has physical properties very well suited for this application (*see* Subheading 2.1) and can be inexpensively processed in a standard laboratory. In addition, beside the already presented soft lithography, we can use a different, more powerful technique to create the micrometric apertures: air molding. This method was introduced by Sigworth's group [48]; it consists of curing the polymer on a heated metal plate with a small aperture through which nitrogen is blown. The airflow creates a bubble-shaped cavity that ends with a micrometric opening. Unlike soft lithography, this technique can be used to fabricate apertures with both concave and convex cross sections, and in particular apertures with smoother internal walls (desirable to improve the quality of the seal with cell membrane) and geometry especially suitable for patch clamping (featuring submicron apertures on the surface, but large bubble-like cavities underneath). The air molding technique is inspired by an industrial production technique applied to large scale production of plastic and can be used for serial production. This method can also be used to cheaply fabricate arrays of micro-apertures—arrays of 16 micro-apertures have been demonstrated (unpublished data). In addition, it can be combined with soft lithography to create apertures with extruded nozzles similar to the tips of the glass pipettes used for patch clamp: planar micropipettes [51]. The planar configuration permits this extension leaving the cells free to connect on a full 2D surface.

The procedure is simple and requires about 10 min to create a complete polymeric PPC chip (*see* Fig. 5), independently from

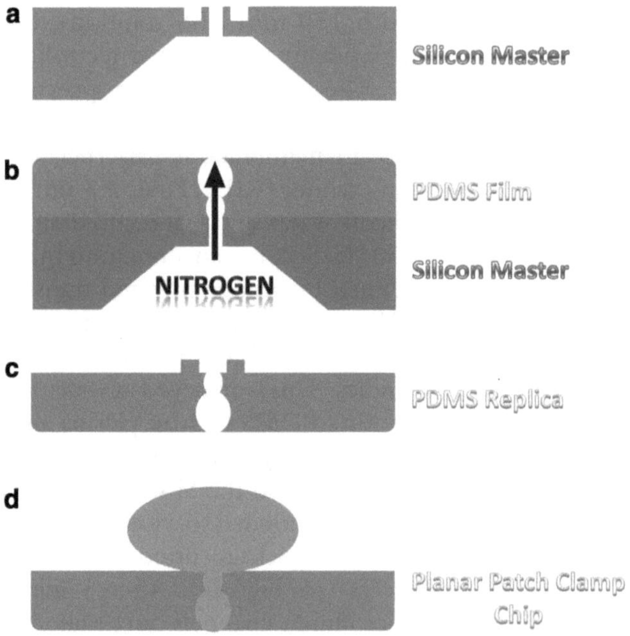

Fig. 5 Extended air molding. (**a**) A micro-structured perforated silicon master is covered with a thin film of liquid polymer. (**b**) Then a small stream of nitrogen, driven through a micrometric channel on the master, opens a cavity in the film. (**c**) Heat curing the polymer while the gas is flowing freezes the bubble-like structure of the air molded channel and then the film can be removed. The morphology of the master around the gas-driving pore is perfectly replicated by the polymer and the so obtained chip can be tuned in its geometry to different purposes and applications, in our case planar patch clamp (**d**)

the number of apertures in the array or the desired geometry. First, PDMS is spin-coated at 4,000 rpm over the master to create a 12 μm layer. Second, a positive pressure is applied to the backside of the membrane through a pressurizable chamber. When the pressure is high enough to overcome the impediment of the aperture and the surface tension of the polymer (about 25–35 psi for a 0.5 μm aperture), it perforates the film, opening a bubble-like cavity through its entire depth. While the gas is flowing and the channel in the polymer is open, the entire sample is heated under a halogen lamp, curing the polymer in about 5 min. The perforated PDMS membrane is then treated in an oxygen plasma (30 W for 60 s), bonded to a thicker support integrating the microfluidics channels, and the polymeric chip carefully peeled from the master.

The masters are usually silicon membranes perforated and micro-structured using a focused ion beam (FIB). This technique is particularly indicated for prototyping because of its flexibility.

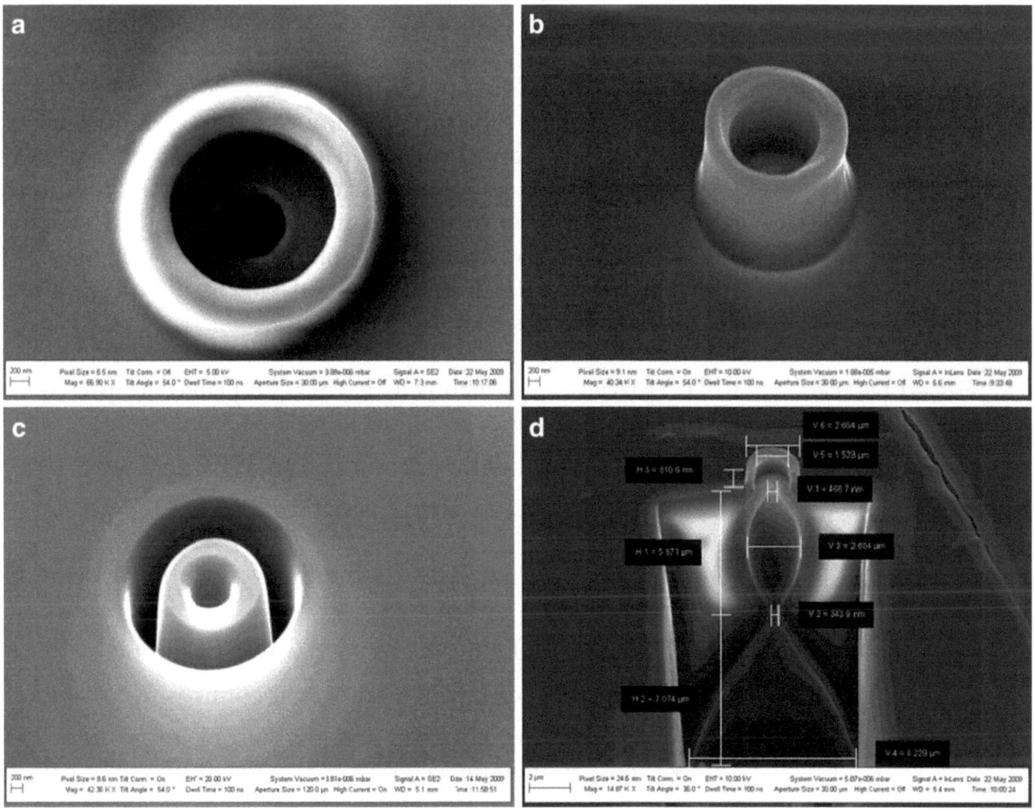

Fig. 6 SEM images of PDMS air molded samples. (**a**, **b**) Replicas obtained from the silicon master reported in (**c**). (**c**) 500 nm pore surrounded by a circular ring. (**d**) FIB cross section of the same sample showing the peculiar bubble-like structure generated by the air molding process

It consists in a direct mask-less fabrication process with great precision and reproducibility. Using FIB, we have realized masters with different morphologies, going from simple pores to complex geometries mimicking the tip of a glass pipette (*see* Fig. 6).

The bubble-like structure of the air molded cavities represents a great advantage for the electrical recordings, especially in whole-cell configuration, because it permits the use of small apertures, which are ideal for small cells like neurons, and does not compromise the quality of the seal while keeping access resistance lows. The slight increase in resistance associated with whole-cell access (due to the invagination of the plasma membrane) is minimized by this particular geometry. In standard PBS, a 0.5 μm aperture in a 12 μm thick PDMS membrane has an access resistance of about 4.8 MΩ, while a 1 μm aperture has a resistance of 1 MΩ. In contract, the shunt capacitance depends on the area of the membrane wetted by the solution. Exposing a 1 mm diameter of a PDMS membrane (0.768 mm^2) will result in a capacitance of about 1.7 pF. This value is low enough to allow for good quality recordings even with single aperture chips that are more adapt to experiments with suspended cells.

Lateral patches have also been demonstrated, whereby the microscopic aperture separates two channels that are in the same plane and therefore molded in the same step [52–54]. Fluxion Biosciences (http://fluxionbio.com/ionflux) has commercialized this idea producing an automated patch-clamp system that allows for recordings from a population of suspended cells positioned over the apertures by suction. In principle this kind of structure could be adapted to perform multiple parallel recordings of cells within a network. However, to our knowledge they have been used only for isolated cells from suspensions.

3.5 Primary Cortical Neurons Culture

3.5.1 Preparation for Dissection

1. The surfaces on which cells have to be plated are pre-coated with Poly-D-Lysine (*see* Subheading 3.6).
2. The plating medium is pre-warmed to 37 °C in a water bath.
3. The dissection hood surface is sterilized with 70 % ethanol.

3.5.2 Dissection

1. A time pregnant rat (E18) is anesthetized with 4 % isoflurane for 3–5 min.
2. The anesthetized rat is decapitated with a guillotine.
3. Using a pair of scissors, the rat belly is cut in a "U" shape.
4. The fetal sacs are removed and transferred into a large dish (145 mm) containing ice-cold PBS and kept on an ice pack.
5. The fetuses are released from the sacs and the heads are cut using scissors. The heads are divided evenly among three 100 mm dishes filled with ice-cold PBS contained on an icepack.
6. The brains are extracted from the skulls using a 20-gauge needle and a forceps (*see* **Note 2**), and then evenly distributed among five 60 mm petri dishes filled with ice-cold PBS on an icepack.
7. The cortices are dissected and transferred into a fresh 60 mm petri dish filled with ice-cold PBS kept on an icepack.
8. The cortices are transferred to a 50 mL conical tube (*see* **Note 3**) using a disposable transfer pipette and centrifuged at 1,000 rpm for 3 min.
9. Under the dissection hood, the supernatant is removed and 5 mL of pre-warmed plating medium added.
10. The cells are dissociated by trituration for ~8 times, ensuring that no large clumps remaining.
11. The 50 mL conical tube is filled with 45 mL of pre-warmed plating medium (final volume should be 50 mL). Allow debris and clumps of cells to settle at the bottom of the tube.
12. 45 mL of the supernatant containing suspended cells is transferred to a new 50 mL conical tube (*see* **Note 4**).

13. The cells are suspended by pipetting up and down 3–4 times. A 50 μL sample of the suspended cells is added to 200 μL of trypan blue in a 1.5 mL Eppendorf vial.
14. The cells are counted using a hemocytometer and then the neurons are ready to be plated on coated Poly-D-lysine (PDL) surfaces.

3.6 Neuronal Cells Placement

Once the chips are assembled, it is necessary to effectively couple the neuronal network with the PPC array. Consequently a strategy to control the positioning of the cells and guiding them on the apertures is needed. Topographical features have been shown to restrict cells and guide the growth of cellular processes [55–58]; however, chemical patterning is the favored method [59] for cell positioning because it improves cell adhesion to the substrate and promotes formation of a high-quality cell to aperture seal. The ability to synthesize networks of cells of almost arbitrary shapes by controlling cell positioning and axonal growth could also be used to investigate the relationship between topology and function. In order to induce placement of neurons on the chips, PDL (a cell adhesion factor commonly used for neuronal cells; *see* Subheading 3.5.1) patterns can be created on the chip's surface using different techniques [60]. Among these, micro-contact printing has been demonstrated to transfer chemicals with high spatial resolution [61, 62]. The main steps of the procedure [63] are summarized in Fig. 7.

Briefly, it consists of printing the chemicals as an ink using a micro-structured PDMS stamp. The stamp can be designed to create chemical features of sizes and spacing comparable to the one of the PPC array (*see* Subheading 3.1). In order to remove any remaining non-polymerized PDMS, stamps should be washed in solvents by reflux [64]. Subsequently, they are mounted on glass slides and washed in 70 % ethanol, deionized (DI) water, and 10 % SDS. Modification of the PDMS surface with SDS has been shown to enhance the transfer of poly-lysine to the substrate [65]. Stamps are then quickly rinsed in DI water, blow-dried with nitrogen and immersed for 30 min in a PDL saline solution (33 mg/mL PDL in standard PBS). After the stamps are properly inked and blow-dried, they are first applied to the chips with a constant and homogeneous pressure (500 g for 1 min) and then left on the chip for an additional minute without pressure. Finally, the chips are sterilized (2 min in 70 % ethanol), rinsed in DI water and dried. In order to align the PDL pattern with the PPC apertures, the stamp has to be aligned carefully with the chip during the printing process.

After PDL functionalization, the chips are ready to be plated with neuronal cells (i.e., primary cortical neurons). Neurons aggregate on the PDL patterns and their processes extend preferentially, though not exclusively, to immediate neighboring sites (*see* Fig. 8). Such control in cell placement is enough to guide the cells on the PPC apertures and permits parallel intracellular recording from connected neurons.

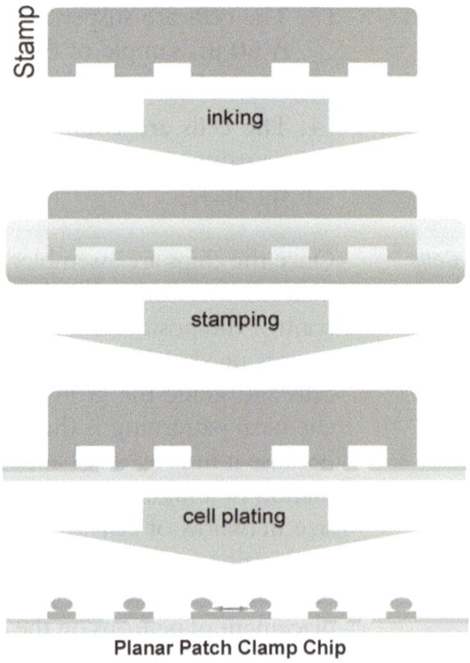

Fig. 7 Sketch summarizing the main steps of the micro-contact printing procedure. A PDMS stamp is inked in PDL saline solution and then stamped on the planar patch clamp chip in order to control the positioning and the connections of the neuronal cells

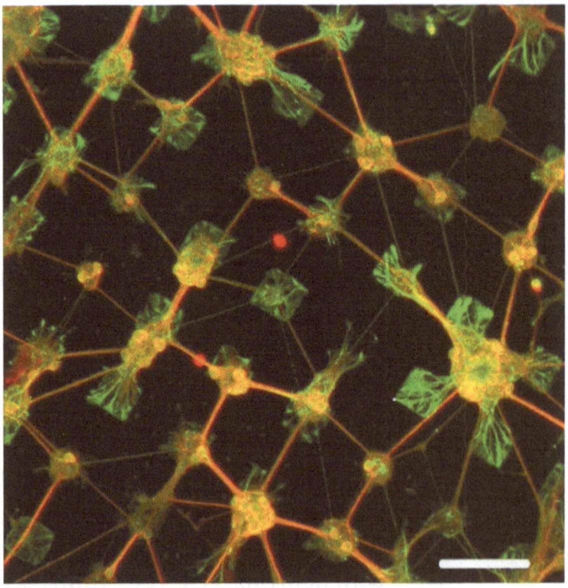

Fig. 8 Immunofluorescence staining of cultured brain cells on chips stamped with 50 μm PDL pattern spaced by 50 μm separation (scale bar: 100 μm). Neurons are stained *red* with MAP2, while astrocytes are stained *green* with GFAP

4 Notes

1. When choosing FBS, it is important to test different lots of FBS in order to ensure optimal growth of cultures. Do not completely deplete the current FBS, but store it in order to have a benchmark to compare to. This is more important for FBS than for Horse Serum.
2. The head is held using a pair of forceps in the eye sockets with and the skin and skull is pierced with a 20-gauge needle down the midline to expose the brain. To remove the brain, squeeze both sides of the head with a pair of forceps.
3. It is suggested to maintain ~5 mL volume with all the cortices collected.
4. It is important to avoid extracting cells from the very bottom of the tube in order to avoid plating debris or large aggregates of cells.

5 Future Perspective

Among the materials discussed above, silicon and quartz are the most compatible with conventional semiconductor manufacturing methods. The incorporation of silicon and quartz chips in microfluidic systems requires integration—this service is offered by a number of microfluidic foundries worldwide (see www.fluidicmems.com). Polyimide and PDMS are also compatible with semiconductor manufacturing methods. They are cheaper and, due to their mechanical flexibility, easier to integrate than silicon and quartz. PDMS is less accepted in mainstream fabrication; however, it is easiest material to use for monolithic integration of the PPC and the microfluidic systems around it. In addition, it presents definite advantages for cell culture such as oxygen and humidity permeation.

Neurodegenerative diseases exert a staggering impact on society by causing human suffering and economic burden, and have been identified by the World Health Organization as an escalating challenge for the aging populations of developed countries. The search for new therapeutic targets and the need for faster, higher-content assessment of drug candidates have created an urgent demand for improved cell-based drug screening technologies. Understanding brain function, how the brain is affected by disease, and how therapeutics can protect or restore the brain, requires detailed understanding of neuronal communication from the network level down to the level of individual synapses and ion channels. Since all brain functions involve networks of cells, studying synaptic communication is fundamental to understanding neuronal function. Ion channels control transmembrane ionic currents and voltage and, in so doing, initiate signal transduction pathways within cells, resulting in propagation of signaling between neuronal synapses. Pathologies often affect specific ion channel types and manifest as synaptic dysfunction. Due to their central role in signal initiation and propagation, ion channels are a primary target for therapeutic intervention. Traditionally, ion channel activity has been monitored in single cells using the glass pipette patch-clamp technique. However, the patch-clamp technique is severely limited by the number of cells which can simultaneously record. Recent advances in robotics have enabled measurements in brain slices from simultaneous pipette patch-clamp electrodes, but these techniques are still very demanding and required highly skilled personal. This does not meet demands of pharmaceutical companies, which require high throughput screening on patch-clamped neuronal cultures in order to evaluate the effect of drugs on synaptic transmission and ion channel activity. Consequently, a PPC chip that combines key benefits of both conventional patch clamp and MEAs on a chip will be ideal for pharmaceutical companies, as well as for neuroscientists to provide insight in neural computation, the nature of neurological diseases, and the treatment of diseases. Neurons that are aligned to the apertures by patterned chemical adhesion or guidance cues can subsequently form synaptic connections; this capability offer the opportunity to develop more physiologically relevant in vitro models, extract unique information on brain function and network communication, and better predictions of therapeutic efficacy, notably as pertaining to synaptic physiology. Additionally, the planar nature of the devices allows for the possibility of integration with other technologies; for instance, combining such a system with PPC electrodes would add electrophysiological control to cell biology experiments, and provide even more detail to synaptic dynamics investigations.

References

1. Sigworth FJ, Klemic KG (2005) Microchip technology in ion-channel research. IEEE Trans Nanobioscience 4:121–127
2. Fertig N, Meyer C, Blick RH et al (2001) Microstructured glass chip for ion-channel electrophysiology. Phys Rev E Stat Nonlin Soft Matter Phys 64(1):040901
3. Behrends JC, Fertig N (2007) Planar patch clamping. In: Walz W (ed) Patch-clamp analysis, advanced techniques. Humana Press, Totowa, NJ, pp 411–433
4. Sondermann M, George M, Fertig N et al (2006) High-resolution electrophysiology on a chip: transient dynamics of alamethicin channel formation. Biochim Biophys Acta 1758(4):545–551
5. Martina M, Luk C, Py C et al (2011) Recordings of cultured neurons and synaptic activity using patch-clamp chips. J Neural Eng 8(3):034002. doi:10.1088/1741-2560/8/3/034002
6. Harms GS, Orr G, Montal M et al (2003) Probing conformational changes of gramicidin ion channels by single-molecule patch-clamp fluorescence microscopy. Biophys J 85: 1826–1838
7. Gullo MR, Akiyama T, Frederix PLTM et al (2005) Towards a planar sample support for in situ experiments in structural biology. Microelectron Eng 78–79:571–574
8. Dunlop J, Bowlby M, Peri R et al (2008) High-throughput electrophysiology: an emerging paradigm for ion-channel screening

and physiology. Nat Rev Drug Discov 7: 358–368
9. Taketani M, Baudry M (2006) Advances in network electrophysiology: using multi-electrode arrays. Springer, New York, NY
10. Jones IL, Livi P, Lewandowska MK et al (2011) The potential of microelectrode arrays and microelectronics for biomedical research and diagnostics. Anal Bioanal Chem 399: 2313–2329
11. Berdondini L, van der Wal PD, Guenat O et al (2005) High-density electrode array for imaging in vitro electrophysiological activity. Biosens Bioelectron 21(1):167–174
12. Maccione A, Garofalo M, Nieus T et al (2012) Multiscale functional connectivity estimation on low-density neuronal cultures recorded by high-density CMOS micro electrode arrays. J Neurosci Methods 207(2):161–171
13. Fromherz P (2006) Three levels of neuroelectronic interfacing: silicon chips with ion channels, nerve cells, and brain tissue. Ann N Y Acad Sci 1093:143–160
14. Patolsky F, Timko BP, Yu G et al (2006) Detection, stimulation, and inhibition of neuronal signals with high-density nanowire transistor arrays. Science 313:1100–1104
15. Frey U, Egert U, Heer F et al (2009) Microelectronic system for high-resolution mapping of extracellular electric fields applied to brain slices. Biosens Bioelectron 24: 2191–2198
16. Berdondini L, Imfeld K, Maccione A et al (2009) Active pixel sensor array for high spatio-temporal resolution electrophysiological recordings from single cell to large scale neuronal networks. Lab Chip 9:2644–2651
17. Ferrea E, Maccione A, Medrihan L et al (2012) Large-scale, high-resolution electrophysiological imaging of field potentials in brain slices with microelectronic multielectrode arrays. Front Neural Circ 6:80. doi:10.3389/fncir.2012.00080
18. Shein M, Greenbaum A, Gabay T et al (2009) Engineered neuronal circuits shaped and interfaced with carbon nanotube microelectrode arrays. Biomed Microdevices 11(2): 495–501
19. Huys R, Braeken D, Micholt L et al (2011) Micro-sized syringes for single-cell fluidic access integrated on a micro-electrode array CMOS chip. Conf Proc IEEE Eng Med Biol Soc 2011:7650–7653. doi:10.1109/IEMBS.2011.6091885
20. Hai A, Shappir J, Spira ME (2010) In-cell recordings by extracellular microelectrodes. Nat Methods 7:200–202
21. Mealing G, Py C (2011) Patch-clamp array neurochips: value in interrogating simple neuronal networks with high resolution. Expert Rev Med Devices 8(1):3–5
22. Mealing G, Bani-Yaghoub M, Tremblay R et al (2005) Application of polymer microstructures with controlled surface chemistries as a platform for creating and interfacing with synthetic neural networks. In: Proceedings of the IEEE international joint conference on neuronal networks, vol 5. IEEE, New York, NY, pp 3115–3116. doi:10.1109/IJCNN.2005.1556425
23. Branch DW, Corey JM, Weyhenmeyer JA et al (1998) Microstamp patterns of biomolecules for high-resolution neuronal networks. Med Biol Eng Comput 36(1):135–141
24. Offenhausser A, Bocker-Meffert S, Decker T et al (2007) Microcontact printing of proteins for neuronal cell guidance. Soft Matter 3: 290–298
25. Park TH, Shuler ML (2003) Integration of cell culture and microfabrication technology. Biotechnol Prog 19(2):243–253
26. Fertig N, Tile A, Blick RH et al (1999) Stable integration of isolated cell membrane patches in a nanomachine aperture: a step towards a novel device for membrane physiology. Appl Phys Lett 77(8):1218–1220
27. Schmidt C (2000) A chip-based biosensor for the functional analysis of single ion channels. Angew Chem Int Ed Engl 39(17): 3137–3140
28. Gad-el-Hak M (2001) The MEMS handbook. CRC Press, Boca Raton
29. Kaul RA, Syed NI, Fromherz P (2004) Neuron-semiconductor chip with chemical synapse between identified neurons. Phys Rev Lett 92(3):038102
30. Lehnert T, Gijs MAM, Netzer R et al (2002) Realization of hollow SiO2 micronozzles for electrical measurements on living cells. Appl Phys Lett 81:5063–5065
31. Stett A, Burkhardt C, Weber U et al (2003) CYTOCENTERING: a novel technique enabling automated cell-by-cell patch clamping with the CYTOPATCH chip. Receptors Channels 9(1):59–66
32. Pantoja R, Nagarah JM, Starace DM et al (2004) Silicon chip-based patch-clamp electrodes integrated with PDMS microfluidics. Biosens Bioelectron 20(3):509–517
33. Matthews B, Judy JW (2006) Design and fabrication of a micromachined planar patch-clamp substrate with integrated microfluidics for single-cell measurements. J Microelectromech Syst 15:214–222

34. Curtis JC, Baldwin K, Dworak BJ et al (2008) Seal formation in silicon planar patch-clamp microstructures. J Microelectromech Syst 17: 974–983
35. Py C, Denhoff MW, Martina M et al (2010) A novel silicon patch-clamp chip permits high-fidelity recording of ion channel activity from functionally defined neurons. Biotechnol Bioeng 107(4):593–600
36. Sordel T, Kermarrec F, Sinquin Y et al (2010) The development of high quality seals for silicon patch-clamp chips. Biomaterials 31:7398–7410
37. Fertig N, Behrends JC, George M et al (2003) Microstructured apertures in planar glass substrates for ion channel research. Receptors Channels 9:29–40
38. Stett A, Bucher V, Burkhardt C et al (2003) Patch-clamping of primary cardiac cells with micro-openings in polyimide films. Med Biol Eng Comput 41:233–240
39. Klemic KG, Klemic JF, Reed MA et al (2002) Micromolded PDMS planar electrode allows patch clamp electrical recordings from cells. Biosens Bioelectron 17(6–7):597–604
40. Chen C-Y, Tu T-Y, Chen C-H et al (2009) Patch clamping on plane glass-fabrication of hourglass aperture and high-yield ion channel recording. Lab Chip 9:2370–2380
41. Nagarah JM, Paek E, Luo Y et al (2010) Batch fabrication of high-performance planar patch-clamp devices in quartz. Adv Mater 22: 4622–4627
42. Metz S, Holzer R, Renaud P (2001) Polyimide-based microfluidic devices. Lab Chip 1:29–34
43. Metz S, Bertsch A, Bertrand D et al (2004) Flexible polyimide probes with microelectrodes and embedded microfluidic channels for simultaneous drug delivery and multichannel monitoring of bioelectric activity. Biosens Bioelectron 19:1309–1318
44. Kristensen BW, Noraberg J, Thiébaud P et al (2001) Biocompatibility of silicon-based arrays of electrodes coupled to organotypic hippocampal brain slice cultures. Brain Res 896: 1–17
45. Xia Y, Whitesides GM (1998) Soft lithography. Angew Chem Int Ed 37:550–575
46. Weibel DB, Diluzio WR, Whitesides GM (2007) Microfabrication meets microbiology. Nat Rev Microbiol 5:209–218
47. Martinez D, Py C, Denhoff MW et al (2010) High-fidelity patch-clamp recordings from neurons cultured on a polymer microchip. Biom Microdev 12:977–985
48. Klemic KG, Klemic JF, Sigworth FJ (2005) An air-molding technique for fabricating PDMS planar patch-clamp electrodes. Pflugers Arch 449:564–572
49. Alberti M, Snakenborg D, Lopacinska J et al (2010) Characterization of a patch-clamp microchannel array towards neuronal networks analysis. Microfluid Nanofluid 9:963–972
50. Py C, Martina M, Monette R, Comas T, Denhoff MW, Luk C et al. (2012) Culturing and electrophysiology of cells on NRCC patch-clamp chips. *J Vis Exp.* (60), e3288, doi: 10.3791/3288
51. Bosca A, Magrassi R, Firpo G et al. (2009) Air molding for planar patch clamp on adherent neuronal networks. IEEE-NANO 2009. 9th IEEE conference on nanotechnology, July 26–30, 2009 Genoa, Italy
52. Lau AY, Hung PJ, Wu AR et al (2006) Open-access microfluidic patch-clamp array with raised lateral cell trapping sites. Lab Chip 6:1510–1515
53. Tang KC, Reboud J, Kwok YL et al (2010) Lateral patch-clamping in a standard 1536-well microplate format. Lab Chip 10:1044–1050
54. Seo J, Ionescu-Zanetti C, Diamond J et al (2004) Integrated multiple patch-clamp array chip via lateral cell trapping junctions. Appl Phys Lett 84:1973–1975
55. Faid K, Voicu R, Bani-Yaghoub M et al (2005) Rapid fabrication and chemical patterning of polymer microstructures and their applications as a platform for cell cultures. Biomed Microdevices 7:179–184
56. Merz M, Fromherz P (2005) Silicon chip interfaced with a geometrically defined net of snail neurons. Adv Funct Mater 15:739–744
57. Sorkin R, Greenbaum A, David-Pur M et al (2009) Process entanglement as a neuronal anchorage mechanism to rough surfaces. Nanotechnology 20(1):015101. doi:10.1088/0957-4484/20/1/015101
58. Maher MP, Pine J, Wright J et al (1999) The neurochip: a new multielectrode device for stimulating and recording from cultured neurons. J Neurosci Meth 87:45–56
59. Wheeler BC, Brewer GJ (2010) Designing neural networks in culture: experiments are described for controlled growth, of nerve cells taken from rats, in predesigned geometrical patterns on laboratory culture dishes. Proc IEEE 98:398–406
60. Petrelli A, Marconi E, Salerno M et al (2013) Nano-volume drop patterning for rapid on-chip neuronal connect-ability assays. Lab Chip 13(22):4419–4429
61. Qin D, Xia Y, Whitesides GM (2010) Soft lithography for micro- and nanoscale patterning. Nat Protoc 5:491–502

62. Marconi E, Nieus T, Maccione A et al (2012) Emergent functional properties of neuronal networks with controlled topology. PLoS One 7(4):e34648. doi:10.1371/journal.pone.0034648
63. Charrier A, Martinez D, Monette R et al (2010) Cell placement and guidance on substrates for neurochip interfaces. Biotechnol Bioeng 105(2):368–373
64. Voicu R, Faid K, Farah AA et al (2007) Nanotemplating for two-dimensional molecular imprinting. Langmuir 23(10):5452–5458
65. Chang JC, Brewer GJ, Wheeler BC (2003) A modified microstamping technique enhances polylysine transfer and neuronal cell patterning. Biomaterials 24:2863–2870
66. Py C, Martina M, Diaz-Quijada GA et al (2011) From understanding cellular function to novel drug discovery: the role of planar patch-clamp array chip technology. Front Pharmacol 2:51. doi:10.3389/fphar.2011.00051
67. Ionescu-Zanetti C, Shaw RM, Seo J et al (2005) Mammalian electrophysiology on a microfluidic platform. Proc Natl Acad Sci U S A 102:9112–9117

Part II

Physiology

Chapter 7

Methods for the Study of Synaptic Receptor Functional Properties

Enrica Maria Petrini and Andrea Barberis

Abstract

The generation of a synaptic current at the postsynaptic element (PSCs) is the result of a dynamic sequence of events including the release of the neurotransmitter, its diffusion in the synaptic cleft, and the activation of neurotransmitter receptors located at the postsynaptic side. It is widely accepted that the amplitude and the duration of PSCs are largely dictated by the gating properties of postsynaptic receptors. However, the knowledge of the properties of postsynaptic receptors is mostly derived from steady-state analysis, a condition that is substantially different from the non-equilibrium activation of synaptic receptors imposed by submillisecond neurotransmitter exposures. Given the technical limitations to reproduce the brief "synaptic-like" agonist pulse durations, the functioning of postsynaptic receptors during synaptic transmission is not fully elucidated and the "on-demand" postsynaptic activation of synapses cannot be easily achieved. In this chapter, we review the diverse approaches to study receptor gating at times relevant for synaptic transmission and novel optical/optogenetic techniques for controlling synaptic activity at the postsynaptic level. In addition, we emphasize the role of non-equilibrium in unmasking specific features of synaptic receptor gating and the recent advances in photonics for the light-control of neuronal activity at the single-receptor level.

Key words Synaptic transmission, Ligand-gated channels, Receptor gating, Neurotransmitter transient, Ultrafast perfusion, Concentration jumps, Neurotransmitter uncaging, Light-gated receptors, Optogenetics, Plasmonic devices

1 Introduction

The elementary information transfer between two neurons is represented by the presynaptic release of a synaptic neurotransmitter vesicle (quantum) that activates neurotransmitter receptors at the postsynaptic side to elicit a unitary postsynaptic current. The amplitude and the duration of such postsynaptic signals play a pivotal role in the temporal and spatial summation of synaptic inputs, a fundamental process that is responsible for the dendritic synaptic filtering, integration, and control of the neuronal firing activity [1]. The importance of the PSCs shape in the control of the neuronal network functioning is highlighted by the fact that

Marzia Martina and Stefano Taverna (eds.), *Patch-Clamp Methods and Protocols*, Methods in Molecular Biology, vol. 1183, DOI 10.1007/978-1-4939-1096-0_7, © Springer Science+Business Media New York 2014

amplitude/timing of PSCs is highly diverse and finely modulated in both different neuronal subtypes and in specific neuronal subregions (axon, soma, different dendritic locations). In addition, alterations of the amplitude and duration of both excitatory and inhibitory postsynaptic currents are believed to underlie important forms of synaptic plasticity [2]. The decay time constant of both excitatory and inhibitory postsynaptic currents may vary from few milliseconds to several tens of milliseconds, thus allowing for a wide-range control of the timing of neuronal signaling. This is exemplified at the GABAergic inhibitory system, where different subclasses of interneurons control the activity of pyramidal neurons with inhibitory postsynaptic currents (IPSCs) showing highly diverse timing and kinetic properties. Such IPSC features are believed to underlie important network properties including the generation of rhythmic activity for the implementation of specific brain states. Traditionally, it has been assumed that, following uniquantal presynaptic release, the concentration of neurotransmitter reached in the cleft would super-saturate the postsynaptic receptors [3], thus implying that the kinetics properties of unitary PSCs were solely dictated by the gating properties of synaptic receptors, independently from the mechanisms of presynaptic release. This idea was also supported by the fact that the kinetics of currents mediated by specific receptors (e.g., AMPA, NMDA, glycine, and GABA receptors) qualitatively matched those of the corresponding synaptic currents [4–7]. This scenario suggested a "division of labor" among different receptor classes and/or subtypes showing specific neurotransmitter affinity and gating properties that results in an extraordinary variety of synaptic current shapes. However, the observation that, after the release of a single synaptic vesicle the neurotransmitter remains in the cleft for ~100 μs [8] added a further level of complexity in the mechanisms of postsynaptic receptor activation. Indeed, such neurotransmitter synaptic exposure is briefer than the time needed by the receptors to reach the fully open state, thus implying that postsynaptic receptors are activated in "non-equilibrium." In these conditions, the amplitude and time course of the synaptic current are dictated by a complex interplay between receptor gating properties and the time course of neurotransmitter concentration in the synaptic cleft. The understanding of synaptic receptor activation in "synaptic conditions" is necessary to define the "transfer function" between presynaptic release and the generation of the postsynaptic current. This relation can be only established by studying the kinetics of currents elicited by "controlled" pulses of exogenous neurotransmitter approaching the duration of synaptic neurotransmitter exposures. To date, the knowledge about the properties of postsynaptic receptors activated in synaptic conditions has been limited by the difficulty to control and reproduce the synaptic neurotransmitter exposures.

To overcome this limitation, a considerable effort has been made over the last two decades to build and optimize devices that allow for the submillisecond perfusion of neurotransmitters to ligand-gated receptors. The use of such "ultrafast perfusion" approaches revealed new features of receptor gating that would have been undetectable in "equilibrium" experiments and which are responsible for the specific characteristics of synaptic currents [9, 10]. In addition, the study of receptors at "synaptic" time exposures showed unexpected interactions between pharmacological agents and receptor functioning, thus emphasizing the concept that the pharmacological action of drugs targeting postsynaptic receptors closely depends on the dynamics of transmitter concentration in the synaptic cleft. In the present review we discuss the methodologies for the delivery of brief submillisecond neurotransmitter pulses, by emphasizing the conceptual implications of receptor activation in "non-equilibrium" conditions. Besides the classical fast-perfusion approaches, we also focus on the recent technological advances exploiting optical/optogenetic tools to achieve high temporal and spatial resolution of the synaptic receptor activation at intact synapses. In addition, we consider novel illumination devices capable of focusing light in sub-diffraction limited spots [11] for the control of optogenetic/optical tools at the single molecule level.

2 Theta Glass-Based UltraFast Perfusion

Postsynaptic currents mediated by fast ionotropic ligand-gated receptors show fast onset and decay kinetics. Unitary excitatory and inhibitory currents, indeed, peak in less than 1 ms and relax back to the baseline in few milliseconds and few tens of millisecond, respectively. Over the last two decades, an enormous effort has been made to build perfusion systems able to match such fast current kinetics by the application of controlled exogenous agonist. Indeed, the possibility to manipulate the concentration of the receptor agonist allows for determining the "dose–response curve," the fundamental input–output function operated by synapses. With this approach it is possible to define the basic properties of receptors functioning, including (1) the receptor affinity/efficacy, (2) the receptor modulation by pharmacological agents and (3) the relationship between receptor structure and function. However, in order to be significant for synaptic transmission the functional link between agonist and receptor opening must be established in conditions approaching those occurring at the synapses. The first perfusion systems aimed at exogenously activating ligand-gated neurotransmitter receptors were based on the replacement of the extracellular medium with a neurotransmitter-containing solution in the recording chamber. In these conditions, the duration of the solution exchange could vary from tens of seconds to minutes, more than five orders of magnitude slower than synaptic neurotransmitter transients. Devices exploiting small diameter pipettes positioned in front of the neuron to be perfused (including multibarrel systems, U-tubes, Y-tubes, etc.) dramatically enhanced the perfusion rate performances by allowing for complete solution exchange around the neuron of interest in hundreds of milliseconds to few tens of milliseconds. The use of these devices has significantly improved the understanding of the basic features of most fast ligand-gated neurotransmitter receptors, by clarifying several aspects of both the structure/function and the pharmacological profile of excitatory and inhibitory receptors. However, even at the best of their performances, such devices yielded a solution exchange time that was critically slower than synaptic neurotransmitter exposures. A major breakthrough in the delivery of millisecond (or even submillisecond) agonist pulses was achieved by (1) creating a perfusion device with a thin interface between the control solution and neurotransmitter-containing solutions to be moved over the sample and (2) limiting the surface of the sample to be perfused [9, 10, 12]. In these conditions, indeed, it is possible to apply (and remove) the agonist-containing solution in tens of microseconds. A thin and stable interface between two solutions can be easily created by exploiting theta-glass tubes (so called because the profile of the capillary mouth recalls the Greek letter "theta"—*see* Fig. 1a). Briefly, a theta barrel forms two independent channels continuously conveying the control solution and the neurotransmitter-containing solution, respectively (Fig. 1a). The patch pipette, with an excised-out patch on its tip, is positioned in front of the control stream. By means of a fast piezo electric translator, the theta-barrel capillary is switched in a way that the patch pipette tip crosses the solution interface and jumps in the neurotransmitter stream. Subsequently, the theta is switched back to the original position bringing the patch pipette tip in the control solution stream (Fig. 1a, inset). Using this device it is possible to obtain solution exchange time of ~25–100 μs, while the shortest pulse duration achievable is typically ~1 ms (Fig. 1b) (but *see* also [13]). Such brief agonist exposure, although still longer than that occurring at the synapse during synaptic transmission, represents a good approximation of the "synaptic conditions," thus allowing for studying postsynaptic receptors activated in a time range relevant for synaptic transmission. In addition, since currents mediated by fast-ligand gated channels display submillisecond rise time (~300 μs), the exchange time of ~25–100 μs is adequate to satisfactory resolve the activation kinetics of postsynaptic receptors (Fig. 1c–e). By exploiting this system, therefore, it is possible to achieve the "concentration clamp" through the application of the neurotransmitter pulses at precise concentration and duration. These systems are usually referred to as *ultrafast perfusion*, *concentration jumps*, *or liquid filaments.*

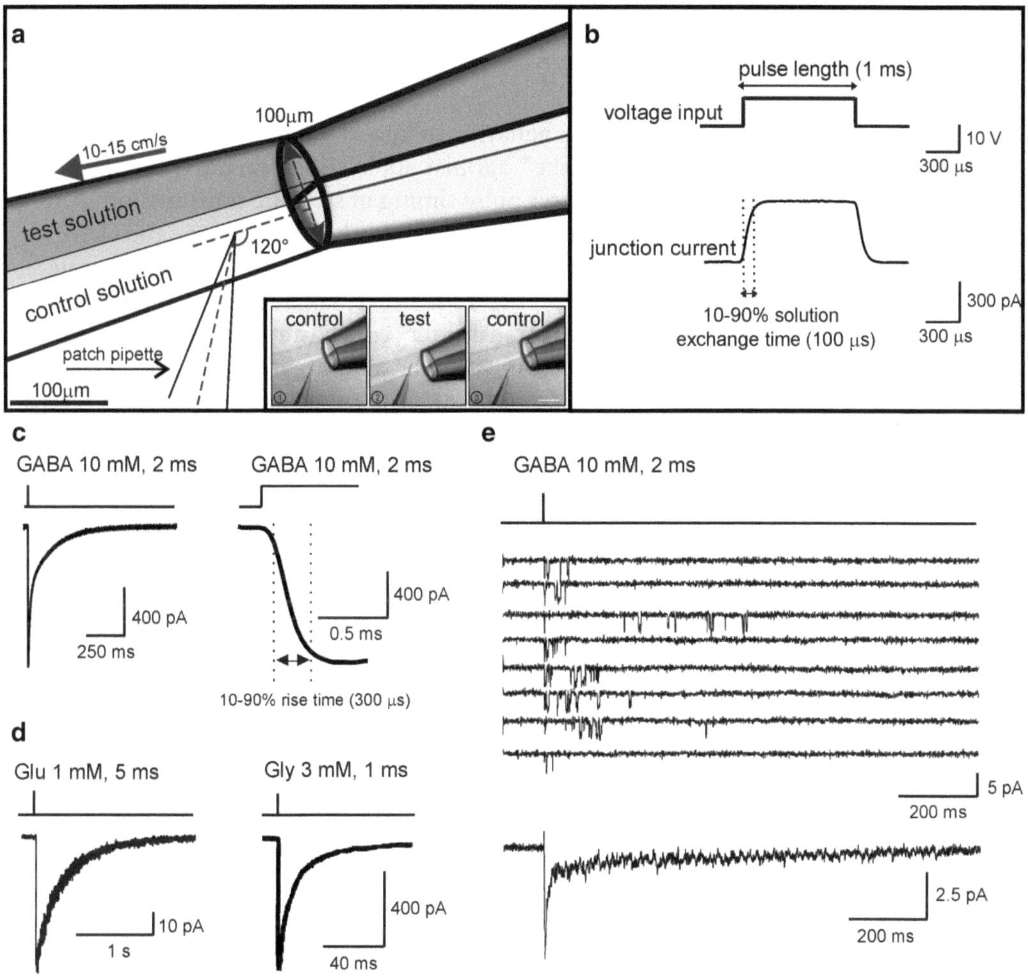

Fig. 1 Receptor activation with the theta-barrel based ultrafast perfusion system. (**a**) Schematic representation of the theta-barrel based ultrafast perfusion system. The control solution and the neurotransmitter-containing solution (test solution) continuously flow in the theta pipette at the speed of 10–15 cm/s generating a thin solution interface. The patch pipette axis forms a 120° angle with the solution stream axis in order to reduce the mechanic stress. *Inset*: Sequence of the theta pipette translations. (*1*) The patch pipette is in the control solution stream; (*2*) the piezo electric translator quickly moves the theta tube so that the patch pipette is perfused with the test solution; (*3*) the theta tube is back to the original position. (**b**) Test of the solution exchange time with open-tip liquid junction current. *Upper panel*: Input voltage pulse applied to the piezo electric translator (1 ms, 10 V). *Lower panel*: profile of the solution exchange measured at the open tip patch pipette following 1 ms piezo translator pulse. (**c**) *Left panel*: $GABA_AR$-mediated currents evoked by brief (2 ms) pulses of saturating GABA (10 mM). Currents were recorded in excised-out patches pulled from HEK293-cells transfected with $\alpha1\beta2\gamma2$ $GABA_AR$ subtype. *Right panel*: same trace as in (**a**) shown in expanded time scale in order to appreciate the fast current onset (10–90 % rise time: 300 μs). (**d**) *Right Panel*: NMDAR-mediated current (GluN1-GluN2C subtype) in response to 1 mM glutamate and 0.5 mM glycine (5 ms pulse). *Left panel*: Currents mediated by GlyRs (1 ms, 3 mM glycine). (**e**) Example of $GABA_AR$-mediated currents evoked by short pulses of saturating GABA resolved at single-channel level. *Upper panel*: eight sweeps showing the stochastic single channel activation. *Bottom panel*: macroscopic current obtained by averaging the eight single channel sweeps shown in the *upper panel*. Modified with permission from [9]

3 Characterization of Synaptic Receptor Properties by Exploiting UltraFast Perfusion

3.1 Synaptic Current Amplitude Depends on Synaptic Neurotransmitter Exposure

The study of the behavior of ligand-gated receptors challenged by brief "synaptic like" agonist applications revealed the importance of neurotransmitter pulse timing in synaptic transmission. After the release of the synaptic vesicle, at both glutamatergic and GABAergic synapses the neurotransmitter concentration in the synaptic cleft reaches 1–3 mM [14, 15]. When applied in equilibrium conditions, such neurotransmitter concentrations are saturating (or near-saturating) for most of the ligand-gated receptors. However it has to be born in mind that, at the synapse, postsynaptic receptors contact the neurotransmitter for only ~100 μs [15]. At such exposure time durations, 1–3 mM GABA and glutamate are no longer saturating. Indeed, in these conditions (defined as "non-equilibrium") the binding reaction is limited by the brief permanence of the neurotransmitter in the synaptic cleft. In addition, the degree of activation of postsynaptic receptors (hence the amplitude of synaptic currents) is highly susceptible to the duration of the "synaptic pulse" in the cleft (Fig. 2a) [16, 17]. The implications of "non-equilibrium" in receptor activation are manifold and require the revision of several assumptions derived from "equilibrium pharmacology" that have been erroneously employed to explain the functioning of ligand-gated receptors at synapses. For instance, according to classical pharmacology, the degree of receptor activation is only dictated by the agonist concentration and is described by dose–response curves. In contrast, in non-equilibrium conditions, receptor activation depends on both the agonist concentration (C) and time of agonist application (t), leading to the effective pulse strength approximating the product $C \times t$ (Fig. 2b). Along the same line, the definition of "affinity" as the ratio between the unbinding rate constant and the binding rate constant of an agonist for its receptor ($K_d = k_{off}/k_{on}$) assumes that the agonist is exposed to the receptor long enough to reach equilibrium ($t \gg 1/(k_{off} + k_{on})$). If the time of agonist exposure is briefer than the time needed to reach the equilibrium, however, the degree of receptor binding will be proportional to the absolute values of the rate constant ($k_{on} + k_{off}$) and will be substantially independent from the equilibrium dissociation constant ($K_d = k_{off}/k_{on}$) (Fig. 2c). Several experimental approaches revealed that the amplitude of postsynaptic currents is extremely sensitive to the duration of the neurotransmitter transient in the synaptic cleft. For instance, one of the strategies to interfere with the neurotransmitter diffusion in the cleft exploited the polymer dextran to prolong the permanence of agonist in the synaptic cleft, by increasing the viscosity of the extracellular medium [18–20]. At both glutamatergic and GABAergic synapses, synaptic currents recorded in the presence of dextran

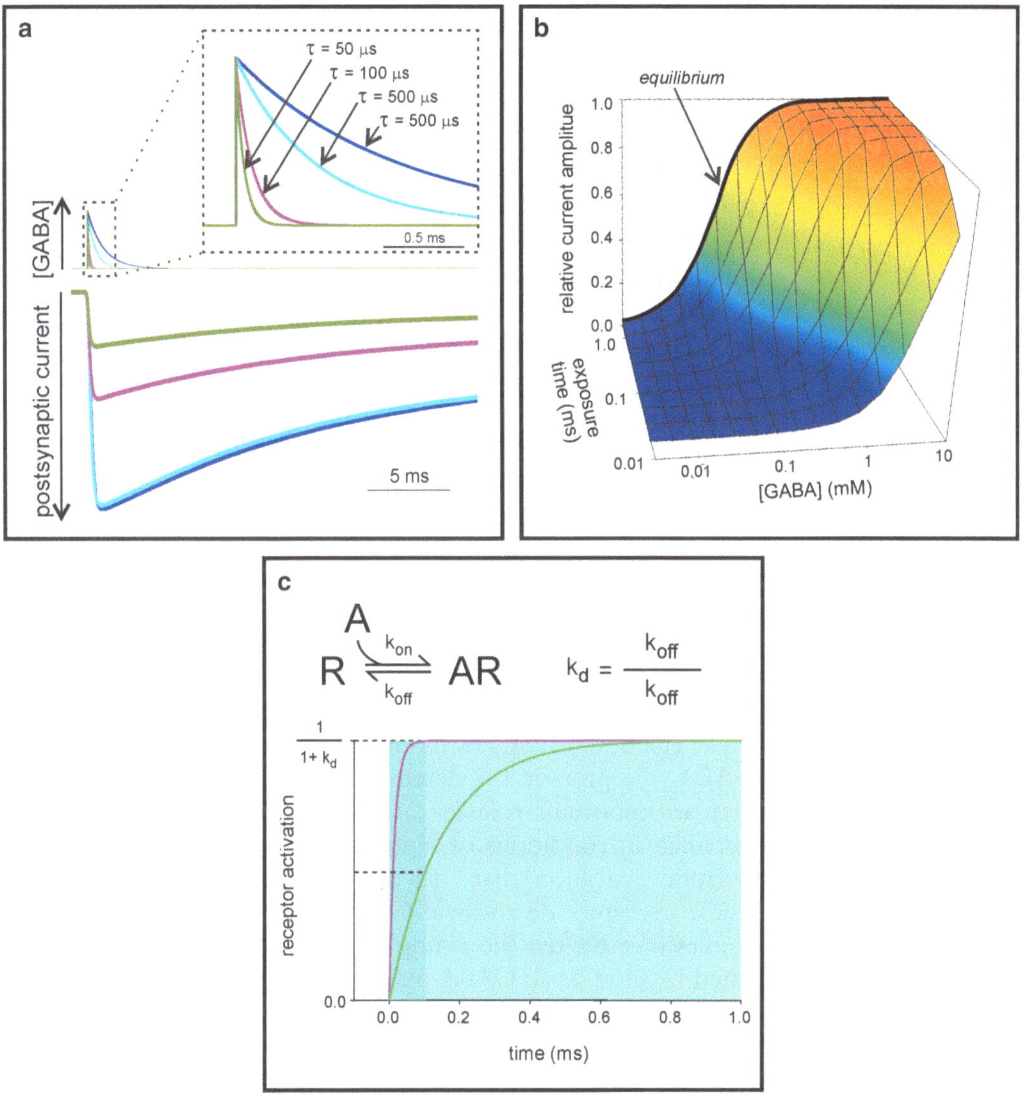

Fig. 2 Non-equilibrium pharmacology. (**a**) Simulated GABAergic inhibitory postsynaptic currents (IPSCs) in responses to (simulated) synaptic transients of GABA concentration at different duration. Please note that currents evoked by fast GABA transient (*green*, $\tau = 50$ μs) are smaller than those evoked by longer duration of GABA transient (*blue*, $\tau = 500$ μs) due to incomplete binding reaction. Modified with permission from [8]. (**b**) Simulated dose–response curve in equilibrium conditions (*thick curve*) and at different GABA exposure times (*thin curves*). Please note, for instance, that GABA 1 mM is saturating when applied for the time needed to reach the peak (equilibrium) but it fails to evoke the maximal currents when exposed for only 100 μs. (**c**) Two simulated curves representing the binding reaction (at constant agonist concentration). Assume that *red curve* corresponds to $k_{on} = 15$ ms^{-1} mM^{-1} and $k_{off} = 15$ ms^{-1}, while the *green curve* corresponds to $k_{on} = 150$ ms^{-1} mM^{-1} and $k_{off} = 150$ ms^{-1}. Although the two reactions show the same equilibrium dissociation constants ($k_{off}/k_{on} = 1$ mM, thus reaching the same equilibrium level), the different absolute values of binding and unbinding rate constants determine markedly different time courses: in case of short agonist exposure (*dark blue*) the activation level of the *green curve* will be significantly lower that the *red* one. This indicates that, in non-equilibrium, the activation level is dictated by the interplay between rate constants absolute values and the duration of agonist application

showed indeed larger amplitudes [18–20] confirming that, at central synapses, the activation of postsynaptic receptors can be limited by the shortness of neurotransmitter exposures.

3.2 Synaptic Current Kinetics Depends on Synaptic Neurotransmitter Exposure

In addition to its influence on current amplitude, the temporal profile of neurotransmitter concentration also contributes to the kinetics of synaptic currents by unmasking specific features of postsynaptic receptor gating. In particular, at both excitatory and inhibitory receptors synapses the influence of the neurotransmitter timing on current kinetics has been shown to mainly rely on the fine tuning of receptor fast desensitization [21–23]. Although desensitization is a non-conductive state, it importantly contributes to shaping current kinetics mediated by ligand-gated receptors [24]. For instance, the deactivation kinetics of NMDA and $GABA_A$ receptors are prolonged by multiple reopenings of receptors previously accumulated in the desensitized state [22, 25]. In this context, the use of ultrafast perfusion has been crucial to demonstrate that, in particular in $GABA_A$ receptors, "synaptic-like" agonist pulses (1 ms) induce a significant receptor accumulation in the desensitized state, thus raising the possibility that fast desensitization could participate in inhibitory synaptic transmission [22, 26]. Indeed, changes in the neurotransmitter exposures may tune the decay kinetics of IPSCs by modulating the relative accumulation of $GABA_A$ receptors in the desensitized state. This dependence may play an important role in case of sustained presynaptic release obtained in conditions of multiquantal release and/or repetitive synaptic activation that may induce a robust accumulation of $GABA_A$ receptor desensitization [27]. Along the same line, several studies investigated the gating features of $GABA_A$ receptors activated by ultra-weak GABA pulses in order to minimize the effect of receptor desensitization. Taking into account that the "effective pulse strength" is $\sim C \times t$ and that the fastest perfusion systems allow for agonist exposures of ~1 ms, it is possible to deliver "synaptic like" GABA pulses by lowering the GABA concentration (C). By using this expedient, it has been demonstrated that GABA currents elicited by ultra-weak pulses of GABA exogenous displayed decay kinetics faster than those evoked by brief, high-concentration GABA pulses [22, 23] (Fig. 3a, b). This kinetic behavior has been also attributed to the fact that, in conditions of "ultra-weak" GABA

Fig. 3 (continued) 100 ms pulses (300 ms gap) show strong induction of desensitization. *Right panel*: the delivery of 1.5 ms pulse pairs (to the same patch as in *left panel*) induces poor receptor desensitization. (**i**) Example of current obtained from patches pulled from ventrobasal (VB, *left panel*) and reticular (RT, *right panel*) thalamic nuclei. (**j**) Currents recorded in patches pulled from ventrobasal and reticular nuclei shown normalized and expanded time scale. Note the slower onset kinetics of patches pulled from reticular nucleus with respect to that of ventrobasal patches. Panels **a** and **b** modified with permission from [23]. Panels **c–f** modified with permission from [8]. Panels **g–j** modified with permission from [9]

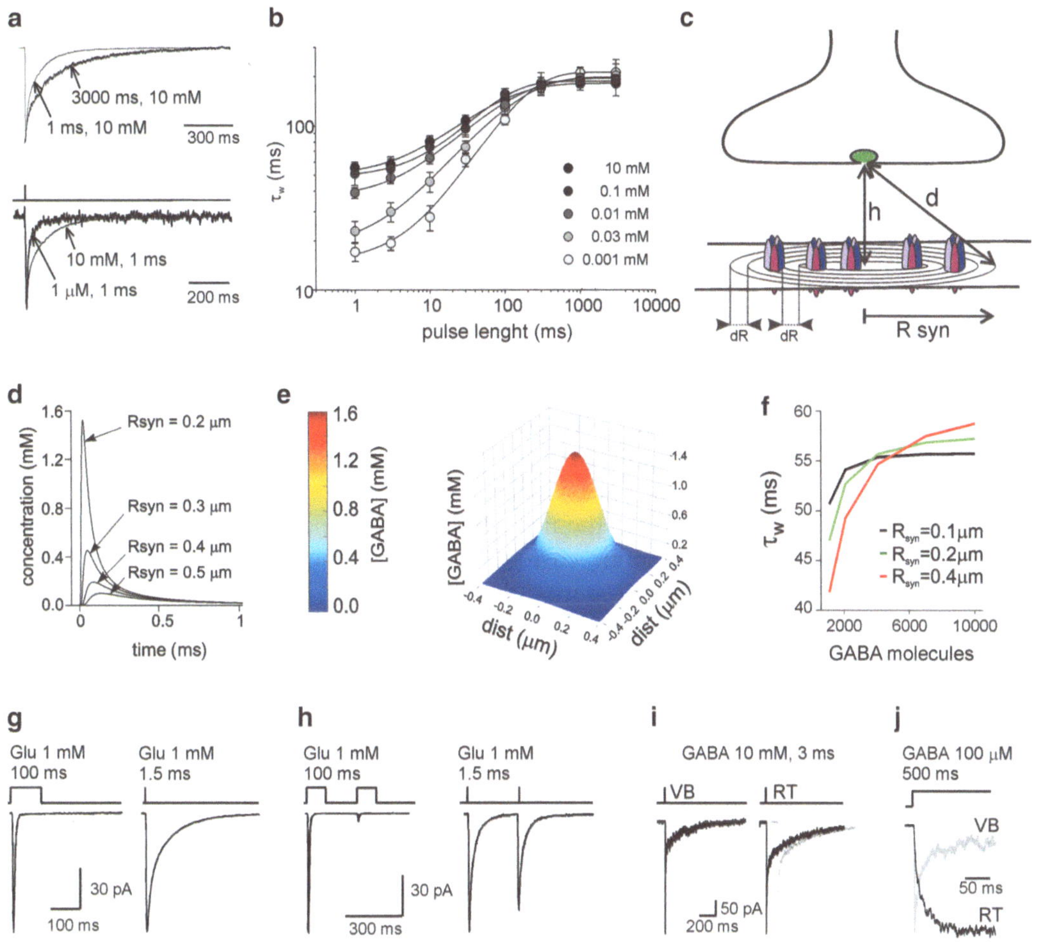

Fig. 3 Dependence of decay/deactivation kinetics on pulse "strength" and neurotransmitter spatial distribution/dynamics. (**a**) *Upper panel*: decay kinetics of $GABA_AR$-mediated currents following an application of either 3,000 or 10 ms. Please note that the current evoked by a 3,000 ms pulse is slower (with respect to that evoked by 1 ms) due to the higher receptor accumulation in the desensitized state(s). *Lower panel*: $GABA_A$-mediated currents evoked by pulses at different "strength" ($C \times t$). The current evoked by lower pulse strength (1 μM, 1 ms) is faster than that evoked by a stronger pulse (10 mM, 1 ms) due to the activation in the monoliganded state that, by showing poor desensitization, uncouples the functional interplay between desensitization and open state(s). (**b**) summary of the dependence of decay kinetics on the pulse duration at different GABA concentration (0.001–10 mM, *see* legend). (**c**) Schematic representation of the GABAergic synapses. The distance from the releasing site to the synaptic site is indicated by *d*, while *R*syn represents the synaptic disk radius. (**d**) Simulated time course of the GABA concentration following release of 4,000 GABA molecules, measured at different distances from the disk center. Please note that in the vicinity of the disk center the GABA concentration peak is markedly higher with respect to the disk periphery. (**e**) Simulated spatial profile of GABA diffusion in the cleft obtained 30 μs after the instantaneous release of 4,000 GABA molecules. (**f**) Dependence of the decay time (τ_w) of simulated IPSCs on the number of GABA molecules released. At low number of GABA molecules released, receptors located at the disk periphery sense a low dose of GABA, activating preferentially in the monoliganded configuration and contributing therefore to IPSCs speed-up. Please note that the IPSCs decay was also influenced by the synapse radius as a result of the different GABA concentrations sensed by receptors at different distance from the simulated vesicle release. (**g**) Short (1.5 ms) and long (100 ms) glutamate pulses applied to recombinant kainate receptors (GluK2/GluK5 subtype). Note that long pulses elicit fast-decaying currents while short ("synaptic-like") pulses unmask slow deactivation kinetics. (**h**) Desensitization properties of GluK2/GluK5 receptors. *Left panel*: two consecutive

exposures, $GABA_A$ receptors would bind only one of the two molecules of GABA needed for full opening. In the "singly bound state," $GABA_A$ receptors poorly desensitize, thus accelerating the current deactivation by reducing the chances of reopenings of receptors accumulated in the desensitized state [22]. The dependence of the decay kinetics on the pulse strength due to the monoliganded activation of $GABA_A$ receptors is also confirmed by single channel experiments. Indeed, single channel openings at lower conductance with respect to the main conductance level (~16 pS vs. ~27 pS) are significantly more frequent at low GABA concentrations (2 μM) and show markedly briefer mean open times (~1 ms and ~11 ms for the 16 pS and the 27 pS conductance levels, respectively) [28]. The short-living opening of $GABA_A$ receptors at lower conductance is a hallmark for low efficacy and is expected to determine the fast decay kinetics observed in Fig. 2a. These data confirm that the neurotransmitter exposure can shape the kinetics of GABAergic currents by tuning the amount of desensitization that, in turn, is modulated by the degree of $GABA_A$ receptor occupancy (singly vs. fully bound). In addition, more recently, the dependence of singly bound activation on the time course of neurotransmitter in the synaptic cleft has been addressed with a computer modelling approach, by studying the influence of presynaptic release on receptor activation. Model simulations revealed that, after the release of the synaptic vesicle, the neurotransmitter concentration profile at the postsynaptic side is not uniform, peaking at high concentrations (1–3 mM) in the center of the synapse and exponentially decreasing at more peripheral areas of the synaptic zone (<100 μM, at 400 nm off-center) (Fig. 3c–e, [23]). In these conditions, $GABA_A$ receptors located at the edges of the synaptic disk "sense" a fast transient of low GABA concentration, thus increasing their probability of activation in the singly bound state thus speeding up the IPSCs decay kinetics (Fig. 3f). Therefore it can be summarized that the spatiotemporal dynamics of neurotransmitter concentration in the synaptic cleft can (1) unmask the gating features of $GABA_A$ receptors (singly bound activation) that contribute to specific kinetics of GABAergic currents and (2) efficiently tune the fraction of receptors activated in either the fully bound or the singly bound state according to the receptor "localization" in the synaptic disk. The latter conclusion emphasizes the fact that not only the gating properties but also the distribution of $GABA_A$ receptors can modulate the kinetics of IPSCs.

Another interesting example of decay kinetics dependence upon the neurotransmitter exposure time is represented by the "synaptic" behavior of kainate receptors (KARs). Kainate receptors are present at several central synapses (including mossy fibers-CA3 synapses) and have been shown to mediate slow decaying synaptic currents (KAR-EPSCs), playing an important role in the refinement of the integration of the synaptic signaling and the neuronal spiking output [29]. The decay kinetics of KAR-EPSCs is extremely slow ($\tau \sim$ 70–100 ms) and strongly differs from the fast kinetics obtained in heterologous systems expressing recombinant kainate receptors [5]. The use of the ultrafast perfusion system revealed an unexpected dependence of the KAR-mediated current decay kinetics on the duration of exogenous glutamate pulses. Indeed, while long and saturating glutamate pulses (100 ms, 10 mM) delivered to recombinant GluK2/GluK5 kainate receptors showed the conventional fast (AMPA-like) deactivation kinetics ($\tau \sim 2$–3 ms), brief "synaptic-like" glutamate exposures (1 ms, 10 mM) elicited slow decaying currents, mimicking the kinetic behavior observed in KAR-mediated synaptic currents [21] (Fig. 3g). This time dependence of the deactivation kinetics has been explained in terms of modulation of desensitization: brief glutamate pulses did not induce accumulation of desensitization (Fig. 3h) thus unmasking slow decay kinetics. In contrast, the massive entry into desensitization elicited by long glutamate exposures accelerated the current decay [21]. More recently, the molecular basis of this mechanism has been provided [30]. Indeed, GluK2/GluK5 kainate receptors possess two types of binding sites for glutamate: one with high affinity and mediating poor desensitization and one with low affinity mediating strong desensitization [30]. Following the brief synaptic exposures, glutamate preferentially activates the high affinity/low desensitizing site(s), providing a plausible mechanism for the slow decay kinetics of KAR-EPSCs. The apparent opposite regulation of the current kinetics by "synaptic like" pulses in $GABA_A$- and KAR-mediated currents (that are accelerated and slowed down by brief agonist exposures, respectively) relies on the fact that. differently from $GABA_AR$, desensitization contributes to fast decay kinetics of KARs-mediated currents. Besides the study of channel biophysics in non-equilibrium conditions, the ultrafast perfusion system can be also suitable for mapping the kinetic properties of receptors expressed both in different brain regions and/or cell type. As shown in Fig. 3i, j, for instance, $GABA_AR$-mediated currents obtained from patches pulled from either ventrobasal or reticular thalamic nuclei show markedly different kinetics, paralleling the different kinetics of IPSCs recorded in the same brain regions [31]. Altogether, this system is suitable for detecting changes in number and/or gating properties of postsynaptic receptors after, for instance, the application of drugs or plasticity protocols. In principle, it would be possible to expand the use of the fast perfusion system in in vivo experiments by pulling patches from for example deep layers of the cortex otherwise difficult to be studied with optogenetic approaches due the poor light penetration.

3.3 Synaptic Current Pharmacological Sensitivity Depends on Synaptic Neurotransmitter Exposure

As considered above, the fact the synaptic receptors are activated in conditions of "non-equilibrium" implies that synaptic currents are extremely sensitive to changes in the duration/concentration of the neurotransmitter in the synaptic cleft. As a corollary of this concept, due to the non-equilibrium synaptic currents would also be extremely susceptible to small alterations of the properties of the postsynaptic receptors. In this framework, the effects of pharmacological agents affecting the functioning of postsynaptic receptors (in particular neurotransmitter binding rate and/or opening rate) are expected to significantly differ when observed in equilibrium conditions (bath application or perfusion system exchanging in hundreds of ms) or during synaptic transmission. In this context, the ultrafast perfusion is an "obligatory" tool to study the mechanisms of the pharmacological modulation of PSCs. It is interesting to note that the determination of the neurotransmitter waveform in the cleft and the mechanisms of actions of drugs on PSCs can be considered two aspects of the same problem. Indeed, the indirect estimations of the transient of neurotransmitter in the synaptic cleft is an important "side product" of most of the studies aimed at understanding the mechanisms of action of drugs in "synaptic conditions." In fact, once the mechanisms of action of a given pharmacological agent are defined with the ultrafast perfusion system, computer modelling allows for the extrapolation (or deconvolution) of the presynaptic waveform responsible for the "effects" of the drug on PSCs observed experimentally. The concept underlying this procedure is described in Fig. 4. The strong dependence of the PSCs pharmacological modulation on the neurotransmitter transient in the cleft is exemplified by the antagonism of Chlorpromazine (CPZ) on GABAergic currents. This widely used neuroleptic drug exerts its therapeutic action by inhibiting dopamine receptors [32], but it has been demonstrated to also block $GABA_A$ receptors, an effect that may be responsible for the induction of seizures when administered at high doses [33]. The antagonistic action of CPZ on $GABA_AR$, surprisingly, was much stronger on mIPSCs with respect to currents elicited by exogenous GABA pulses. The analysis of CPZ effects on currents evoked by ultrafast perfusion system revealed that this drug selectively decreases the binding rate constant of GABA for $GABA_A$ receptors (k_{on}) [15]. This effect was demonstrated by a marked slowdown of the current onset in the presence of CPZ at lower GABA concentrations (100 μM–1 mM) that was progressively "overcome" at supersaturating GABA concentrations (10 mM) (Fig. 5a, b). Indeed, the GABA "rate of binding" for $GABA_A$ receptors is given by $k_{on} \times [GABA]$. In the presence of CPZ, the lower k_{on} can be compensated by the increase of the GABA concentration. In synaptic conditions (brief GABA exposures), however, due the lower k_{on}, the binding reaction cannot be completed, thus leading to severe reduction of the current amplitude in CPZ. In contrast, currents elicited by longer pulse durations (e.g., exogenous pulses) will be less sensitive to the CPZ block due to the possibility for the binding reaction to progress until reaching its equilibrium Fig. 5c [15]. This concept is analogous to that described in Fig. 2c. A more complex example of interplay between the neurotransmitter transient in the synaptic cleft and the receptor functioning is provided by the modulation of GABAergic currents by protons (extracellular pH). Acidic pH, indeed, not only reduces the binding rate constant of

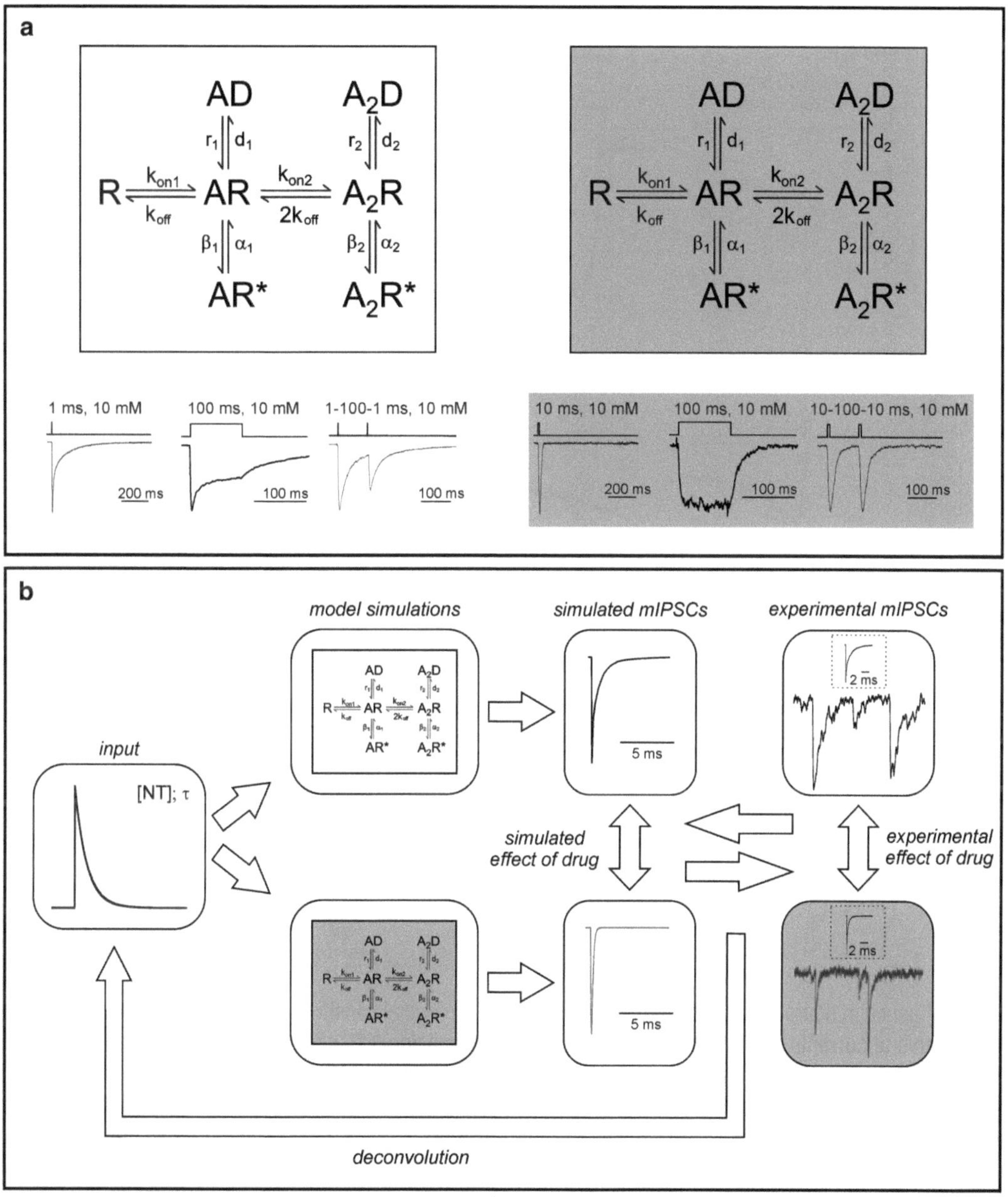

Fig. 4 Schematic representation of the "deconvolution" procedure. (**a**) The rate constants of the kinetic scheme (we show here the Jones and Westbrook model, used to simulate GABAergic currents) are defined by the formal fit that best reproduces the waveform of $GABA_AR$-mediated currents experimentally obtained by receptor activation with several experimental protocols including short pulses (1 ms), long pulses (100 ms), and paired pulses. The same procedure is repeated in the presence of the drug to be studied (*green background*) in order to define the alterations of the rate constant reproducing the effect of the drug experimentally observed. (**b**) After having defined the two sets of rate constants values reproducing the current behavior both in control and in the presence of the drug, respectively, the simulated drug effect is compared with the experimental one. In particular, the deconvolution process consists in "finding" the best values of the neurotransmitter (NT) profile in the cleft (input: [NT], τ) that minimize the differences between the simulated and the experimental drug effect. The rationale of this is that the drug effect strictly depends on the "input function" (neurotransmitter concentration temporal profile)

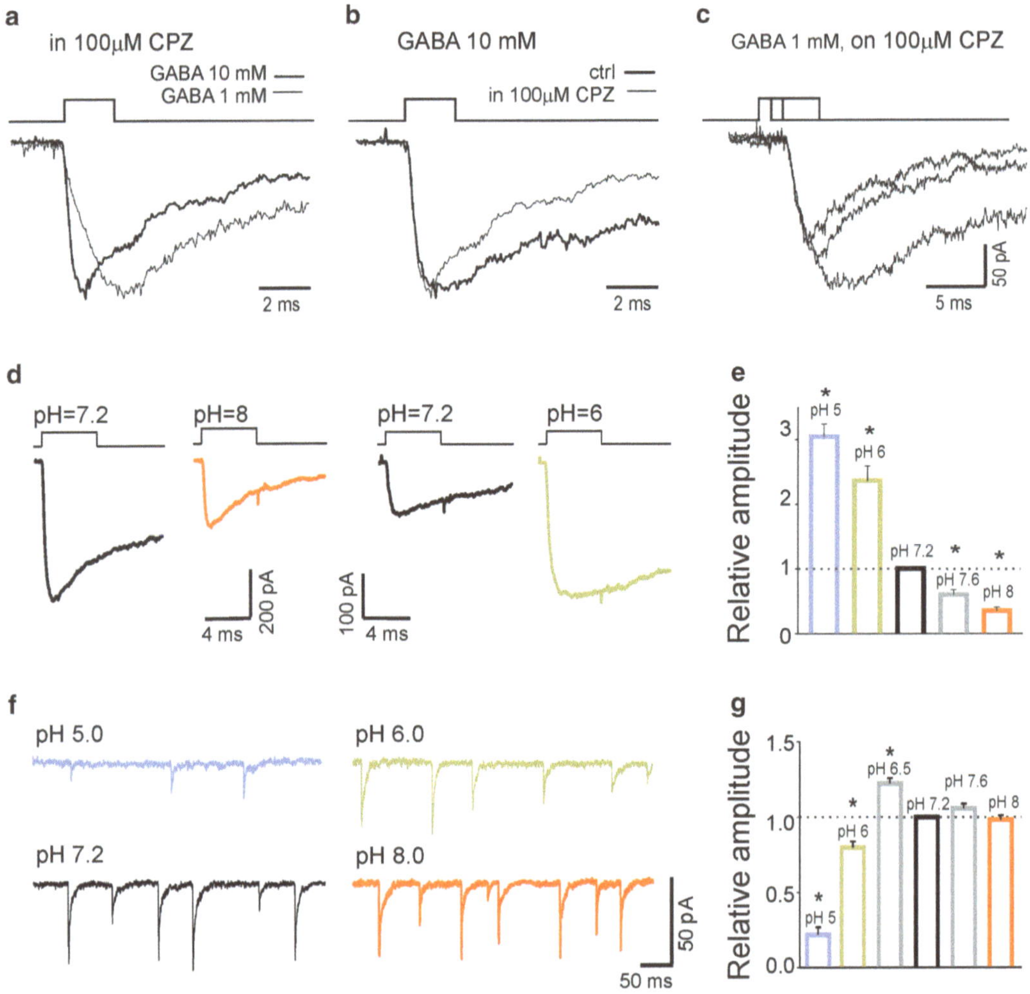

Fig. 5 Influence of neurotransmitter "timing" in the pharmacological sensitivity of synaptic GABAergic currents. (**a**) Example of currents elicited in the presence of Chlorpromazine (CPZ) by short GABA pulses (2 ms) at both 1 and 10 mM GABA. (**b**) Currents evoked by 2 ms pulses of GABA 10 mM in the presence (*thin line*) and in the absence (*thick line*) of CPZ. Please note that high GABA concentrations (10 mM) "overcome" the CPZ effect. (**c**) Currents evoked by 1 mM GABA pulses at different time duration (1, 2, and 5 ms) in the presence of CPZ. Please note that lower k_{on} induced by CPZ reveals that the pulse length represents the "rate limiting" for receptor activation. (**d**) Current responses to short application of saturating GABA pulses at different pH values. (**e**) Summary of relative amplitudes dependence on extracellular pH (normalized to the current amplitude recorded at pH = 7.2). (**f**) mIPSCs recordings at various pH values. (**g**) Summary of the dependence of mIPSC averaged amplitudes on extracellular pH. Please note the different amplitude dependence of both mIPSCs and current evoked by brief saturating GABA pulses, due to the different activation condition (*see* text). Panels **a–c** modified with permission from [15]. Panels **d–g** modified with permission from [8]

$GABA_AR$ (k_{on}) but also interferes with the gating properties of $GABA_AR$ by reducing the rate of entry into the desensitized state (*d2*) [34]. As observed for chlorpromazine, the block exerted by pH in fast-perfusion-evoked currents was substantially different from that observed in mIPSCs. The acidification of extracellular pH led to a monotonic amplitude increase of currents evoked by fast-perfusion saturating GABA pulses (10–30 mM, 2 ms) (Fig. 5d, e). In contrast, mIPSCs amplitude displayed biphasic pH dependence, with a mild increase a slight acidic pH, (6.5) and a steep current inhibition at pH < 6.5 (Fig. 5f, g). Such different pH dependence, again, can be explained in terms of different timing of neurotransmitter exposure to $GABA_A$ receptors. In currents evoked by exogenous fast-perfusion GABA pulses, similarly to the chlorpromazine effect, the decreased k_{on} at acidic pH is counter balanced by the application of super-saturating GABA pulses (10–30 mM, 2 ms) that maintain the $k_{on} \times [GABA]$ product constant. However, the reduction of receptor desensitization "reroute" more receptors to the open state (instead of entering into desensitized state), thus determining the massive current amplitude increase observed in Fig. 5e. In contrast, in mIPSCs (elicited by brief, synaptic GABA pulses) the lower desensitization induces an increase of mIPSCs current at pH ~ 6.5, while at pH values <6.5 the decrease of k_{on} becomes overwhelming with respect to the decrease of desensitization, thus determining a marked reduction of receptor occupancy with consequent inhibition of mIPSCs amplitude (Fig. 5g) [34]. The mechanisms of PSCs drug sensitivity described above emphasize the concept that the conditions of receptor activation are entitled to be considered an important part of the "pharmacological profile" of a given receptor.

4 Control of Receptor Activation/Deactivation by Light

Whereas the ultrafast perfusion system has been invaluable to understand the basic rules governing the receptor functioning during synaptic transmission, several limitations prevent the use of this approach for the study of the synaptic function in a more "integrated" environment. Indeed, the major drawback of the ultrafast perfusion system is that receptors are activated in the excised outside-out patches. In this configuration, (1) it is impossible to control the receptor subtypes included in the patch, (2) the intracellular factors/interactors are rapidly washed out and (3) the interactions with the cytoskeleton are completely disrupted. In addition, it is impossible to observe the impact of synaptic activation at the level of the neuron/network. In order to overcome these limitations, the behavior of synaptic receptors should be studied in conditions by which "synaptic-like" neurotransmitter pulses are delivered to postsynaptic receptors at intact synapses. The investigation of the synaptic receptor behavior at individual synapses represents a major technological challenge. Indeed, not only the neurotransmitter has to be applied in the submillisecond time scale, but also it has to be spatially restricted at the nanometer range, being the postsynaptic size ~100–200 nm. The advent of the "uncaging technique," which combines photochemistry and optical techniques to photorelease neurotransmitters, has represented a major improvement in the control of synaptic receptor activation at intact synapses, approximating the temporal and spatial characteristics of synaptic neurotransmitter release. Neurotransmitter (in particular glutamate or GABA) is covalently bound to a specimen (cage) that hinders the neurotransmitter receptor binding to its receptor. After brief illumination with UV light, the bounds between the neurotransmitter and the "cage" are readily broken, leaving the neurotransmitter free to bind the receptor and to exert its physiological action. In the best experimental conditions, light can be radially restricted to ~0.4 μm (roughly on par with the *X*, *Y* synapse dimensions) and the duration of illumination can be controlled in the submillisecond time range. In addition, the two-photon illumination allows for reducing the volume of photorelease (with respect to that obtained with one-photon excitation) by limiting the axial illumination (*Z* axis) to 1–2 μm (Fig. 6a, left panel). The use of light to photorelease caged neurotransmitter allows for substantial experimental versatility. Indeed, it is possible to elicit "synaptic like" responses at individual synapses located in different sub-neuronal regions and/or to study the response of several synapses activated at specific spatial/temporal patterns, a valuable tool to investigate the dendritic integration of synaptic inputs. Besides these enormous advantages for the control of the synaptic receptor activation, this method also shows some drawbacks. As mentioned above, indeed, an ideal perfusion system should be able to control both the delivery and the removal of the agonist. While the photorelease reaction is very fast and could satisfactory mimic the "timing" of vesicle release, the neurotransmitter removal (hence the "pulse" duration) cannot be controlled. The main problem is that, although the photorelease occurs in a "known" volume, after uncaging the neurotransmitter diffuses in the neighboring synaptic area, thus significantly exceeding both the time and the space constraints imposed by UV illumination (Fig. 6a, right panel). It could be objected that the diffusion process also governs the dynamics of neurotransmitter after the vesicle release in the synaptic cleft, thus mimicking the "physiological conditions." However, the conditions in the two systems are substantially different: at the synapse, the neurotransmitter diffuses from a 20 nm diameter vesicle at concentration 100–150 mM. During the photorelease process, the "effective" neurotransmitter concentration is 0.6–6 mM [35, 36] diffusing from a ~0.4 μm (*x*–*y*) and 1–2 μm (*z*) "spot." The latter conditions, inevitably, favor the activation of extrasynaptic receptors, thus weakening the equivalence between unitary synaptic currents (mPSCs) and uncaging currents (uPSCs). The neurotransmitter "spillover" after UV photorelease, however, seems to mainly affect responses obtained at GABAergic synapses with respect to the glutamatergic. Indeed, while uEPSCs satisfactorily match the kinetics

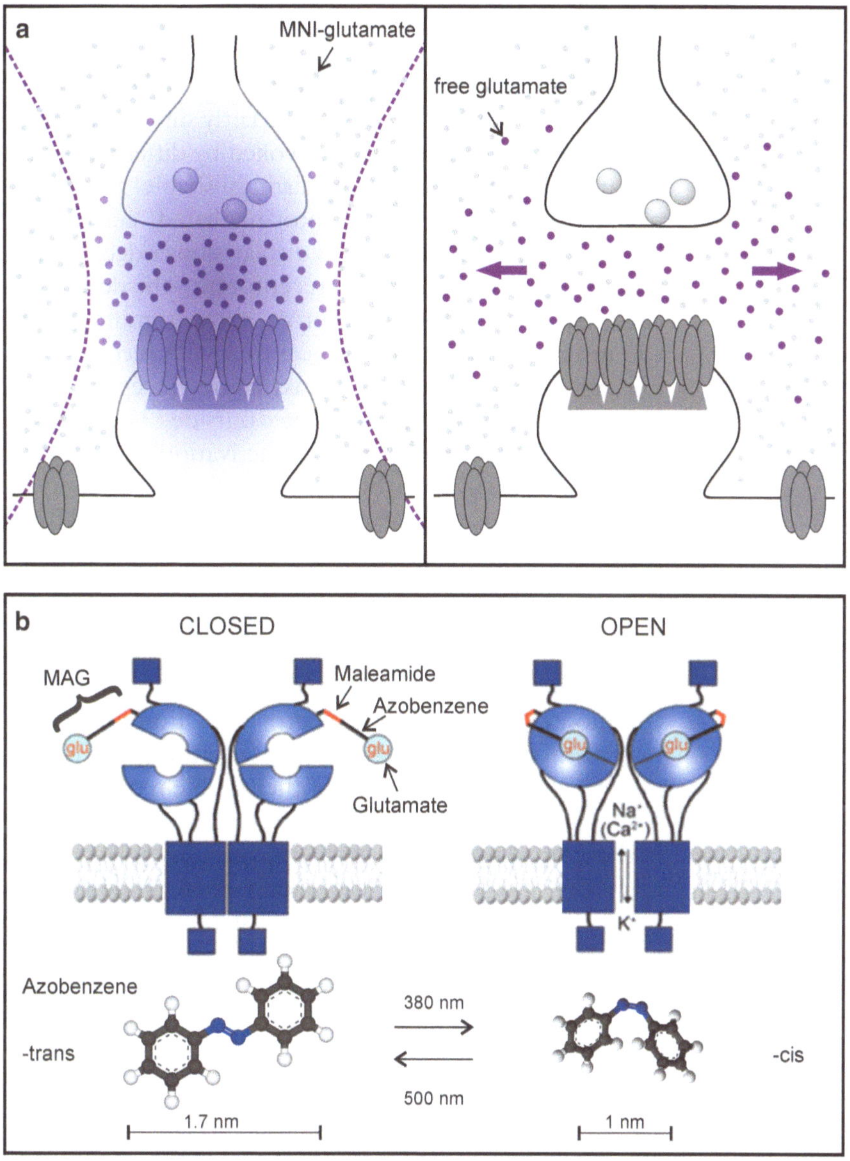

Fig. 6 Activation of ligand-gated receptor with light. (**a**) Uncaging technique. Caged neurotransmitter molecules (e.g., MNI-glutamate, *gray spots*) are photoreleased in the excitation volume that, using 2P stimulation, can be restricted radially to ~0.4 μm and axially to ~1–2 μm. Please note that although the light spot radial dimension roughly approximate the synapse size (*left panel*), the photoreleased neurotransmitter molecules (e.g., free glutamate, *violet spots*) diffuse outside the synapse, potentially activating extrasynaptic receptors (*right panel*). (**b**) Light control of ligand-gated receptor by photoswitchable tethered ligands (PTLs). Glutamate receptor GluK2 binds MAG (Maleimide–Azobenzene–Glutamate) by the reaction between maleimide and an engineered cysteine located in a specific position on the GluK2 protein. The schematic molecular structure of the azobenzene (the photoswitchable part of MAG) is represented (*bottom panel*) in both the trans- and cis-isomerization (*left* and *right*, respectively). These conformations are controlled by light at different wavelengths: UV (cis-) or >400 nm (trans-), respectively. The conformational change between the two isomerization states determines the binding of glutamate to the receptor binding site. Please note that GluK2 is closed in the trans-configuration (*left panel*) while readily opens after the cis-photoisomerization (*right panel*). Modified with permission from [45]

of mEPSCs (but *see* also [35]), uIPSCs are substantially slower than their synaptic counterparts [37]. These differences could be ascribed to several factors. First, the desensitization and deactivation kinetics of AMPA receptor are fairly similar, [5] indicating that the decay of current responses evoked by different duration of glutamate pulses could be hardly distinguishable. In contrast, as mentioned above, $GABA_A$ receptor-mediated currents evoked by longer GABA applications will display slower decay kinetics, due to the stronger accumulation of receptor into the desensitized state(s) [22]. Second, $GABA_A$ receptors show significant higher neurotransmitter affinity with respect to AMPA receptors thus increasing the probability of extrasynaptic receptor activation by photoreleased GABA diffusing outside GABAergic synapses (assuming "equilibrium" receptor activation) [26, 38]. Concerning the receptor activation by spillover of photoreleased neurotransmitter, the presence of extrasynaptic "tonic" $GABA_A$ receptors with very high GABA affinity further increases the possibility of uIPSCs "contamination" by activation of non-synaptic receptors [39]. Third, at glutamatergic synapses the difference of density between synaptic and extrasynaptic receptors is significantly higher than at GABAergic synapses [40], leading to a lower relative contribution of extrasynaptic receptor activation in the uEPSCs current with respect to uIPSCs. In addition to the uncontrolled neurotransmitter diffusion, another problem of the uncaging technique is that photoreleasable compounds are not completely inert in their caged form. This problem, again, is more critical at $GABA_A$ receptors. For instance, the widely used glutamate caged compound (MNI-glutamate) exerts a strong antagonistic action on $GABA_A$ receptors [41]. This may represent a problem if, for instance, the experimental conditions require (or assume) functional GABAergic inhibition. This problem may be overcome by the use of the Rubi-glutamate that, nominally, does not block $GABA_ARs$ [42]. A different (and more problematic) issue concerning "GABA photorelease" is the fact that GABA-caged compounds are efficient competitive blockers of $GABA_A$ receptor itself [37]. Although newer GABA caged compounds (e.g., DPNI-GABA) are designed to minimize this unwanted effect, it is not clear how the competitive action of these compounds could impact on the amplitude and kinetics of uIPSCs. Interestingly, the caged compounds based on the Ruthenium photochemistry (Rubi-GABA or Rubi-glutamate) are photoreleased following illumination with light at visible wavelengths, allowing for the combined uncaging of two different caged compounds, typically performed in the UV and visible range, respectively [43]. While the use of caged compounds is still an excellent approach for the study of synaptic receptors, new technologies for the light-control of receptor activity are aimed at overcoming the disadvantages caused by the diffusion of the photoreleasable neurotransmitter. To this concern, the Isacoff

Lab has developed a novel approach for the light-controlled activation and deactivation of ionotropic glutamate receptors, by engineering kainate receptor (GluK2) in order to bind photoswitchable tethered ligands (PTLs) (Fig. 6b) [44, 45]. The PTL developed by the Isacoff Lab is represented by the Maleimide–Azobenzene–Glutamate molecule (MAG): the ligand (glutamate) is linked to the photoswitch (azobenzene) that, in turn, is anchored to a mutated cysteine on the receptor (GluK2) by means of a cysteine-reactive group (maleimide) (Fig. 6b). The length of the azobenzene molecule is significantly different in the *cis* and *trans* configurations, being 1 nm and 1.7 nm, respectively (Fig. 6b, bottom). The conformational change induced by the photoisomerization from the *trans*-to-*cis* configuration, shortens the azobenzene molecule length by 0.7 nm and consequently presents the glutamate molecule to the receptor binding site, thus triggering the receptor opening (Fig. 6b). In contrast, the *cis*-to-*trans* isomerization removes the glutamate molecule from the receptor binding site, determining the receptor deactivation (Fig. 6b). Interestingly, the *trans*-to-*cis* and *cis*-to-*trans* isomerizations can be efficiently operated by light at two different wavelengths (380 nm and >400 nm, respectively), allowing for the precise light-control of both the activation and the deactivation reactions (Fig. 6b, bottom). As mentioned above, MAG binds a mutated cysteine on the engineered GluK2 receptor. Although many cysteine residues could virtually compete with the mutated one for the binding of MAG, it is possible to "direct" the maleimide (on MAG molecule) in the vicinity of the cysteine of interest by exploiting the relatively high affinity binding between the glutamate (on MAG molecule) and its binding site. Clearly, during the MAG binding process, (also called photo-affinity labeling) azobenzene has to be kept in the *cis* isomerization, since, in this configuration, the length of MAG is optimal for the simultaneous binding of the cysteine of interest and the receptor binding site. It has to be noted that the PTLs approach has been mainly developed to control the neuronal excitability as an alternative to the use of the more popular opsins [46]. However, in addition to the study of the role of controlled receptor activation in the synapse/network functioning, PTLs-based regulation of receptor activity turns out also to provide an excellent opportunity for the study of the functioning of ligand-gated receptors. The advantages of the use of these light-gated channels (LiGluK2) are manifold. By using PTLs, the agonist is directly linked to the receptor of interest, thus avoiding uncontrolled neurotransmitter diffusion. In keeping with this, differently from another limitation of the uncaging technique, it is possible to control the deactivation of receptors, thus allowing for precise control of the "stimulus duration." These features, together with the possibility to activate the receptors exclusively in the illuminated area(s), represent the ideal conditions for the control of ligand-gated receptor activity. In addition, LiGluK2 are trafficked and clustered at synapses as "normal" recombinant glutamate receptors, allowing for the control of synapses at unprecedented spatial and temporal resolution (Barberis et al., unpublished observations). Moreover, the stability of the photoisomerization reaction makes the LiGluK2 remaining in the activated or inactivated state even in the dark. As a consequence, in experimental conditions in which long exposures are needed, after a short 380 nm pulse triggering the *trans*-to-*cis* photoisomerization, the receptor will stay open until another short >400 nm pulse will close the channel, with considerable lower phototoxicity as compared to the uncaging technique that requires constant illumination for the whole duration of the pulse. Besides the extraordinary possibility to independently control ligand-gated receptors activation/deactivation with light, the PTLs technique at its present stage also shows some weaknesses. First, the shape of currents mediated by LiGluK2 is not perfectly matching that obtained with the "conventional" glutamate applications. The reason for such different kinetic behavior could be the altered receptor gating by the photoswitch tethering and/or the heteromerization of the LiGluK2 subunit with unknown proteins already expressed in neurons. In support of the latter hypothesis, the kinetics of LiGluK2 light-evoked responses are significantly different when observed in neurons or HEK293 cells, being the responses in HEK293 cell significantly closer to the "conventional" GluK2-mediated currents (Barberis et al., unpublished observations). Second, in order to obtain light-gated currents with LiGluK2 it is necessary to transfect the GluK2 receptors bearing the cysteine point-mutation that allows for the binding of the MAG molecule, therefore limiting the analysis to currents evoked by GluK2 recombinant receptors. To this concern, the possibility to target endogenous cysteine residues by specific PTLs would greatly improve the versatility of this technique, by enabling the light-control of native receptors. In order to achieve this important amelioration of the PTLs technique, however, it will be important to obtain detailed crystallographic maps of the receptor to be targeted, in order to study the relative distances and spatial organization of the cysteine residues in relation to the receptor binding site. So far, few receptor/channels have been engineered for PTLs-based light-control. These include the acetylcholine receptors (AChR) [47] and the voltage-gated potassium channels [48, 49] that, differently from the ionotropic GluK2 receptors, are inactivated by light, since the "ligand" used for this specific PTL is the tetraethylammonium (TEA), a specific potassium channel blocker. It is plausible to believe that, in the mid-long term, several receptors channels, enzymes and seconds messengers will be engineered with the PTLs technology providing an unprecedented control of the membrane excitability, intracellular biochemical pathways and synaptic function. Concerning the study of the synapse, so far, only LiGluK2 are suitable for the control of synaptic receptors in a "synaptic" time scale. It has to be remarked, however, that although postsynaptic kainate receptors contribute to the tuning and the refinement of excitatory signals, they do not play a primary role in excitatory synaptic transmission. In this concern, the development of light-gated AMPA, NMDA, and $GABA_A$R receptors would dramatically improve the control of both excitatory and inhibitory synaptic transmission, significantly increasing the possibilities for the study of dendritic synaptic signal integration and its implication at the level of network functioning. In addition, the development of new light-sensitive channels/receptors should be paralleled by the design of novel PTLs with specific spectral properties, in order to allow for the simultaneous "multicolor" control of diverse light-sensitive effectors.

4.1 Focusing Light Below the Diffraction Limit

The advent of the use of natural or engineered proteins capable to change the neuronal electrophysiological activity in response to light determined a revolutionary transformation of the approach for the study of the CNS functioning. Indeed, the development of light-sensitive ion channels allowed to exploit the "power of light" to control the neuronal activity with unprecedented precision and selectivity. In fact, light (1) minimally interferes with the biological system (unless for the photodamage); (2) allows for ultra-brief time exposures and the control of multiple light-sensitive devices at different wavelengths; (3) allows for fast or simultaneous "multi-spot" illumination by exploiting "sequential scanning" or "structured light," respectively; (4) can be shaped in complex spatial configurations including "light sheets" or holographic (3D) shapes and, (5) can achieve a considerable tissue penetration with minimal scattering and photodamage by the use of two-photon (2P) excitation. In addition to the advantages listed above (and their different possible combinations), the use of light is also relatively inexpensive. The application of these illumination techniques in combination with the neuronal expression of the bacterial opsins capable to either favor (channelrhodopsin-2, ChR2) or prevent (halorhodopsin, NpHR) the generation of action potential determined an impressive advancement in the comprehension of the brain microcircuits functioning [50]. In addition, the "in vivo" activation/inhibition of specific neuronal subpopulations allowed for the first time the real-time study of the relation between neuronal activity and behavior [51]. While opsins efficiently control the neuronal excitability at "single cell level," the PTLs-based optogenetic approach mentioned above, by targeting neuronal synaptic receptors, offers the possibility to control the activation of individual synapses at postsynaptic level. In order to reach this goal, however, it is essential to restrict the illumination at nanometric level. Although the possibility to shape light seems to be limitless, using conventional technology, it is still impossible to focus light below the "diffraction limit" imposed by the Abbe's barrier ($1.22\lambda/NA$, where λ is the light wavelength and NA is the objective numerical aperture). This means that focusing a laser beam (e.g., 488 nm) using 100× objective with 1.4 numerical aperture, the diffraction limited spot size will be ~425 nm. Even by assuming ideal conditions of "zero scattering" at the sample, such spot dimension is still considerably larger than synapses (~200 nm) [52] precluding the selective illumination of the synaptic area without the contamination of the synaptic response by extrasynaptic receptors (or neighboring synapses in the close vicinity). The aforementioned spatial limit, in addition, would completely prevent the possibility to study the functional properties of receptors compartmentalized in subregions of the synaptic area (synaptic nanodomains), as recently shown by superresolution studies [52–54]. Taking into account these limitations, light would appear inadequate to control the activation of "optogenetic" or "light sensitive" tools at the nanoscopic scale. However it has to be noted that the Abbe's barrier concerns "regular" optical systems. Recent advancements in photonics, indeed, show that metallic materials acquire unusual properties when arranged in one-dimensional structures (such as nanowires), providing new perspectives for the manipulation of light [55]. As shown by De Angelis et al. [11, 56], it is possible to control the field enhancement at the end of a nanowire (or a tapered tip) to obtain a spot of light with nanometer dimension: indeed, the spatial confinement of light in the near field is comparable to the radius of curvature of the tapered or linear nanowires, which, for the tapered case, lies between 2.5 and 5 nm. This approach, based on plasmon polarities technology [57], exploits the increment of the localized electric field created by the interaction of a focused laser beam with the metallic surface of nanostructures. Exploiting this technology, it is possible to direct tapered tips to the nano-region of interest by using nano-positioning devices such as piezo actuators or AFM scanning probes obtaining sub-diffraction limited illumination spots in the range of 10–30 nm. The use if this technology enables restricting the illumination to individual synapses, and, in specific experimental conditions, even to single molecules. The possibility to control the activation of receptors with nanometric and submillisecond precision, in combination with the use of nanoscopy/superesolution techniques like STED, PALM, or STORM, will help answer many open questions about the synaptic function, including the relation between the receptor gating and the receptor localization in membrane micro- or nano-domains and the interaction with specific proteins/interactors. In addition, the possibility to precisely control of the activation/deactivation of whole synapses will substantially improve the resolution for the study of dendritic integration of excitatory and inhibitory synaptic inputs, a key mechanism for neuronal information processing. Finally, the possibility to excite/inhibit individual receptors (or small number of receptors) will open a new level of investigation on the role of ligand-gated receptors in the neuronal computation. In fact, the electric signals carried by receptors may absolve to different physiological role(s) when activated individually or simultaneously with other receptors. Indeed, the activation of an individual AMPA receptor would mainly contribute to the "dendritic noise" while the activation of 50–100 AMPA receptors generates an EPSP: both signals however will importantly contribute to the neuronal output. The combination of optogenetics with novel illumination techniques will allow for the study of the multilevel effects of ligand-gated receptor activation and their role in the neuronal signaling.

Acknowledgments

This work was supported by the Telethon grant GGP11043 to A.B.

References

1. Branco T, Hausser M (2010) The single dendritic branch as a fundamental functional unit in the nervous system. Curr Opin Neurobiol 20(4):494–502. doi:10.1016/j.conb.2010.07.009
2. Klausberger T, Somogyi P (2008) Neuronal diversity and temporal dynamics: the unity of hippocampal circuit operations. Science 321(5885):53–57. doi:10.1126/science.1149381, 321/5885/53 [pii]
3. Tang CM, Margulis M, Shi QY, Fielding A (1994) Saturation of postsynaptic glutamate receptors after quantal release of transmitter. Neuron 13(6):1385–1393
4. Barberis A, Petrini EM, Cherubini E, Mozrzymas JW (2002) Allosteric interaction of zinc with recombinant alpha(1)beta(2) gamma(2) and alpha(1)beta(2) GABA(A) receptors. Neuropharmacology 43(4):607–618, S0028390802001090 [pii]
5. Bowie D, Lange GD (2002) Functional stoichiometry of glutamate receptor desensitization. J Neurosci 22(9):3392–3403, 20026333
6. Dravid SM, Prakash A, Traynelis SF (2008) Activation of recombinant NR1/NR2C NMDA receptors. J Physiol 586(Pt 18):4425–4439. doi:10.1113/jphysiol.2008.158634
7. Pitt SJ, Sivilotti LG, Beato M (2008) High intracellular chloride slows the decay of glycinergic currents. J Neurosci 28(45):11454–11467. doi:10.1523/JNEUROSCI.3890-08.2008
8. Barberis A, Petrini EM, Mozrzymas JW (2011) Impact of synaptic neurotransmitter concentration time course on the kinetics and pharmacological modulation of inhibitory synaptic currents. Front Cell Neurosci 5:6. doi:10.3389/fncel.2011.00006
9. Barberis A (2012) Fast perfusion methods for the study of ligand-gated ion channels. In: Fellin THM (ed) Neuronal network analysis, vol 67. Springer, New York, NY, pp 173–187. doi:10.1007/7657_2011_20
10. Jonas P (1995) Fast application of agonist to isolated membrane patches. In: Sakmann B, Neher E (eds) Single channel recordings. Plenum Press, New York, NY
11. De Angelis F, Das G, Candeloro P et al (2010) Nanoscale chemical mapping using three-dimensional adiabatic compression of surface plasmon polaritons. Nat Nanotechnol 5(1):67–72. doi:10.1038/nnano.2009.348
12. Franke C, Hatt H, Dudel J (1987) Liquid filament switch for ultra-fast exchanges of solutions at excised patches of synaptic membrane of crayfish muscle. Neurosci Lett 77(2):199–204
13. Moffatt L, Hume RI (2007) Responses of rat P2X2 receptors to ultrashort pulses of ATP provide insights into ATP binding and channel gating. J Gen Physiol 130(2):183–201. doi:10.1085/jgp.200709779
14. Clements JD, Lester RA, Tong G et al (1992) The time course of glutamate in the synaptic cleft. Science 258(5087):1498–1501
15. Mozrzymas JW, Barberis A, Michalak K, Cherubini E (1999) Chlorpromazine inhibits

miniature GABAergic currents by reducing the binding and by increasing the unbinding rate of GABAA receptors. J Neurosci 19(7): 2474–2488
16. He L, Wu XS, Mohan R, Wu LG (2006) Two modes of fusion pore opening revealed by cell-attached recordings at a synapse. Nature 444(7115):102–105. doi:10.1038/nature05250
17. Popescu G, Robert A, Howe JR, Auerbach A (2004) Reaction mechanism determines NMDA receptor response to repetitive stimulation. Nature 430(7001):790–793. doi:10.1038/nature02775
18. Barberis A, Petrini EM, Cherubini E (2004) Presynaptic source of quantal size variability at GABAergic synapses in rat hippocampal neurons in culture. Eur J Neurosci 20(7):1803–1810. doi:10.1111/j.1460-9568.2004.03624.x, EJN3624 [pii]
19. Min MY, Rusakov DA, Kullmann DM (1998) Activation of AMPA, kainate, and metabotropic receptors at hippocampal mossy fiber synapses: role of glutamate diffusion. Neuron 21(3):561–570, S0896-6273(00)80566-8 [pii]
20. Perrais D, Ropert N (2000) Altering the concentration of GABA in the synaptic cleft potentiates miniature IPSCs in rat occipital cortex. Eur J Neurosci 12(1):400–404, ejn957 [pii]
21. Barberis A, Sachidhanandam S, Mulle C (2008) GluR6/KA2 kainate receptors mediate slow-deactivating currents. J Neurosci 28(25):6402–6406. doi:10.1523/JNEUROSCI.1204-08.2008, 28/25/6402 [pii]
22. Jones MV, Westbrook GL (1995) Desensitized states prolong GABAA channel responses to brief agonist pulses. Neuron 15(1):181–191, 0896-6273(95)90075-6 [pii]
23. Petrini EM, Nieus T, Ravasenga T et al (2011) Influence of GABAAR monoliganded states on GABAergic responses. J Neurosci 31(5):1752–1761. doi:10.1523/JNEUROSCI.1453-10.2011
24. Mozrzymas JW, Barberis A, Mercik K, Zarnowska ED (2003) Binding sites, singly bound states, and conformation coupling shape GABA-evoked currents. J Neurophysiol 89(2):871–883. doi:10.1152/jn.00951.2002
25. Lester RA, Jahr CE (1992) NMDA channel behavior depends on agonist affinity. J Neurosci 12(2):635–643
26. Barberis A, Mozrzymas JW, Ortinski PI, Vicini S (2007) Desensitization and binding properties determine distinct alpha1beta2gamma2 and alpha3beta2gamma2 GABA(A) receptor-channel kinetic behavior. Eur J Neurosci 25(9):2726–2740. doi:10.1111/j.1460-9568.2007.05530.x, EJN5530 [pii]
27. Bragina L, Marchionni I, Omrani A et al (2008) GAT-1 regulates both tonic and phasic GABA(A) receptor-mediated inhibition in the cerebral cortex. J Neurochem 105(5):1781–1793.doi:10.1111/j.1471-4159.2008.05273.x
28. Macdonald RL, Rogers CJ, Twyman RE (1989) Kinetic properties of the GABAA receptor main conductance state of mouse spinal cord neurones in culture. J Physiol 410:479–499
29. Sachidhanandam S, Blanchet C, Jeantet Y et al (2009) Kainate receptors act as conditional amplifiers of spike transmission at hippocampal mossy fiber synapses. J Neurosci 29(15):5000–5008. doi:10.1523/JNEUROSCI.5807-08.2009
30. Mott DD, Rojas A, Fisher JL, Dingledine RJ, Benveniste M (2010) Subunit-specific desensitization of heteromeric kainate receptors. J Physiol 588(Pt 4):683–700, jphysiol.2009.185207 [pii]
31. Mozrzymas JW, Barberis A, Vicini S (2007) GABAergic currents in RT and VB thalamic nuclei follow kinetic pattern of alpha3- and alpha1-subunit-containing GABAA receptors. Eur J Neurosci 26(3):657–665. doi:10.1111/j.1460-9568.2007.05693.x, EJN5693 [pii]
32. Seeman P (1980) Brain dopamine receptors. Pharmacol Rev 32(3):229–313
33. Toone BK, Fenton GW (1977) Epileptic seizures induced by psychotropic drugs. Psychol Med 7(2):265–270
34. Mozrzymas JW, Zarnowska ED, Pytel M, Mercik K (2003) Modulation of GABA(A) receptors by hydrogen ions reveals synaptic GABA transient and a crucial role of the desensitization process. J Neurosci 23(22):7981–7992, 23/22/7981 [pii]
35. DiGregorio DA, Rothman JS, Nielsen TA, Silver RA (2007) Desensitization properties of AMPA receptors at the cerebellar mossy fiber granule cell synapse. J Neurosci 27(31):8344–8357. doi:10.1523/JNEUROSCI.2399-07.2007
36. Matsuzaki M, Ellis-Davies GC, Nemoto T et al (2001) Dendritic spine geometry is critical for AMPA receptor expression in hippocampal CA1 pyramidal neurons. Nat Neurosci 4(11):1086–1092. doi:10.1038/nn736
37. Trigo FF, Papageorgiou G, Corrie JE, Ogden D (2009) Laser photolysis of DPNI-GABA, a tool for investigating the properties and distribution of GABA receptors and for silencing neurons in situ. J Neurosci Methods 181(2):159–169. doi:10.1016/j.jneumeth.2009.04.022

38. Patneau DK, Mayer ML (1990) Structure-activity relationships for amino acid transmitter candidates acting at N-methyl-D-aspartate and quisqualate receptors. J Neurosci 10(7):2385–2399
39. Glykys J, Mody I (2007) Activation of GABAA receptors: views from outside the synaptic cleft. Neuron 56(5):763–770. doi:10.1016/j.neuron.2007.11.002
40. Levi S, Triller A (2006) Neurotransmitter dynamics. In: Kittler JT, Moss SJ (eds) The dynamic synapse: molecular methods in ionotropic receptor biology, Frontiers in neuroscience. CRC Press, Boca Raton, FL
41. Fino E, Araya R, Peterka DS et al (2009) RuBi-glutamate: two-photon and visible-light photoactivation of neurons and dendritic spines. Front Neural Circ 3:2. doi:10.3389/neuro.04.002.2009
42. Fino E, Yuste R (2011) Dense inhibitory connectivity in neocortex. Neuron 69(6):1188–1203. doi:10.1016/j.neuron.2011.02.025
43. Hayama T, Noguchi J, Watanabe S et al (2013) GABA promotes the competitive selection of dendritic spines by controlling local Ca2+ signaling. Nat Neurosci 16(10):1409–1416. doi:10.1038/nn.3496
44. Gorostiza P, Isacoff EY (2008) Optical switches for remote and noninvasive control of cell signaling. Science 322(5900):395–399. doi:10.1126/science.1166022
45. Szobota S, Gorostiza P, Del Bene F et al (2007) Remote control of neuronal activity with a light-gated glutamate receptor. Neuron 54(4):535–545. doi:10.1016/j.neuron.2007.05.010
46. Fenno L, Yizhar O, Deisseroth K (2011) The development and application of optogenetics. Annu Rev Neurosci 34:389–412. doi:10.1146/annurev-neuro-061010-113817
47. Bartels E, Wassermann NH, Erlanger BF (1971) Photochromic activators of the acetylcholine receptor. Proc Natl Acad Sci U S A 68(8):1820–1823
48. Banghart M, Borges K, Isacoff E et al (2004) Light-activated ion channels for remote control of neuronal firing. Nat Neurosci 7(12):1381–1386. doi:10.1038/nn1356
49. Chambers JJ, Banghart MR, Trauner D, Kramer RH (2006) Light-induced depolarization of neurons using a modified Shaker K(+) channel and a molecular photoswitch. J Neurophysiol 96(5):2792–2796. doi:10.1152/jn.00318.2006
50. Mathur BN, Tanahira C, Tamamaki N, Lovinger DM (2013) Voltage drives diverse endocannabinoid signals to mediate striatal microcircuit-specific plasticity. Nat Neurosci 16(9):1275–1283. doi:10.1038/nn.3478
51. Tye KM, Deisseroth K (2012) Optogenetic investigation of neural circuits underlying brain disease in animal models. Nat Rev Neurosci 13(4):251–266. doi:10.1038/nrn3171
52. Specht CG, Izeddin I, Rodriguez PC et al (2013) Quantitative nanoscopy of inhibitory synapses: counting gephyrin molecules and receptor binding sites. Neuron 79(2):308–321. doi:10.1016/j.neuron.2013.05.013
53. MacGillavry HD, Song Y, Raghavachari S, Blanpied TA (2013) Nanoscale scaffolding domains within the postsynaptic density concentrate synaptic AMPA receptors. Neuron 78(4):615–622. doi:10.1016/j.neuron.2013.03.009
54. Nair D, Hosy E, Petersen JD et al (2013) Super-resolution imaging reveals that AMPA receptors inside synapses are dynamically organized in nanodomains regulated by PSD95. J Neurosci 33(32):13204–13224. doi:10.1523/JNEUROSCI.2381-12.2013
55. Deshpande VV, Bockrath M, Glazman LI, Yacoby A (2010) Electron liquids and solids in one dimension. Nature 464(7286):209–216. doi:10.1038/nature08918
56. De Angelis F, Patrini M, Das G et al (2008) A hybrid plasmonic-photonic nanodevice for label-free detection of a few molecules. Nano Lett 8(8):2321–2327. doi:10.1021/nl801112e
57. Raether H (1988) Surface plasmons. Springer, Berlin

Chapter 8

Single-Cell RT-PCR, a Technique to Decipher the Electrical, Anatomical, and Genetic Determinants of Neuronal Diversity

Maria Toledo-Rodriguez and Henry Markram

Abstract

The patch-clamp technique has allowed for detailed studies on the electrical properties of neurons. Dye loading through patch pipettes enabled characterizing the morphological properties of the neurons. In addition, the patch-clamp technique also allows for harvesting mRNA from single cells to study gene expression at the single cell level (known as single-cell RT-PCR). The combination of these three approaches makes possible the study of the GEM profile of neurons (gene expression, electrophysiology, and morphology) using a single patch pipette and patch-clamp recording. This combination provides a powerful technique to investigate and correlate the neuron's gene expression with its phenotype (electrical behavior and morphology). The harvesting and amplification of single cell mRNA for gene expression studies is a challenging task, especially for researchers with sparse or no training in molecular biology (*see* **Notes 1, 2** and **5**). Here we describe in detail the GEM profiling approach with special attention to the gene expression profiling.

Key words Single-cell RT-PCR, Multiplex PCR, Ion channel, Calcium binding protein, Neuropeptide, Morphology, Electrophysiology

1 Introduction

Neurons, the building blocks of the nervous system, largely differ in terms of their anatomical, electrophysiological, and gene expression properties, indicating unique functional roles for any given cell type. In recent years there has been an explosive increase in the amount of electrical, morphological, and gene expression data available from neurons. Nevertheless, despite this increase, the principles of diversification linking different data sets have not been clearly defined. Two key obstacles have impeded the derivation of these principles of diversification: (a) technical: it was and still is technically very challenging to study single neurons simultaneously at the electrical, morphological and gene expression level; (b) data set size: due to the technical difficulties too few neurons have been studied, preventing proper statistical analysis (*see* **Note 3**). In order

Marzia Martina and Stefano Taverna (eds.), *Patch-Clamp Methods and Protocols*, Methods in Molecular Biology, vol. 1183,
DOI 10.1007/978-1-4939-1096-0_8,

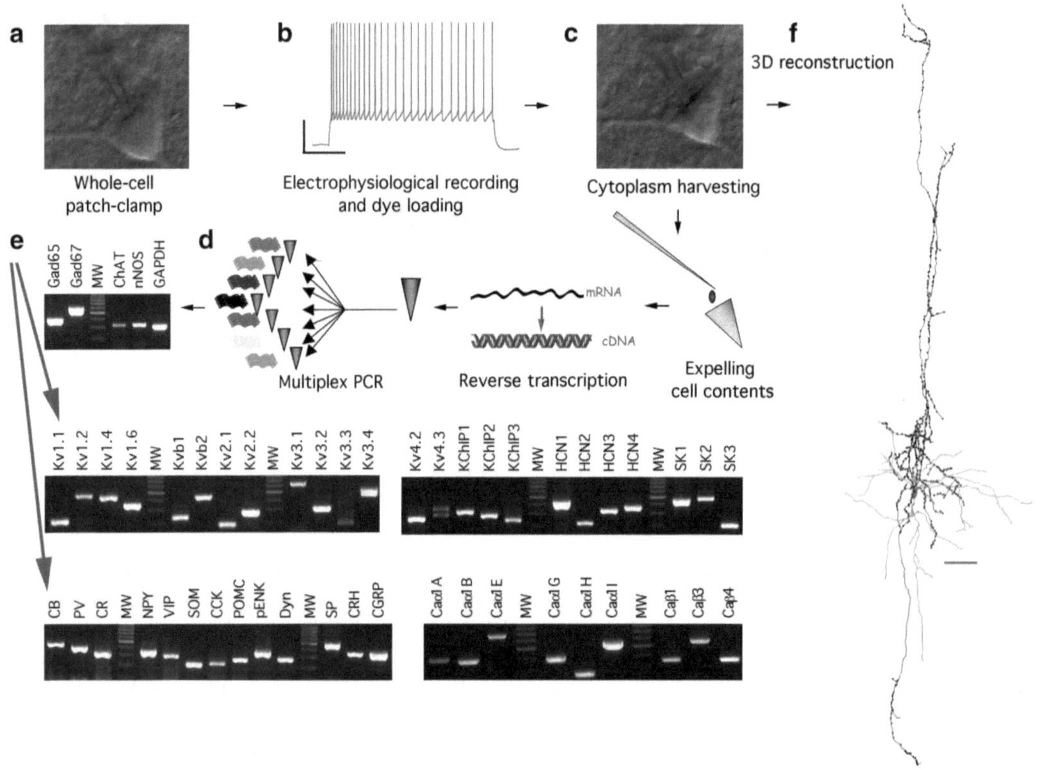

Fig. 1 The step-by-step single cell multiplex RT-PCR technique. (**a**) Whole-cell patch-clamp recordings from different classes of neocortical neurons are obtained. (**b**) A complex set of stimulation protocols are employed. Scale bars: current 20 mV, time 500 ms. (**c**) At the end of each recording the neuron's cytoplasm is extracted for subsequent reverse transcription. (**d**) Multiplex RT-PCR is performed in order to simultaneously detect the expression of 51 mRNAs from each neuron. (**e**) Agarose gels showing the 51 mRNA species simultaneously amplified from 250 pg of total brain mRNA. These include: the voltage activated K^+ channels (Kv1.1/2/4/6, Kvβ1/2, Kv2.1/2, Kv3.1/2/3/4, Kv4.2/3, KchIP1/2/3); the K^+/Na^+ permeable hyperpolarization activated channels (HCN1/2/3/4); the calcium activated K^+ channels (SK1/2/3); the voltage activated Ca^{2+} channels (Caα1A/B/E/G/E/I, Caβ1/3/4); the calcium-binding proteins (Calbindin (CB), Parvalbumin (PV), Calretinin (CR)); the neuropeptides [Neuropeptide Y (NPY), Vasoactive Intestinal Peptide (VIP), Somatostatin (SOM), Cholecystokinin (CCK), Proenkephalin (pENK), Proopiomelanocortin (POMC), Dynorphin (Dyn), Substance P (SP), Corticotropin Releasing Hormone (CRH), and Calcitonin Gene Related Peptide (CGRP)]; the enzymes [Glutamic Acid Decarboxylase (Gad 65 and Gad 67), Choline Acetyltransferase (ChAT), Nitric Oxide Synthase (nNOS), and Glyceraldehyde-3-Phosphate Dehydrogenase (GAPDH)]. *See* Table 1 for the size of the PCR product as predicted by its mRNA sequence. MW: 100 bp ladder molecular weight marker. (**f**) During recording neurons are also loaded with a dye allowing subsequent morphological classification and 3D computer anatomical reconstruction. Soma and dendrites in *light grey* and axon in *black*. Scale bar: 100 μm

to explore the principles that guide neuronal diversity, we and other groups developed a multidisciplinary approach, which allowed for studying single neurons simultaneously at the morphological, electrophysiological, and gene expression level [1–7]. For this purpose whole-cell patch-clamp recordings are combined with single-cell reverse transcription, multiplex PCR, and morphological identification (Figs. 1 and 2).

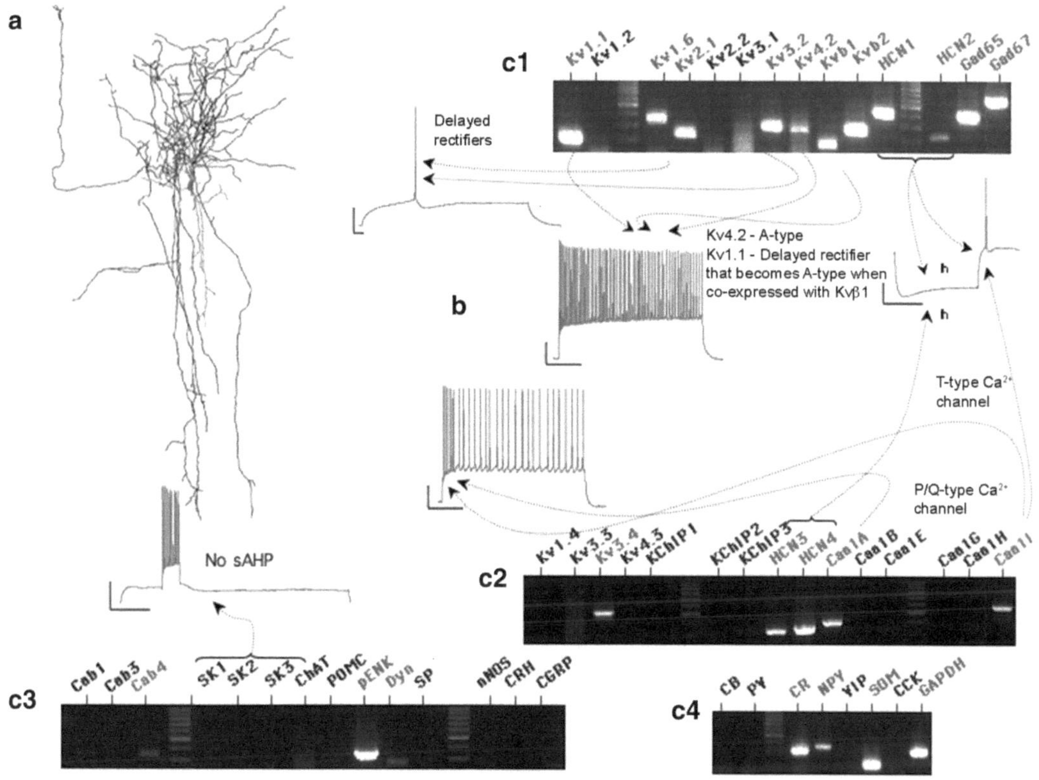

Fig. 2 A representative example of the GEM profiling of a layer 3 bitufted cell. (**a**) 3D computer reconstruction. Scale bar: 100 μm. (**b**) Representative responses to step-current injections. Scale bars: current 20 mV, time 500 ms. (**c1–4**) Agarose gels showing the mRNA expression profile of 51 K^+ and Ca^{2+} channels, Ca^{2+} binding proteins, neuropeptides, and enzymes. For abbreviations *see* Fig. 1 legend. Soma and dendrites in *light grey* and axon in *black*

2 Materials

2.1 Solutions

1. Extracellular solution: 125 mM NaCl, 2.5 mM KCl, 25 mM glucose, 25 mM $NaHCO_3$, 1.25 mM NaH_2PO_4, 2 mM $CaCl_2$, and 1 mM $MgCl_2$.
2. RNase-free intracellular solution containing biocytin: 100 mM potassium gluconate, 20 mM KCl, 4 mM ATP-Mg, 10 mM phosphocreatine, 0.3 mM GTP, 10 mM Hepes (pH 7.3, 310 mOsmol, adjusted with sucrose), and 0.5 % biocytin. The intracellular solution is prepared under RNAse-free conditions: RNAse-free water; glassware and pH meter are cleaned with NaOH (10 N), and chemicals are opened from the first time using gloves and RNAse-free tools. After preparation, the intracellular solution is filtered, tested for RNAse contamination by incubating the solution overnight at 37 °C with RNA

followed by agarose gel electrophoresis. The intracellular solution is confirmed RNAse-free if the two sharp bands of the ribosomal RNA appear in the gel. Finally the solution is aliquoted in single-use vials and stored at –20 °C.

3. Fixative: 0.1 M phosphate buffer (2.65 g/l $NaH_2PO_4.H_2O$, 14 g/l K_2HPO_4; pH 7.4) containing 2 % paraformaldehyde, 1 % glutaraldehyde and 0.3 % saturated picric acid.
4. ABC-Triton X-100 solution: 20 μl A+20 μl B+10 μl 10 % Triton X-100 in 1 ml Phosphate Buffer (PB). Prepare 30 min before use.
5. DAB solution: mix 17.5 mg DAB (1 ml) 200 μl 1 % cobalt chloride, 150 μl 1 % ammonium nickel sulfate and 0.1 M PB to 25 ml. Then add 0.03 % H_2O_2 (10 μl 30 % H_2O_2 in 1 ml PB) 1~2 min. Note: DAB reaction needs to be done under covering.
6. Aqueous mounting medium.
7. RT mix 1: 0.4 μl of RNAsin (40 U/μl) 4 μl oligo-dT primer (25 ng/μl) and add RNAse-free water up to 12 μl.
8. RT mix 2: 4 μl 5× Superscript III buffer, 1 μl 0.1 M DTT, 2 μl 5 mM dNTP, 0.4 μl RNAsin (40 U/μl), and 0.5 μl Superscript III (200 U/μl).
9. First PCR mix: 100 nM of each of the primers, 200 μM of each dNTP, 20 μl solution Q, and 5 U of HotStarTaq DNA Polymerase in a final volume of 100 μl.
10. Second PCR mix: 1 μM of its specific primers, 2 μl of the first PCR product (template), 200 μM of each dNTP, 1 M Betaine, and 1 U of TaqZol DNA Polymerase, in a final volume of 20 μl.

2.2 RNAse-Free Tools

1. RNAse-free pipettes: bake capillaries before pulling at 200 °C overnight. Clean puller with RNAse-Zap according to the manufacturer's instructions. Use gloves and autoclaved tweezers to handle the capillaries. Store the pooled capillaries in an RNAse-free container.
2. RNAse-free electrode wire: re-chlorinate the electrode at the beginning of each recording session or when suspected RNAse contamination.

3 Methods

3.1 Sample Preparation and Patch-Clamp Recordings

1. Rapidly decapitate the rat (use anesthesia when working with adult animals), extract the brain, and cut the two hemispheres apart.
2. Section neocortical slices (sagittal; 300 mm thick) on a vibratome filled with iced extracellular solution.

3. After sectioning leave the slices in extracellular solution with oxygenation at 34 °C for 30 min (incubate the chamber in a water bath). Afterwards remove the chamber to room temperature.
4. Transfer one slice to the patch-clamp setup for recording.
5. Identify neurons using IR-DIC microscopy [8].
6. Perform somatic whole-cell recordings using RNAse-free pipettes—resistance 3 MΩ—and 2–4 μl intracellular solution (containing biocytin). For a good intracellular filling stay in the neuron at least 20 min.

3.2 Histological Procedures

1. After recording fix the slices for 24 h in cold fixative.
2. Rinse the slices several times (10 min each) in PB.
3. Transfer the slices into PB containing 3 % H_2O_2 for 30 min.
4. Rinse five to six times in PB (10 min each).
5. Incubate the slices overnight at 4 °C in ABC-Triton X-100 solution.
6. Wash sections several times in PB (10 min each).
7. Develop with DAB solution under visual control using a bright-field microscope until all processes of the cells appeared clearly visible (usually 2–4 min).
8. Stop the reaction by transferring the sections into PB.
9. After washing in the same buffer, mount the slices in mounting.

3.3 Single-Cell RT-Multiplex PCR

1. At the end of the recording aspirate the cell's cytoplasm into the recording pipette under visual control by applying gentle negative pressure. Best results are obtained with several [3–8] gentle suction pulses. The strength and number of aspiration pulses has to be calibrated for each cell type based on the soma size. You should visually see a slight decrease in cell volume. Only process cells whose the seal was intact throughout the recording, and whose nucleus was not harvested (harvesting the nucleus will result in genomic DNA contamination and death of the cell, and therefore compromising the morphology). To avoid harvesting the nucleus, place the pipette parallel to the cell membrane (i.e., not facing directly the center of the soma).
2. Once the negative pressure pulses end, slowly withdraw the electrode from the cell to form an outside-out patch to prevent contamination of the pipette (you may see the neuron's membrane stretching as the pipette withdraws). If this happens gently tap on the electrode holder so that the vibrations will break the membrane bridge. This will prevent over-stretching and

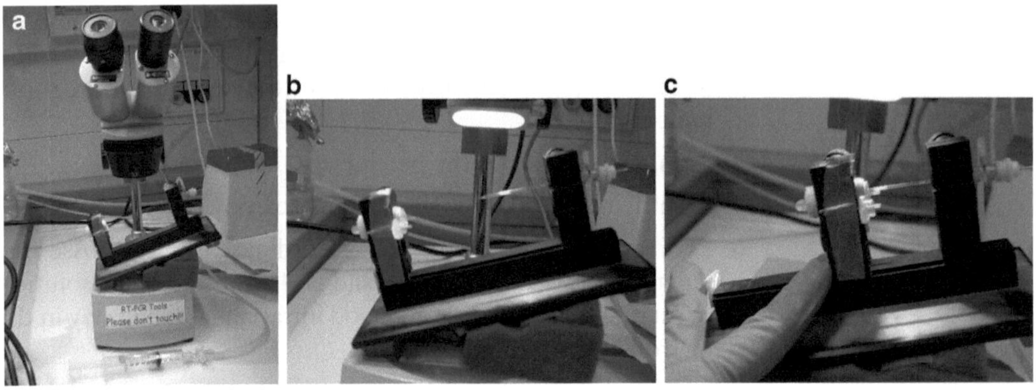

Fig. 3 Expelling the single cell contents in the patch pipette, device and procedure. (**a**) Device to expel the single cell contents after harvesting into the patch pipette. (**b**) Patch pipette secured to the pipette holder. (**c**) Breaking the pipette tip against the microcentrifuge tube bottom

breaking the cell's membrane leading to cell death. If you suspect the neuron has been damaged place the slide immediately in fixative (otherwise biocytin will leak through the damaged membrane diminishing the quality of the morphology).

3. Secure the pipette in a pipette holder (Fig. 3a, b), apply positive pressure. Under visual control bring the pipette tip close to the bottom of an RNAse-free microcentrifuge tube and break the tip against the tube bottom (Fig. 3c). In some cases breaking the pipette tip may reduce the RNA yield. Therefore, if possible, expel the pipette contents by applying strong positive pressure.
4. To the single cell contents add RT mix 1. Incubate at 70 °C for 5 min (*see* **Notes 2** and **3**).
5. Cool on ice for 1 min and briefly centrifuge to bring the contents to the tube bottom.
6. Add 8 μl of RT mix 2 and incubate for 50 min at 50 °C.
7. Heat-inactivate the reverse transcriptase by incubating at 70 °C for 10 min. At this step the cDNA can be frozen and stored at −20 °C before further processing.
8. Optimize each multiplex PCR conditions using total RNA purified from rat brain, so that a PCR product could be detected from 250 pg to 1 ng of total RNA without contamination caused by unspecific amplification. Table 1 lists the primer pairs we included in the different multiplexes, the name and accession number of the genes amplified, and the length of the PCR product. In this case, four different multi-

Table 1
List of all primers used in this study

mRNA	GeneBank #	Primers from 5′ to 3′		Amplicon size (bp)
GAPDH[a]	M17701	Sense Antisense	GCCATCAACGACCCCTTCAT TTCACACCCATCACAAACAT	315
Gad65[b]	M72422	Sense Antisense	TCTTTTCTCCTGGTGGTGCC CCCCAAGCAGCATCCACAT [Gad down]	391
Gad67[b]	M76177	Sense Antisense	TACGGGGTTCGCACAGGTC [Gad down]	600
ChAT[c]	M88488	Sense Antisense	GGCCATTGACAACCATCTTCTG CTTGAACTGCAGAGGTCTCTCAT	321
nNOS	U67309	Sense Antisense	CTTCCGAAGCTTCTGGCAAC GCTGGATGGCTTTGAGGACAT	328
CB[b]	M27839	Sense Antisense	AGGCACGAAAGAAGGCTGGAT TCCCACACATTTTGATTCCCTG	432
PV[b]	M12725	Sense Antisense	AAGAGTGCGGATGATGTGAAGA ATTGTTTCTCCAGCATTTTCCAG	389
CR[b]	X66974	Sense Antisense	CTGGAGAAGGCAAGGAAAGGT AGGTTCATCATAGGGACGGTTG	311
NPY[b]	M15880	Sense Antisense	GCCCAGAGCAGAGCACCC CAAGTTTCATTTCCCATCACCA	362
VIP[b]	X02341	Sense Antisense	TGCCTTAGCGGAGAATGACA CCTCACTGCTCCTCTTCCCA	290
Som[b]	K02248	Sense Antisense	ATCGTCCTGGCTTTGGGC GCCTCATCTCGTCCTGCTCA	208
CCK[b]	K01259	Sense Antisense	CGCACTGCTAGCCCGATACA TTTCTCATTCCGCCTCCTCC	216
SP[d]	M15191	Sense Antisense	GAGCATCTTCTTCAGAGAATCGC TCGCTGGCAAACTTGTACAACTC	468/513
POMC	K01878 J00759	Sense Antisense	ATAGACGTGTGGAGCTGGTG (in exon 2) TTCCTCCGCACGCCTCTGC (in exon 3)	253
PENK[d]	M28263	Sense Antisense	CAAACAGGATGAGAGCCACTTGC GCTTCTGCAGCTCCTTTGCTTC	370
Dyn[e]	M32781 M32783	Sense Antisense	GCCATAGGGGGATTTGGTAGC AACCTCAGAGGGGATCACAAG	266
CRH	M54987	Sense Antisense	AACTCAGAGCCCAAGTACGTTGA TCACCCATGCGGATCAGAATC	335
CGRP	L29188 L00111	Sense Antisense	GGTCGGGAGGTGTGGTGAAG (in exon 5) TCGACAGGGTGGTTTATGGGG (in exon 6)	420

(continued)

Table 1 (continued)

mRNA	GeneBank #	Primers from 5′ to 3′		Amplicon size (bp)
Kv1.1[f]	M26161	Sense Antisense	CCGCCGCAGCTCCTCTACT CAAGGGTTTTGTTTGGGGGCTTTT	209
Kv1.2[f]	X16003	Sense Antisense	GAAAAGTAGAAGTGCCTCTACCATAA TTGATATGGTGTGGGGGCTATGA	458
Kv1.4[f]	X16002	Sense Antisense	CTGGGGGACAAGTCAGAGTATCTA ACTCTCCTCGGGACCACCT	434
Kv1.6	X17621	Sense Antisense	GGGAACGGCGGTCCAGCTA GTGCATCTCATTCACGTGACTGAT	351
Kv2.1[g]	X16476	Sense Antisense	CAACTTCGAGGCGGGAGTC TCCAGTCAACCCTTCTGAGGAGTA	229
Kv2.2[g]	M77482	Sense Antisense	ACCAGGAGGTTAGCCAAAAAGACT AGGCCCCTTATCTCTGCTTAGTGT	446
Kv3.1[g]	X62840	Sense Antisense	CCAACAAGGTGGAGTTCATCAAG TGGTGTGGAGAGTTTACGACAGATT	640
Kv3.2[g]	X62839	Sense Antisense	ACCTAATGATCCCTCAGCGAGTGA CAAAATGTAGGTGAGCTTGCCAGAG	302
Kv3.3	M84211	Sense Antisense	GAGACCCCCGTCCCAATG CGGGGGAAGGGGCATAGTC	179
Kv3.4[f]	X62841	Sense Antisense	TCAGGCACACGGGACAGAAAC GGGCAGAGGACTTGGGAGACATA	418/522
Kv4.2[f]	S64320	Sense Antisense	CCGAATCCCAAATGCCAATGTG CCTGACGATGTTTCCTCCCGAATA	265
Kv4.3	U42975	Sense Antisense	GGGCAAGACCACGTCACTCA CTGCCCTGGATGTGGATGGT	296/386
Kvb1[f]	X70662	Sense Antisense	AAGGGAGAAAACAGCAAAACAAGC TGGCACCAAGGTTTTCAATGAGTT	170
Kvb2	X76724	Sense Antisense	ACAGTGGCATCCCACCCTACT GTGGACGATGGAGGACGACAAT	283
KChIP1	AB046443	Sense Antisense	AAAGGCGACCCTCCAAAGATAAG GGACAGTTCCTCTCAGCAAAATCG [KChIP1 *long*]	330
KChIP2	AF269283	Sense Antisense	RCARKYCCTKTACCGAGGCTTCA [KChIP up] CAAGCATTTCCTCCTTTGTGATAC	289
KChIP3	AB043892	Sense Antisense	[KChIP up] GTCGTAGAGATTGAAGGCCCACT	249
HCN1	AF247450	Sense Antisense	CCTCAAATGACAGCCCTGAATTG TCGGTGTGGAACTACCAGGTGT	405

(continued)

Table 1
(continued)

mRNA	GeneBank #	Primers from 5′ to 3′		Amplicon size (bp)
HCN2	AF247451	Sense Antisense	CTCTCCGGCAACGCGTGTG AGTCCCTGCGGTCCGGACT	211
HCN3	AF247452	Sense Antisense	TGCCCCTCTCCCCTGATTC TTCCAGAGCCTTTGCGCCTA	335
HCN4	AF247453	Sense Antisense	AACCTGGGGGCTGGACAGA CTGGGCAGCCTGTGGAGAG	462
SK1	U69885	Sense Antisense	GCATGTGCACAACTTCATGATGGA [SK up] TGGGCGGCTGTGGTCAGGTG	449
SK2	U69882	Sense Antisense	[SK up] CGCTCAGCATTGTAGGTGACATG	461
SK3	U69884	Sense Antisense	[SK up] CATCTTGACACCCCTCAGTTGG	192
Caa1A[e]	M64373	Sense Antisense	GAGCGGCTGGATGACACAGAAC CTGGCGACTCACCCTGGATGTC	420
Caa1B	M92905	Sense Antisense	TTGGCTCCTTCTTCATGCTCAAC GATAAGGAACCGGAACATCTTCTC	409
Caa1E[e]	L15453	Sense Antisense	AGACTGTGGTGACTTTTGAGGACC GAGCTATGGGGCACCATGGCTT	693
Caa1G	AF027984	Sense Antisense	TGGGCTCCTTCTTCATGATCAAC [CaT up] GGAACTCTGAGCGTCCCATTAC	407
Caa1H	AF290213	Sense Antisense	[CaT up] CTGCTGGGATCCACCTTCTTAC	271
Caa1I	AF086827	Sense Antisense	[CaT up] AGGTCCGAGGAGACCCCATC	556
Cab1	X61394	Sense Antisense	CCCTAAACTGCTGTGGGTGGA CCCAGCTCTGCTCCCCAAAG	359
Cab3	M88751	Sense Antisense	ACTGACCACCTCCTGCCCTAC GTCCTGCCTCACCTGCACTG	555
Cab4	L02315	Sense Antisense	GCTATGGTATTTGTTTGCTGGAAG GACTGCAGAAGGAACAACACCTC	351

For abbreviations *see* Fig. 1 legend. R = A or G; K = G or T; Y = C or T

[a] [17]

[b] [16]

[c] [22]

[d] [21]

[e] [18]

[f] [19]

[g] [20]

plex PCRs reactions were performed to test the expression of 51 mRNA species from each cell. Each multiplex PCR reaction is performed using a different amount of c-DNA depending on the multiplex complexity and the number of mRNA molecules suspected to be expressed by the cell. During calibration, different combinations of genes are distributed between the pools and different primer pairs are tested until an even amplification of all genes in the pool was obtained (*see* **Notes 1** and **4**). The genes co-amplified in each multiplex PCR reaction were (for abbreviations *see* Fig. 1 legend; for primers and amplicon length *see* Table 1):

- Pool I → (CB, PV, CR, NPY, VIP, SOM, CCK, and GAPDH).
- Pool II → (SP, ChAT, POMC, pENK, Dyn, nNOS, CRH, CGRP, and Caα1E).
- Pool III → (Kv1.1, Kv1.2, Kv1.6, Kv2.1, Kv2.2, Kv3.1, Kv3.2, Kv4.2, Kvβ1, Kvβ2, HCN1, HCN2, Gad65, and Gad67).
- Pool IV → (Kv1.4, Kv3.3, Kv3.4, Kv4.3, KChIP1, KChIP2, KChIP3, HCN3, HCN4, Caα1A, Caα1B, Caα1G, Caα1H, Caα1I, Caβ1, Caβ3, Caβ4, SK1, SK2, and SK3).

9. The first PCR amplification round consists of 10 min hot start at 95 °C followed by 25 cycles [94 °C for 40 s, 56 °C (pools I, II and IV) or 58 °C (pool III) for 40 s and 72 °C for 1 min] performed with a programmable thermocycler. For each pool all genes are simultaneously amplified in a single tube containing 1/10 (pools I and IV) or 2/5 (pools II and III) of the RT product. (Please notice, in this example we had to divide the sample during the first amplification due to nonspecific interactions between the primers. If the primers do not interact, confirmed using total brain cDNA, then all primers should be included in a single re-amplification mix containing the entire RT mix).
10. The second PCR consists of 40 cycles [94 °C for 40 s, 56 °C (pools I, II, and IV) or 58 °C (pool III) for 40 s, and 72 °C for 1 min]. In this case, each gene is individually amplified in a separate test tube.
11. The products of the second PCR are analyzed in 1.5 % agarose gels using ethidium bromide (*see* **Note 3**).
12. Amplification specificity can be randomly verified by restriction analysis.

3.4 Controls for the Single-Cell RT-Multiplex PCR

1. For each PCR amplification, controls for contaminating artifacts are performed using sterile water instead of cDNA.
2. A control for nonspecific harvesting of surrounding tissue components is randomly employed by advancing pipettes into the slice and retrieving without seal formation and suction.
3. Amplification of genomic DNA can also be avoided by designing primers overspanning introns and/or never harvesting the cell nucleus.
4. Controls in which the RT is omitted should give negative results (no bands).

3.5 Conclusion

1. The combination of patch-clamp recordings with morphological identification and gene expression characterization provides a novel systems approach to relate gene expression profiles, neuronal morphology and electrical behavior. Using these techniques, it is possible to derive precise information about (a) the correlation and developmental changes between gene expression and behavior; (b) the alterations leading to pathologies in developmental, degenerative, and psychological brain disorders.

4 Notes

1. The detection of specific mRNA transcripts from a single cell is a technical challenge due to the very low amount of mRNA contained in a single neuron (approximately 10^6 molecules transcribed from about 15,000 different genes [9]). Besides, this very small number of mRNA molecules may be (a) lost in one of the numerous steps from the harvesting of the neuron's cytoplasm to the final PCR amplification; or (b) destroyed by endogenous or contaminating RNAses. Depending on the number of genes to be studied the investigator can choose between different approaches.

   (a) *One gene per cell* [2].

   - In this case the entire contents harvested from the cell are reverse-transcribed and PCR-amplified for one single gene.
   - Advantages: is the simplest approach to set (requires only designing one primer pair) and has the lowest chance to obtain false negatives.
   - Disadvantages: it is not possible to study coexpression of two or more genes and if the gene is not detected it is impossible to know if the gene is not expressed in that specific cell or there was a failure in the harvesting or amplification (false negative).

(b) *Multiple genes from a single cell.* Depending on the way that the cytoplasmic contents are processed:
Splitting the cell's cytoplasm into as many reactions as genes to be tested and then amplifying each gene independently [10].

- Advantages: (1) Still is a relative simple and straightforward approach (each primer pair is designed independently). (2) Because an internal control can be included (housekeeping gene, a gene always expressed in every cell), it is possible to control for false negatives.
- Disadvantages: As stated before, the number of mRNA molecules contained by a single neuron is small. If the cell contents are spilt into too many fractions it may happen that, although the gene was expressed, the aliquot for testing the expression of that gene did not contain any of its mRNA molecules (giving rise to false negatives).

(c) *Multiplex PCR* [11]. In this case the genes whose expression is going to be tested are pre-amplified in a common PCR reaction before being split (into as many reactions as genes to be tested) and subsequently they are amplified independently (as described in the above sample).

- Advantages: (1) The pre-amplification step ensures that, regardless of the number of fractions, any aliquot may contain at least a few copies of each mRNA species expressed by the neuron. (2) Because an internal control can be included (housekeeping gene, a gene always expressed in every cell) it is possible to control for false negatives.
- Disadvantages: The primer design becomes complex as it is required to avoid cross-interactions between the different primer pairs and amplification products. The number of genes which can be included in the multiplex PCR is limited as the probability to encounter false negatives due to amplification failures rises as the number of primer pairs included in the multiplex pool due to: (1) competition for reagents; (2) interference between the multiple primers (unspecific annealing of one primer with others or with the PCR products from other genes); (3) difficulty in designing primers that will optimally melt and anneal in the same conditions (temperature, salt concentrations, times, etc.) [12].

(d) *Degenerated Primers* [13]. In the particular case where the genes to amplify show a high degree of similarity in their sequence (as in ion channel or receptor families), a single pair of degenerated primers can be designed for the first PCR (located in regions identical or nearly identical for all the genes). Subsequently the identity of each individual gene can be determined by: (1) differences in the size of the PCR products (if the regions amplified have a different size for different members of the family); (2) using restriction enzyme analysis where each selected enzyme specifically cuts the cDNA of only one gene; (3) performing second PCR reaction using nested primers (primers located inside the previously amplified region) specific for each gene.

- Advantages: a relatively simple and straightforward primer design (only one primer pair has to be designed for the first PCR, while for the second PCR each primer pair is designed independently).
- Disadvantages: depending on the sequence similarity of the different genes to be amplified, many times the degenerated primers will not amplify all the genes with the same efficiency (Fig. 4 comparison between the multiplex and degenerated protocols to amplify the members for the Kv1 family).

2. In many instances better yields are obtained with smaller reverse transcription volumes. Therefore, when possible, try to reduce the RT mixture volume to 10 μl (3.85 μl intracellular solution containing cell's contents, 0.2 μl of RNAsin (40 U/μl) 2 μl oligo-dT primer (25 ng/μl, Invitrogen), 2 μl 5× Superscript III buffer, 0.5 μl 0.1 M DTT, 1 μl 5 mM dNTP, 0.2 μl RNAsin (40 U/μl), and 0.25 μl Superscript III (200 U/μl)).
3. If the final goal is to quantify relative gene expression using real time PCR reverse transcription could be performed using a RT kit optimized for downstream real time PCR applications such as High Capacity RNA-to-cDNA Kit (Invitrogen). The use of real time PCR has advantages and disadvantages over the protocol described here. The main advantage is that by knowing the Ct the investigator can normalize the expression of the genes (using one or several housekeeping genes) enabling comparison of expression levels across cells. A disadvantage of real time PCR is that it requires a more specialized equipment and knowledge. Moreover, many of the commercially available real time PCR primer kits may not be suitable for multiplex amplification.

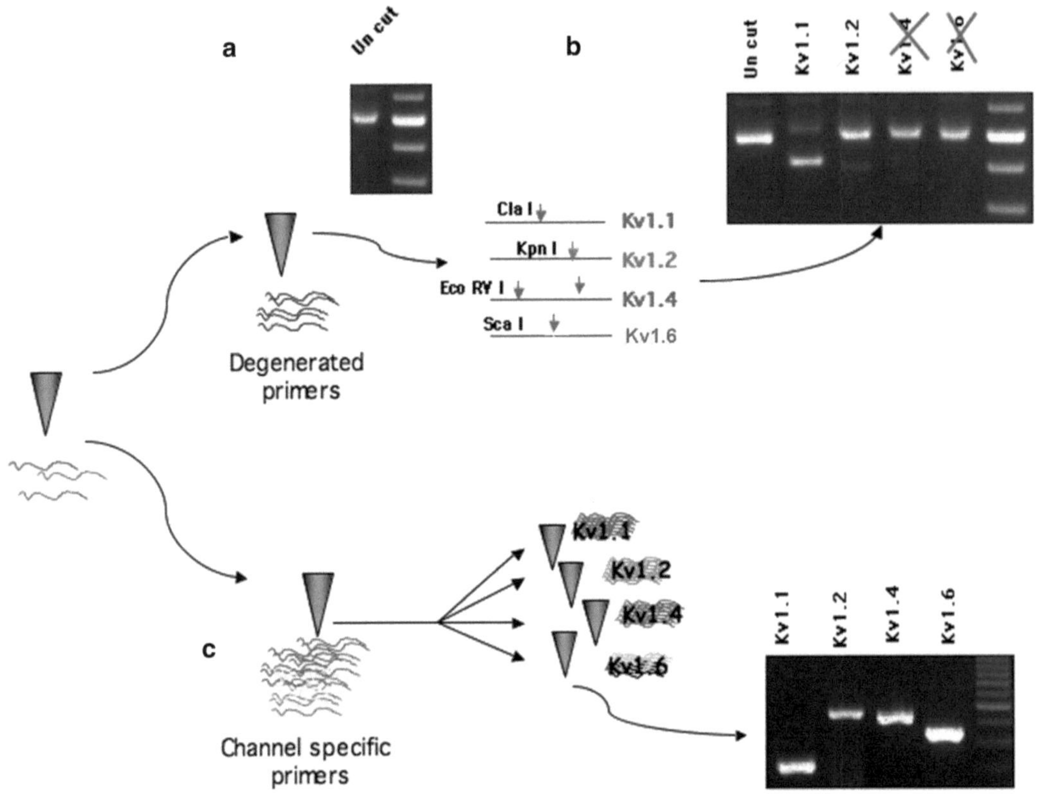

Fig. 4 Kv1 family members amplified using degenerated primers versus channel specific primers. (**a**) Simultaneous amplification of four members of the Kv1 family using degenerated primers. (**b**) Determination of the identity of each Kv1 family member by restriction enzyme analysis. (**c**) Simultaneous amplification of the Kv1 family members Kv1.1, Kv1.2, Kv1.4, and Kv1.6 using gene specific primers

4. As it can be seen, the higher the number of genes to be tested, the higher the (a) challenge designing the experiment; (b) number of manipulations and experiment duration; (c) number of false negatives. For beginners (in particular beginners with low or no experience in molecular biology) we recommend to start by the easiest (one gene per cell), then move to testing multiple genes from a single cell using published primer sequences and only then launch to design their own multiplex mixtures.
5. The analysis of single-cell GEM data presents special opportunities but requires consideration of several sources of systematic and random error, in particular testing for the effects of false negatives in the gene expression. This can be overcome by the use of statistics. Moreover, the cell-wise correlation of GEM information allows efficient application of modern statistical classification techniques with fewer samples and lower noise than might be required from mixed-cellular tissue

extracts. The current best set of statistical techniques is distributed for the free by The Institute for Genomics Research Multi-experiment Viewer [14]. Examples of application of these techniques can be found in the following refs. 5, 6, 15.

Acknowledgements

The authors would like to thank the late Prof. Phil Goodman for helpful insights on the statistical approaches to the single-cell RT-PCR data analysis. We would like to thank Shaoling Ma, Claudia Herzberg, Raya Eilam, and Tal Hetzroni for their technical assistance and Dr Jesper Ryger for his useful comments. This work was supported by the National Alliance for Autism Research and a European Union grant.

References

1. Eberwine J, Yeh H, Miyarisho KCY (1992) Analysis of gene expression in single live neurons. Proc Natl Acad Sci U S A 89: 3010–3014
2. Lambolez B, Audinat E, Bochet P et al (1992) AMPA receptor subunits expressed by single Purkinje cells. Neuron 9:247–258
3. Sucher NJ, Deitcher DL (1995) PCR and patch-clamp analysis of single neurons. Neuron 14:1095–1100
4. Ceranik K, Bender R, Geiger JR et al (1997) A novel type of GABAergic interneuron connecting the input and the output regions of the hippocampus. J Neurosci 17:5380–5394
5. Toledo-Rodriguez M, Goodman P, Illic M et al (2005) Neuropeptide and calcium binding protein gene expression profiles predict neuronal anatomical type in the juvenile rat. J Physiol 567:401–413
6. Toledo-Rodriguez M, Blumenfeld B, Wu C et al (2004) Correlation maps allow neuronal electrical properties to be predicted from single-cell gene expression profiles in rat neocortex. Cereb Cortex 14:1310–1327
7. Wang Y, Gupta A, Toledo-Rodriguez M et al (2002) Anatomical, physiological, molecular and circuit properties of nest basket cells in the developing somatosensory cortex. Cereb Cortex 12:395–410
8. Dodt HU, Zieglgansberger W (1990) Visualizing unstained neurons in living brain slices by infrared DIC-videomicroscopy. Brain Res 537:333–336
9. Sargent TD (1987) Isolation of differentially expressed genes. Methods Enzymol 157: 423–432
10. Foehring RC, Mermelstein PG, Song WJ et al (2000) Unique properties of R-type calcium currents in neocortical and neostriatal neurons. J Neurophysiol 84:2225–2236
11. Ruano D, Lambolez B, Rossier J et al (1995) Kainate receptor subunits expressed in single cultured hippocampal neurons: molecular and functional variants by RNA editing. Neuron 14:1009–1017
12. Edwards MC, Gibbs RA (1994) Multiplex PCR: advantages, development, and applications. PCR Methods Appl 3:S65–S75
13. Plant TD, Schirra C, Katz E et al (1998) Single-cell RT-PCR and functional characterization of Ca^{2+} channels in motoneurons of the rat facial nucleus. J Neurosci 18:9573–9584
14. Saeed AI, Sharov V, White J et al (2003) TM4: a free, open-source system for microarray data management and analysis. Biotechniques 34: 374–378
15. Cauli B, Porter JT, Tsuzuki K et al (2000) Classification of fusiform neocortical interneurons based on unsupervised clustering. Proc Natl Acad Sci U S A 97:6144–6149
16. Cauli B, Audinat E, Lambolez B et al (1997) Molecular and physiological diversity of cortical nonpyramidal cells. J Neurosci 17:3894–3906
17. Aranda-Abreu G, Behar L, Chung S et al (1999) Embryonic lethal abnormal vision-like RNA-binding proteins regulate neurite

outgrowth and tau expression in PC12 cells. J Neurosci 19:6907–6917

18. Glasgow E, Kusano K, Chin H et al (1999) Single cell reverse transcription-polymerase chain reaction analysis of rat supraoptic magnocellular neurons: neuropeptide phenotypes and high voltage-gated calcium channel subtypes. Endocrinology 140:5391–5401
19. Song WJ, Tkatch T, Baranauskas G et al (1998) Somatodendritic depolarization-activated potassium currents in rat neostriatal cholinergic interneurons are predominantly of the A type and attributable to coexpression of Kv4.2 and Kv4.1 subunits. J Neurosci 18:3124–3137
20. Baranauskas G, Tkatch T, Surmeier DJ (1999) Delayed rectifier currents in rat globus pallidus neurons are attributable to Kv2.1 and Kv3.1/3.2 K(+) channels. J Neurosci 19:6394–6404
21. Surmeier DJ, Song WJ, Yan Z (1996) Coordinated expression of dopamine receptors in neostriatal medium spiny neurons. J Neurosci 16:6579–6591
22. Yan Z, Surmeier DJ (1996) Muscarinic (m2/m4) receptors reduce N- and P-type Ca^{2+} currents in rat neostriatal cholinergic interneurons through a fast, membrane-delimited, G-protein pathway. J Neurosci 16:2592–2604

Chapter 9

Whole-Cell Voltage Clamp on Skeletal Muscle Fibers with the Silicone-Clamp Technique

Romain Lefebvre, Sandrine Pouvreau, Claude Collet, Bruno Allard, and Vincent Jacquemond

Abstract

Control of membrane voltage and membrane current measurements are of critical importance for the study of numerous aspects of skeletal muscle physiology and pathophysiology. The silicone-clamp technique makes use of a conventional patch-clamp apparatus to achieve whole-cell voltage clamp of a restricted portion of a fully differentiated adult skeletal muscle fiber. The major part of an isolated muscle fiber is insulated from the extracellular medium with silicone grease and the tip of a single microelectrode connected to the amplifier is then inserted within the fiber through the silicone layer. The method is extremely easy to implement. It represents an alternative to the traditional vaseline-gap isolation and two or three microelectrodes voltage-clamp techniques. The present chapter reviews the benefits of the silicone-clamp technique and provides updated detailed insights into its practical implementation.

Key words Skeletal muscle, Voltage clamp, Silicone grease, Mammalian muscle, Ion channels, Excitation–contraction coupling

1 Introduction

Whole-cell voltage clamp on isolated adult skeletal muscle fibers was first achieved about 50 years ago [1] and has since then been mastered by several groups with either various vaseline-gap isolation methods [2–18] or with two or three microelectrodes techniques [19–25]. Of course each method has its own advantages and drawbacks, but overall none of the microelectrode or gap methods can be readily qualified as easy to implement and use. Furthermore, some specific difficulties inherent to the two above lines of approach were worth circumvent; for instance microelectrodes suffer from their invasive features while gap isolation techniques necessitate a specific experimental chamber within which the transfer and mounting of a single isolated fiber can be a tricky demanding task. There was thus room for a more accessible way to do voltage clamp on skeletal muscle, especially when having in

Marzia Martina and Stefano Taverna (eds.), *Patch-Clamp Methods and Protocols*, Methods in Molecular Biology, vol. 1183, DOI 10.1007/978-1-4939-1096-0_9, © Springer Science+Business Media New York 2014

mind the relative simplicity and easiness of practice of the whole-cell patch-clamp technique. Furthermore, there was also room for avoiding the quite challenging microdissection step to use instead enzymatically isolated fibers [26]. Still, due to their large size and to the extent of the transverse tubular membrane system skeletal muscle fibers have to be left off the list of the cell types readily suitable for conventional whole-cell patch clamping; at least whole-cell patch-clamp should be restricted to muscle fibers of small size for measurements of charge movement and calcium transients under conditions where ion conductances are blocked to increase the membrane resistance and improve the clamp [27]. We initially had the idea that whole-cell voltage clamp could then be achieved by sealing the tip of a patch-clamp pipette on an electrically insulated restricted portion of a muscle fiber, sort of a combination between a gap isolation system and the patch-clamp technique. While attempting to work this out, we found that silicone grease could be used as a non-traumatic, electrically insulating agent on fibers isolated by collagenase treatment of a muscle, opening up the possibility to work on a short portion of such an isolated intact fiber. We also found that the inside of a glass micropipette could be given access to the cytoplasmic compartment of the fiber through the silicone insulating material. These two small tricks have had important practical outcomes. First, enzymatically isolated muscle fibers could, for the first time, be voltage-clamped by a classical gap-isolation method using silicone instead of vaseline [13, 28]. Second, voltage clamp could also be achieved with a patch-clamp amplifier on a short silicone-free end portion of a fiber, simply with the tip of a micropipette inserted within the fiber through the silicone layer [29]; this is what we have called the silicone voltage-clamp technique or silicone-clamp technique. This configuration is much more stable and less damageable to the preparation than sealing the tip of a patch pipette on the silicone-free portion of fiber. Finally, either microinjection or intracellular dialysis through the tip of a pipette impaled within the silicone embedded portion of fiber can be achieved without fearing the detrimental consequences of membrane damage. This makes intracellular loading of a dye, a buffer, or any other compound of interest easy. The simplicity of the silicone voltage-clamp technique makes it readily combinable with an extracellular perfusion system, with fluorescence detection under either conventional or confocal microscopy and with single channel activity measurements using an additional pipette in cell-attached mode [30]. The technique is easy to learn: it takes no more than half a day of practice for a moderately motivated student to be successful and less than a week to be completely confident. The different types of membranous and intracellular signals that can be routinely collected under silicone voltage-clamp conditions were reviewed by Collet et al. [31].

2 Materials

All solutions are prepared with deionized water and stored at 4 °C until use, unless otherwise specified.

2.1 Isolation of the Muscle Fibers

1. Fibers are isolated from the hindlimb *flexor digitorum brevis* and/or interosseus muscles from mouse.
2. Tyrode solution: 140 mM NaCl, 5 mM KCl, 2.5 mM $CaCl_2$, 1 mM $MgCl_2$, 10 mM Hepes, pH 7.2.
3. Collagenase (type 1).

2.2 Preparation of a Muscle Fiber for the Voltage Clamp

1. Silicone grease (*see* **Note 1**).
2. Culture medium supplemented with fetal bovine serum (FBS 10 %) (*see* **Note 2**).

2.3 Electrophysiology

1. The technique makes use of a conventional patch-clamp setup.
2. The composition of the extracellular solution is designed according to the aim of the experiment; care should be taken to avoid conditions generating large currents that may produce membrane voltage escape. For routine measurements of voltage-activated intracellular calcium transients in the silicone-free end portion of fiber, we use the following: 140 mM tetraethylammonium-methanesulfonate, 2.5 mM $CaCl_2$, 2 mM $MgCl_2$, 10 mM Hepes, 1 mM 4-aminopyridine, 2 μM tetrodotoxin, pH 7.2; for measurements of intramembrane charge movement, the $CaCl_2$ concentration is lowered to 0.25 mM and the solution also contains 0.5 mM $CdCl_2$ and 0.3 mM $LaCl_3$ to eliminate the calcium current.
3. Internal solution: 120 mM K-glutamate, 5 mM Na_2ATP, 5 mM sodium phosphocreatine, 5.5 mM $MgCl_2$, 5 mM glucose, 10 mM Hepes, pH 7.2. Aliquots of 1 ml of this solution are stored at −20 °C.

3 Methods

3.1 Isolation of the Muscle Fibers

Intact isolated muscle fibers can be easily obtained by enzymatic treatment of small muscles from mouse. The procedure is particularly efficient with either the *flexor digitorum brevis* or the interosseal muscles. Notice that this procedure also works beautifully with frog toe muscles.

1. Mice are killed by cervical dislocation.
2. Muscles are removed and incubated at 37 °C for 60 min in typically 3 ml of Tyrode solution containing 6 mg of collagenase (*see* **Note 3**).

3. Muscles are then rinsed with collagenase-free Tyrode and can be stored at 4 °C until use. We use them up to 24 h following the dissociation.
4. Isolated fibers are obtained by gently sucking a collagenase-treated muscle a few times in and out of the cut disposable blue tip of a Pipetman (or of the tip of a Pasteur pipette), within the experimental chamber (see next). This process yields tens of viable, healthy fibers from a single round of trituration of one muscle.

3.2 Preparation of a Muscle Fiber for the Voltage Clamp

1. For simple voltage-clamp experiments we use a standard 35 mm culture dish as experimental chamber. For experiments combining voltage-clamp and fluorescence measurements we either use a home-made chamber with a glass coverslip bottom or a 50 mm wide culture μ-Dish. In all cases the bottom of the chamber is first coated with a thin layer of silicone (*see* **Note 4**).
2. The chamber is then filled with culture medium (supplemented with FBS, *see* Subheading 2). This solution is necessary to preserve the integrity of the fibers upon first contact with the silicone (*see* **Note 5**). Notice that isolated fibers do not adhere to the silicone material.
3. The next step consists in embedding the major part of a fiber with silicone so as to leave only a short (typically 50–100 μm long) emerging portion in contact with the extracellular solution. Obviously this is a critical step of the process; although on the few first attempts it may seem hazardous, with practice it gets extremely trivial and can be achieved within 10 s, almost eyes shut. Different ways have been and are still used in the lab to do it (*see* **Note 6**). One simplest way is achieved as follows:
4. A yellow pipette tip is heated and pulled over the flame of a lighter so as to make a ~5–10 cm long thin tube. The end of it is cut with a scalpel blade; the cut opening diameter should be approximately 50–200 μm.
5. A 5 ml standard plastic syringe is filled with silicone.
6. The back opening of the yellow tip is cut with a scalpel blade so that it can be tightened around the end of the 5 ml syringe. Pressure is applied until silicone reaches the end of the yellow capillary.
7. An inverted microscope with either a ×10 or ×20 objective is used. The syringe is held so that the extremity of the yellow tube is positioned close to the median portion of a given fiber, in touch with the silicone layer covering the bottom of the chamber (Fig. 1b).

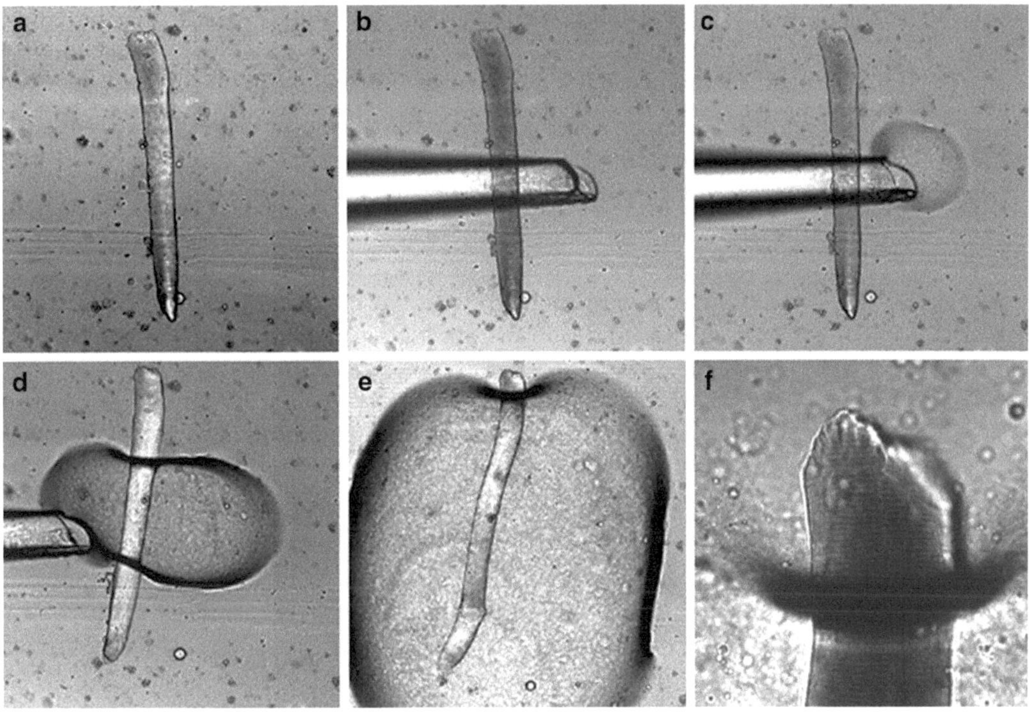

Fig. 1 Silicone embedding of a single isolated mouse muscle fiber. Images of a muscle fiber were taken during the consecutive steps of the process. Here the silicone was pressure-ejected through the broken tip of a glass micropipette. In **a**–**e** and **f**, image side size is 709 μm and 113 μm, respectively. See text for details. (Republished with permission from Springer)

8. Pressure is applied in the syringe so that the silicone flows out of the yellow tube end, at a rate sufficiently low to allow for control of the process.
9. Once the flowing out silicone has merged with the layer on the bottom of the chamber (Fig. 1c), the tube end is moved up then laterally above the fiber and then down on the opposite side of the fiber so as to make a bridge of silicone covering the fiber (Fig. 1d). This is repeated several times along the fiber until only a short portion remains out of the silicone (Fig. 1e, f). Once this is completed, any compound of interest can be microinjected within the portion of fiber embedded in silicone (*see* **Note** 7).
10. We usually embed several fibers with silicone in a same chamber. The extracellular FBS supplemented solution (*see* **Notes 5** and **6**) is then replaced by the experimental extracellular solution of interest before starting the electrophysiological procedure.

3.3 Electrophysiology

1. The voltage-clamp pipettes are pulled so as to yield a typical resistance of 2–3 MΩ when filled with the internal solution (*see* **Note 8**).
2. Positive pressure is maintained in the pipette until the tip is inserted within the silicone layer coating the fiber. This makes the resistance go way up (hundreds of MΩ to GΩ); pressure is then released and the desired holding voltage value is applied.
3. The pipette is then moved further down until the tip gets in touch with the membrane of the fiber. In order for the tip to go through the membrane a single or couple of brief taps on the supporting table is usually sufficient. Access can in many cases be further improved by moving the tip of the pipette within the fiber and/or by giving a few more gentle taps on the table once the tip is in.
4. Following the initial jump in current produced upon insertion of the tip, the holding current usually tends to decrease towards a lower value with a time constant of a few minutes; we thus routinely wait ~10 min for the current to stabilize, adjust the series resistance compensation and start the experiment. The impalement site is usually located at a distance of 100–150 μm from the silicone-free portion of fiber.
5. If intracellular dialysis is to be used, the tip of the pipette is inserted as close as possible to the silicone-free end of fiber. Once in the fiber, the tip is crushed by moving it back and forth in a non-gentle manner towards the bottom of the chamber. Breaking of the tip is followed by the effect of the decreased series resistance on the capacitive current transient elicited by a small voltage pulse. Once this is done, a period of time of approximately 30 min is sufficient to allow for equilibration of the pipette solution into the voltage-clamped end portion of fiber, as can be assessed from images of a fluorescent dye.

Figure 2 shows illustrative examples of membrane current traces obtained from silicone voltage-clamped fibers under various experimental conditions. Figure 2a shows L-type calcium current traces obtained in response to depolarizing pulses of increasing amplitude with a 2 mV increment. Figure 2b shows intramembrane charge movement traces obtained in response to depolarizations to the indicated voltages. Linear capacitive and resistive components of the current have been subtracted. Figure 2c shows membrane current traces recorded from a fiber, the silicone-free portion of which included the endplate area. Inward currents were elicited by puffing carbamylcholine (CCh) onto the endplate region for the indicated periods of time.

Figure 3 shows examples of intracellular Ca^{2+} signals triggered by voltage-clamp depolarization in muscle fibers.

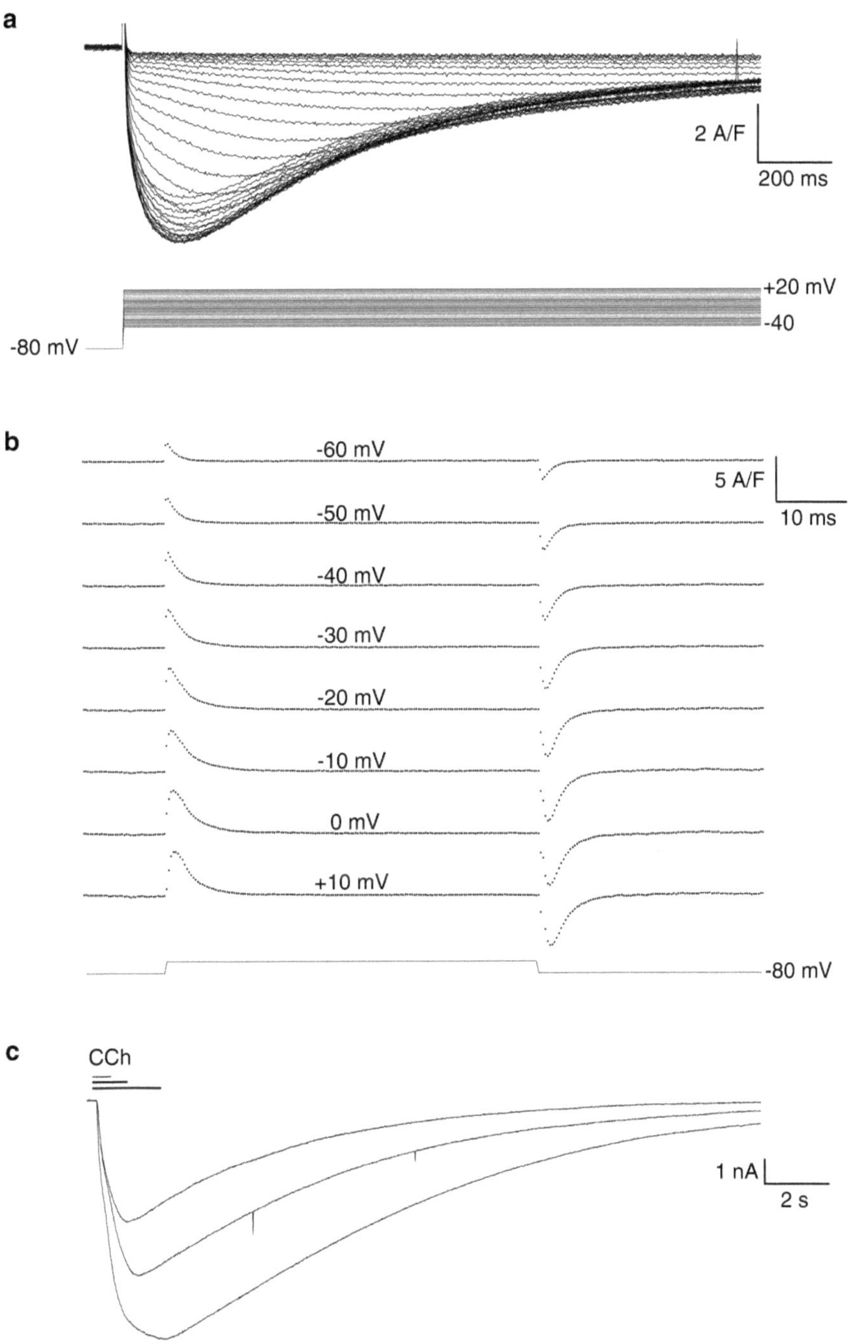

Fig. 2 Examples of membrane current records from silicone voltage-clamped mouse muscle fibers under different conditions. (**a**) Calcium current records obtained in response to depolarizing pulses to the indicated values; the linear resistive component of the current has been subtracted. (**b**) Intramembrane charge movement records in response to depolarizing pulses to the indicated values; linear resistive and capacitive components of the current have been subtracted. (**c**) Membrane currents through endplate nicotinic receptors at −80 mV. The silicone-free portion of fiber included the endplate area; CCh was puffed onto the endplate region using a pipette containing 10 mM CCh diluted in the extracellular solution. The pipette was connected to the Picospritzer apparatus; its tip was placed in the vicinity of the endplate and pressure pulses of the indicated duration were delivered. The extracellular solution contained 130 mM $MgSO_4$, 10 mM $MgCl_2$, 100 mM mannitol, 10 mM PIPES, pH 7.20. (Republished with permission from Springer)

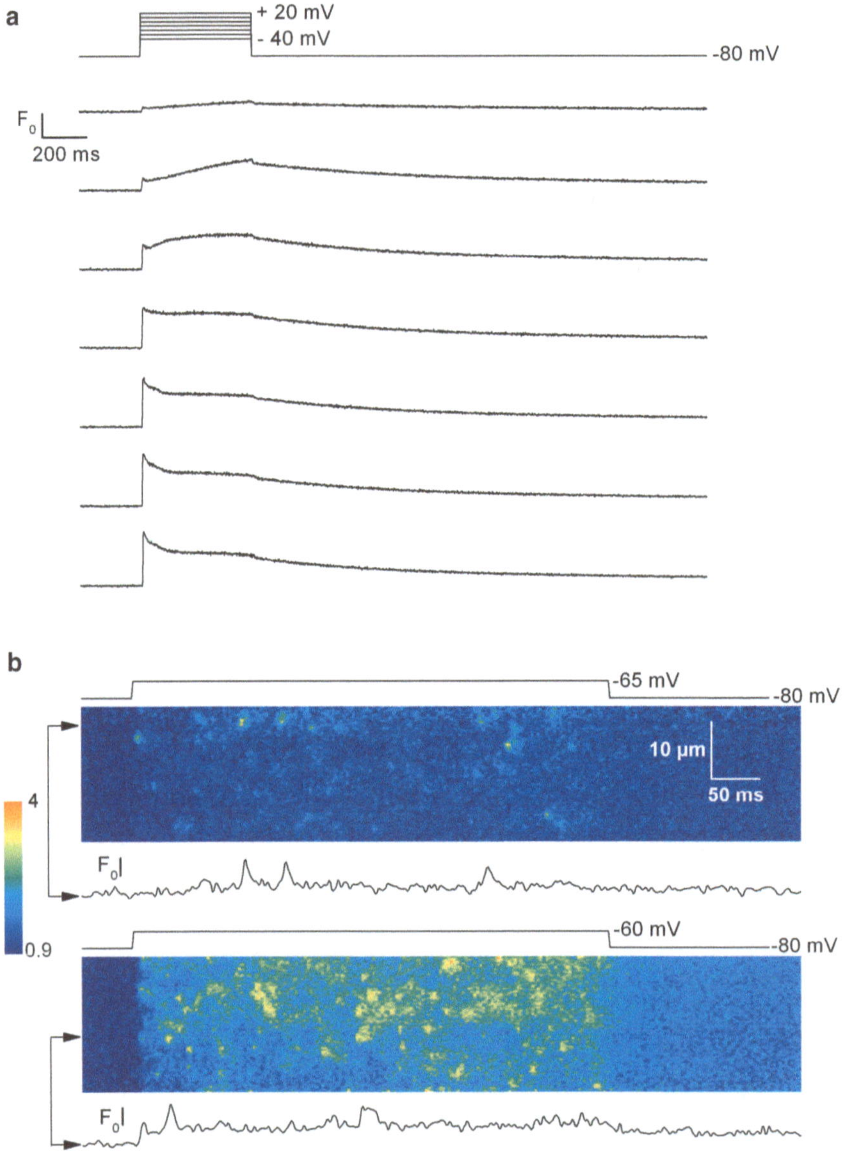

Fig. 3 Examples of intracellular Ca^{2+} signals triggered by depolarizing pulses in muscle fibers under silicone voltage-clamp conditions. (**a**) Changes in the fluorescence of the Ca^{2+} dye rhod-2 in a mouse muscle fiber depolarized by 500 ms-long depolarizing pulses from −80 mV to values ranging between −40 mV and +10 mV (from *top* to *bottom*) with a 10 mV increment. The intracellular medium was dialyzed through the voltage-clamp pipette with a solution containing (in mM) 120 K-glutamate, 5 Na_2-ATP, 5 Na_2-phosphocreatine, 5.5 $MgCl_2$, 5 glucose, 0.1 rhod-2, 20 EGTA, 8 $CaCl_2$, and 5 HEPES. (**b**) Elementary calcium release events (calcium sparks) detected in a frog toe muscle fiber depolarized by 500 ms-long pulse from −80 mV to the indicated values. The fluorescence of the Ca^{2+} dye Fluo-3 was recorded using the line-scan mode of a Zeiss LSM 510 confocal microscope, with a time resolution of 1.54 ms per line. The trace below each line-scan image shows the time course of the fluorescence at the indicated line position. Fluo-3 was pressure microinjected in the fiber prior to voltage clamp

4 Notes

1. A compound called Chemplex 825 (NFO Technologies, KS, USA) has also been used with success. The quality of the silicone is extremely important. Special attention has to be paid to its viscosity. If the silicone is too viscous it can be very hard to get a good insulation of the silicone-embedded main portion of fiber. On the contrary, if the silicone is too fluid the silicone-treated fibers end up spontaneously pop out of the silicone after a few minutes. Although we always purchased the same silicone product (Pâte 7048, *see* Subheading 3) we occasionally had a clear difference in viscosity in certain batches. Also one batch exhibited a surprisingly elevated level of intrinsic fluorescence upon UV excitation, hardly compatible with intracellular calcium measurements with UV excitable dyes. We have recently started working successfully with a remarkably clean and translucent grease called Siliconia (grade 3, Graisse Bellevue, Mont d'Origny, France).
2. Actually, the fetal bovine serum is the key component that prevents fiber damage upon contact with the silicone. FBS can simply be used in standard Tyrode solution or in any experimental solution.
3. The period of time during which the muscles are incubated within the collagenase-containing solution can be more precisely adjusted (within a range of 45–90 min) so as to achieve the best quality of the cells according to the following gross criteria. If the duration of incubation is too short, fibers hardly if any dissociate upon mechanical trituration of the muscles. Conversely if the incubation lasts too long, fibers dissociate very easily but are then usually more fragile and may be more likely to die during the subsequent steps of the process.
4. For this a small amount of the material is first spread all over the bottom of the chamber with a plastic or glass rod. This silicone layer is then made as thin as possible and uniform by sliding either the edge of a square coverslip or a flexible thin plastic rod back and forth against and along the bottom of the chamber.
5. If one uses a standard extracellular solution as for instance Tyrode, the majority of fibers exhibit a dramatic and irreversible contracture upon first contact with the silicone. Why silicone is toxic to the fibers and how this is prevented by the presence of FBS is not entirely clear. Notice that FBS was also reported to protect enzymatically isolated muscle fibers mounted in a double silicone-gap device [28]. When filling the silicone-coated chamber with the FBS supplemented solution, care should be taken that the solution is well equilibrated at room temperature, otherwise air bubbles tend to form on the top of the silicone layer.

6. This step should be carried out in the presence of the FBS supplemented culture medium. The process can be achieved using any crude manually controlled pressure delivery system. Some people in the lab were very happy with a system that uses a glass micropipette with broken tip (~50–100 μm wide), back-filled with silicone by applying pressure through a crude device consisting of a thin plastic capillary tube inserted within the lumen of the pipette, and linked on the other end to a standard 5 ml medical syringe filled with silicone. The silicone-filled pipette is then mounted on a standard patch-clamp holder, the air-pressure input of which is connected to a 50 ml syringe with plastic tubing. Holding the pipette with one hand while controlling pressure with the other hand allows to nicely paint the fibers with silicone. Although we find it easier to leave one extremity of the fiber free of silicone, the procedure can be implemented so that any portion of the fiber length (for instance specifically including the endplate) is left free of silicone.
7. There is no specific trick to be successful with the microinjection under these conditions. The tip of the pipette is inserted through the silicone within the embedded end portion of fiber. Pressure is raised until local swelling is perceptible at the impalement site. In some cases the injection can be facilitated by gently moving the pipette tip back and forth within the fiber. Once the injection is complete pressure is turned off and the pipette removed. A period of at least 30 min is then left to allow for equilibration of the injected solution within the entire myoplasmic volume. For best control of the injection and of the equilibration of the injected compound within the myoplasm, a colorimetric or fluorescent dye can be used in the injected solution (see for instance [32, 33]).
8. In early experiments [13] the voltage-clamp micropipettes had a much smaller tip; they were filled with a solution containing 3 M K-acetate and 20 mM KCl and thus exhibited a similar resistance (1–3 MΩ) as the ones used later. One advantage was then that the insertion of the tip within the silicone-insulated part of fiber was much easier and immediate. However, we found that during long-lasting experiments performed under these conditions the voltage-clamped end portion of the fiber tended to end up swelling, which is the reason why this procedure was abandoned.

Acknowledgements

This work was supported by the Centre National de la Recherche Scientifique, the Université Claude Bernard Lyon 1, and the Association Française contre les Myopathies.

References

1. Adrian RH, Chandler WK, Hodgkin AL (1966) Voltage clamp experiments in skeletal muscle fibres. J Physiol 186:51P–52P
2. Ildefonse M, Rougier O (1972) Voltage-clamp analysis of the early current in frog skeletal muscle fibre using the double sucrose-gap method. J Physiol 222:373–395
3. Hille B, Campbell DT (1976) An improved vaseline gap voltage clamp for skeletal muscle fibers. J Gen Physiol 67:265–293
4. Kovacs L, Schneider MF (1978) Contractile activation by voltage clamp depolarization of cut skeletal muscle fibres. J Physiol 277:483–506
5. Duval A, Léoty C (1980) Comparison between the delayed outward current in slow and fast twitch skeletal muscle in the rat. J Physiol 307:43–57
6. García J, Amador M, Stefani E (1989) Relationship between myoplasmic calcium transients and calcium currents in frog skeletal muscle. J Gen Physiol 94:973–986
7. Jacquemond V, Csernoch L, Klein MG, Schneider MF (1991) Voltage-gated and calcium-gated calcium release during depolarization of skeletal muscle fibers. Biophys J 60:867–873
8. Jong DS, Pape PC, Chandler WK, Baylor SM (1993) Reduction of calcium inactivation of sarcoplasmic reticulum calcium release by fura-2 in voltage-clamped cut twitch fibers from frog muscle. J Gen Physiol 102:333–370
9. Sanchez JA, Vergara J (1994) Modulation of Ca^{2+} transients by photorelease of caged nucleotides in frog skeletal muscle fibers. Am J Physiol 266:C1291–C1300
10. Shvinka N, Caffier G (1994) Cation conductance and efflux induced by polyene antibiotics in the membrane of skeletal muscle fiber. Biophys J 67:143–152
11. Delbono O (1995) Ca^{2+} modulation of sarcoplasmic reticulum Ca^{2+} release in rat skeletal muscle fibers. J Membr Biol 146:91–99
12. Gonzalez A, Caputo C (1996) Ryanodine interferes with charge movement repriming in amphibian skeletal muscle fibers. Biophys J 70:376–382
13. Szentesi P, Jacquemond V, Kovács L, Csernoch L (1997) Intramembrane charge movement and sarcoplasmic calcium release in enzymatically isolated mammalian skeletal muscle fibres. J Physiol 505:371–384
14. Hui CS (1998) A slow calcium-dependent component of charge movement in Rana temporaria cut twitch fibres. J Physiol 509:869–885
15. Struk A, Lehmann-Horn F, Melzer W (1998) Voltage-dependent calcium release in human malignant hyperthermia muscle fibers. Biophys J 75:2402–2410
16. Shirokova N, García J, Ríos E (1998) Local calcium release in mammalian skeletal muscle. J Physiol 512:377–384
17. Pape PC, Carrier N (2002) Calcium release and intramembranous charge movement in frog skeletal muscle fibres with reduced (<250 microM) calcium content. J Physiol 539:253–266
18. Squecco R, Bencini C, Piperio C, Francini F (2004) L-type Ca^{2+} channel and ryanodine receptor cross-talk in frog skeletal muscle. J Physiol 555:137–152
19. Adrian RH, Chandler WK, Hodgkin AL (1970) Voltage clamp experiments in striated muscle fibres. J Physiol 208:607–644
20. Chandler WK, Rakowski RF, Schneider MF (1976) A non-linear voltage dependent charge movement in frog skeletal muscle. J Physiol 254:245–283
21. Heistracher P, Hunt CC (1969) The relation of membrane changes to contraction in twitch muscle fibres. J Physiol 201:589–611
22. Caputo C, Bolaños P, Gonzalez A (1993) Effects of sulfhydryl inhibitors on depolarizations-contraction coupling in frog skeletal muscle fibers. J Gen Physiol 101:411–424
23. Friedrich O, Ehmer T, Fink RH (1999) Calcium currents during contraction and shortening in enzymatically isolated murine skeletal muscle fibres. J Physiol 517:757–770
24. Gómez J, Neco P, Di Franco M, Vergara JL (2006) Calcium release domains in mammalian skeletal muscle studied with two-photon imaging and spot detection techniques. J Gen Physiol 127:623–637
25. Andronache Z, Hamilton SL, Dirksen RT, Melzer W (2009) A retrograde signal from RyR1 alters DHP receptor inactivation and limits window Ca^{2+} release in muscle fibers of Y522S RyR1 knock-in mice. Proc Natl Acad Sci U S A 106:4531–4536
26. Bekoff A, Betz WJ (1977) Physiological properties of dissociated muscle fibres obtained from innervated and denervated adult rat muscle. J Physiol 271:25–40
27. Hernández-Ochoa EO, Schneider MF (2012) Voltage clamp methods for the study of membrane currents and SR Ca^{2+} release in adult skeletal muscle fibres. Prog Biophys Mol Biol 108:98–118
28. Woods CE, Novo D, DiFranco M, Vergara JL (2004) The action potential-evoked sarcoplasmic reticulum calcium release is impaired in mdx mouse muscle fibres. J Physiol 557:59–75

29. Jacquemond V (1997) Indo-1 fluorescence signals elicited by membrane depolarization in enzymatically isolated mouse skeletal muscle fibers. Biophys J 73:920–928
30. Jacquemond V, Allard B (1998) Activation of Ca^{2+}-activated K^{+} channels by an increase in intracellular Ca^{2+} induced by depolarization of mouse skeletal muscle fibres. J Physiol 509:93–102
31. Collet C, Pouvreau S, Csernoch L et al (2004) Calcium signaling in isolated skeletal muscle fibers investigated under "Silicone Voltage-Clamp" conditions. Cell Biochem Biophys 40:225–236
32. Csernoch L, Bernengo JC, Szentesi P, Jacquemond V (1998) Measurements of intracellular Mg^{2+} concentration in mouse skeletal muscle fibers with the fluorescent indicator mag-indo-1. Biophys J 75:957–967
33. Bernengo JC, Collet C, Jacquemond V (2001) Intracellular Mg^{2+} diffusion within isolated rat skeletal muscle fibers. Biophys Chem 89:35–51

Chapter 10

Multiphoton Imaging Approaches for Studying Striatal Dendritic Excitability

Joshua L. Plotkin and D. James Surmeier

Abstract

As the main input nucleus to the basal ganglia, the striatum is responsible for receiving and integrating highly convergent afferents to ultimately guide action selection and movement initiation. Although the majority of this synaptic integration occurs in the dendrites of striatal projection neurons (SPNs), their thin diameter makes them inaccessible with traditional recording electrodes. Recent advances in optical imaging technologies have allowed us and others to start lifting the veil on the mechanisms governing synaptic integration in the striatum by enabling direct dendritic measurements and manipulations. Here we describe how our lab has approached combining 2-photon imaging and photolysis with electrophysiological recordings to study dendritic excitability and synaptic integration in the striatum.

Key words 2-Photon imaging, 2-Photon uncaging, Calcium imaging, Striatum, Dendritic morphology

1 Introduction

Multiple anatomical parameters have conspired against the interrogation of striatal dendritic integration: (1) SPNs have fine diameter dendrites that are difficult to see with the IR-DIC optics typically used for patch clamp recordings and far too thin to patch with electrodes, (2) SPN dendritic trees are highly branching and nonlaminar in orientation, and (3) SPNs receive intermingled convergent excitatory afferents from multiple brain regions, with no obvious fiber tract within the striatum composed of only one afferent type with a defined postsynaptic target [1–5]. Recently, these restrictions have been circumvented with the use of multi-photon optical approaches. By loading SPNs with fluorescent anatomical dyes and calcium indicators via a somatic patch clamp electrode (1) living SPN dendrites in ex vivo slices can be visualized with 2-photon laser scanning microscopy (2PLSM), (2) calcium imaging can be performed at identified dendritic loci as an indirect measure of voltage changes in response to synaptically or somatically delivered

Marzia Martina and Stefano Taverna (eds.), *Patch-Clamp Methods and Protocols*, Methods in Molecular Biology, vol. 1183, DOI 10.1007/978-1-4939-1096-0_10,

stimuli, and (3) individual dendritic spines can be activated by 2-photon photolysis of caged glutamate or single photon optogenetic stimulation of an apposed presynaptic terminal.

2PLSM is based on the nonlinear excitation of fluorophores by an ultrashort pulse laser (typically between 70 and 140 fs). Mode-locking the laser constrains the delivery of photons in time, leading to increased laser power at the sample plane. By focusing these bursts of high photon concentration through the objective lens, the probability of a single fluorophore experiencing 2-photon excitation is specifically increased at the focal volume [6–8]. This excitation volume can be rapidly moved through the *x–y* plane with a pair of mirrors (galvanometers) in an automated fashion to construct a planar image of the sample. Images can then be taken at multiple focal planes to reconstruct the entire neuron or dendritic region of interest in three dimensions. There are several benefits of the multiphoton excitation approach. First, longer excitation wavelengths are delivered to the tissue (in theory twice as long as the single photon wavelength required for the fluorophore of interest), allowing for deeper penetration of the exciting photons. Second, unlike the case with confocal imaging where a pinhole is required to filter out photons originating from points outside the focal plane, multiphoton excitation only excites fluorophores at the focal point, meaning all emitted photons reaching the photomultiplier tubes (PMTs) can be collected as data, greatly increasing signal detection. Third, the ability to limit excitation to a discrete volume allows for local photolysis of caged compounds such a glutamate, since regions above and below the desired focal regions will not experience the 2-photon excitation events required for photolysis. This methods review focuses on multiphoton imaging of SPNs in acute mouse brain slices.

2 Materials

2.1 Solutions

1. Internal recording solutions. Standard potassium: 135 mM $KMeSO_4$, 5 mM KCl, 10 mM HEPES, 2 mM Mg-ATP, 0.2 mM Na-GTP, 5 mM Tris-phosphocreatine, 5 mM Na-phosphocreatine. Standard cesium: 120 mM $CsMeSO_3$, 15 mM CsCl, 8 mM NaCl, 10 mM HEPES, 3 mM Mg-ATP, 0.3 mM Na-GTP, 10 mM TEA, 5 mM QX-314. Internal recording solutions can be used for current or voltage clamp experiments, respectively. The main salt, ionic and carrier components will be determined by the experimental question. The important components specific to 2PLSM imaging are: (1) an anatomical dye (such as Alexa 568). The exact concentration will depend primarily on the sensitivity of the hardware being used, but we find 50 μM to be more than sufficient (*see* **Note 1**). (2) If calcium imaging is being performed, calcium and calcium

buffer (typically EGTA or BAPTA) should be omitted and a fluorescent calcium indicator included, which will act as both reporter and buffer. The precise indicator and concentration used will be determined by the calcium signal being probed and PMT sensitivity (*see* **Note 2**).

2. Recording solution: 124 mM NaCl, 3 mM KCl, 2 mM $CaCl_2$, 1 mM $MgCl_2$, 26 mM $NaHCO_3$, 1 mM NaH_2PO_4, 16.66 mM glucose. This is continuously bubbled with 95 % O_2 and 5 % CO_2.
3. Caged glutamate: 5 mM MNI-glutamate, 124 mM NaCl, 3 mM KCl, 2 mM $CaCl_2$, 1 mM $MgCl_2$, 26 mM HEPES, 1 mM NaH_2PO_4, 16.66 mM glucose, pH 7.2–7.4, 290–310 mOsm. We typically perfuse small quantities of "caged" 4-methoxy-7-nitroindolinyl (MNI)-glutamate onto the slice locally using a syringe controlled by a fixed rate pump—for this reason buffering the pH with HEPES rather than O_2/CO_2 is more practical.

2.2 Lasers

The most common laser for 2PLSM utilizes a solid state 532 nm pump with a Ti:Sapphire based crystal. If photolysis will also be performed, a second such laser is required, set at the appropriate (and different) wavelength.

2.3 Other Hardware

The remaining hardware is necessary to direct the laser to the appropriate sample volume, adjust the power of the laser at the sample volume, and collect the resulting photons emitted by the excited sample fluorophore. This includes:

1. An achromatic wave plate and beam splitter (if the laser will be split to two microscopes).
2. Pockels cells (electro-optical attenuators; our lab's strategy utilizes three Pockels cells).
3. Two sets of galvanometers (used to guide the lasers to targeted coordinates).
4. Dichroic mirror (directing fluorophore-emitted photons to the PMTs).
5. Red and green (or experimentally appropriate for the fluorophores being used) filter cubes.
6. Two PMT/channel detectors and a transmission detector (if Dodt contrast optics will be employed for identifying and patching neurons).
7. An upright microscope with a high magnification water-dipping objective lens with at least 2 mm working distance.

Equipment is installed on a large floating table to minimize vibration. Commercially available systems that accomplish the above tasks are available, such as the Prairie Ultima (Prairie Technologies), which is the system used by our lab.

2.4 Electrophysiology

Standard equipment used for patch clamping is needed, including an amplifier, digitizer, micromanipulators, and electrode head stage. Software is the largest application-specific concern. Our solution has been to use the freely available software package WinFluor (John Dempster, University of Strathclyde) to perform and coordinate electrophysiological recordings with Prairie Technology's proprietary imaging software Prairie View (*see* **Note 3**).

3 Methods

Self-contained 2PLSM systems have recently become commercially available (though at significantly higher costs than custom-built systems). As such we briefly describe the basic light path involved, but the methodology discussed in this chapter focuses more on the end-use procedures involved in a typical 2PLSM experiment, rather than hardware specifics, which can be found elsewhere [6–10].

3.1 The Light Path

A 2PLSM system, whether custom built or commercially bought, guides laser flow and sample photon collection in a meaningful way (Fig. 1). The general sequence of events that must occur is as follows:

1. A mode-locked laser beam is tuned to the appropriate wavelength for the fluorophore to be excited. Since the laser is the most expensive component of the system, and most appropriate commercially available lasers produce far more power than is needed for a single system (i.e., 5 W or more), it is common to split the laser beam to two or more microscopes.
2. The laser beam passes through an electro-optical attenuator (i.e., Pockels cell), allowing for laser power at the sample plane to be controlled.
3. The laser beam is guided into the scan head where a pair of galvanometers directs the beam in the X/Y plane.
4. The laser beam passes through a dichroic mirror into the objective, which defines the focal volume of excitation.
5. Photons emitted by the excited fluorophore at the sample plane are collected (either through the objective lens, the condenser or both) by PMTs behind appropriate filter cubes.

3.2 Loading a Neuron

1. Backfill a thick-walled borosilicate glass electrode (pulled to have a tip resistance of 3.5–6 MΩ) with internal solution containing an appropriate anatomical dye (such as the red dye

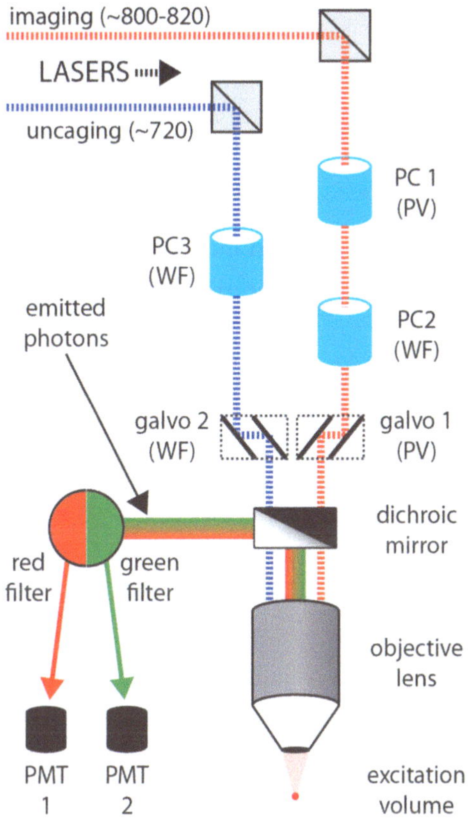

Fig. 1 Diagram of the light path for 2PLSM and 2P photolysis. *Parentheses* indicate the software used to control each component, either Prairie View (PV) or WinFluor (WF). *PC* Pockels cell, *galvo* galvanometer pair, *PMT* photomultiplier tube

Alexa Fluor 568, 50 μM) and a calcium indicator (such as the green dye Fluo-4, ≤100 μM) if warranted. Identify and patch-clamp the neuron of interest, using standard patch clamping procedures.

2. After rupturing the membrane to enter the whole-cell patch-clamp mode, allow the neuron to "fill" with the internal solution for at least 10 min. Adjust the laser power by tuning the Pockels cell so red (Alexa) intensity is bright but not saturated. Typically it is best to keep PMT gain constant (~70 % max).
3. As the anatomical dye loads, distal dendrites will become increasingly visible. When the region of interest is fully visible, and the dye intensity of the corresponding channel stabilizes, data collection can commence.

3.3 Taking a z-Series

1. Load the SPN with an Alexa dye as described above.
2. Determine the region of interest (i.e., the entire SPN or a segment of a dendrite) and adjust the digital zoom to a value

appropriate for the field of view. For an image of an entire SPN (using a 63× objective lens), a digital zoom of 1 is standard (giving a pixel size of ~0.36 μm). However, for higher magnification images of dendrites (using the same objective lens) it is necessary to increase the digital zoom (we typically use a digital zoom of 5, giving a pixel size of ~0.072 μm), to increase the portion of the field of view occupied by the region of interest and increase the number of pixels per unit area of the image.

3. Determine the *z*-direction step size. This is a trade-off between having a small enough step to detect fine details (like spines), but not so small that it unnecessarily increases the sampling time, which can introduce artifacts including drift and diminishing neuron health. This value may vary depending on the final use of the image (*see* **Note 4**).
4. Adjust the dwell time (the amount of time the laser remains on each acquired pixel). We typically use a dwell time of 4 μs during live imaging to identify and focus on regions of interest, and 10 μs when acquiring a z-series.
5. Adjust the laser power (and if necessary the PMT gain) to maximize the signal-to-noise ratio. Signal-to-noise can also be improved by averaging several (usually two to four) images in each plane taken for the z-series.
6. Define the upper and lower limits of the z-series (the top and bottom of the SPN dendritic field or region of interest). The specifics of how the z-series is acquired will be determined by the software used, but ultimately a series of sequential images at varying focal planes are acquired according to the parameters defined above [11, 12].

3.4 Performing a Line Scan

Line scanning is a simple way to acquire calcium signals with a fast acquisition rate. As opposed to whole field imaging, only one row of pixels is scanned with the laser: for example, if the image area is 512×512 pixels (a commonly used resolution), a line scan will allow for an acquisition rate 512 faster than full field imaging (*see* **Note 5**).

1. Patch clamp an SPN with an electrode filled with internal solution containing a red Alexa dye and an appropriate green calcium indicator dye (*see* **Note 2**) as described above.
2. Select a dendrite of interest and center the focus on the specific region to be imaged.
3. Take a reference image of this plane.
4. Determine the scan rate of the line scan to be performed. This is determined by the product of the dwell time (we typically use 10 μs) and the number of pixels in the line. Adding the dead time (time required for the galvanometers to return to

the starting pixel) gives the total time of each individual line scan (L). Dividing the total duration of the protocol to be performed by L will determine the number (repetitions) of line scans that must be performed during the protocol.

5. The next step is to initiate the simultaneous electrophysiology and imaging protocols. The precise method for doing this will differ depending on the software used. We use Prairie View software to set up line scan parameters, and the freely available WinFluor program to write and acquire physiology protocols. WinFluor initiates the physiology protocol, triggers the line scan protocol established in Prairie View and acquires and displays both physiology and imaging data within a single file. We also use WinFluor to control a second Pockels cell in series with the primary imaging Pockels cell for further gain control (*see* **Note 6**).
6. Once the line scan image data is collected, several option exist for analysis. One option is to normalize the change in green channel fluorescence (ΔF) to the baseline green fluorescence (F_o). Other options include normalizing the change in green fluorescence to the red anatomical fluorescence ($\Delta G/R$), to account for dye loading and photobleaching. Variations on these methods take into account fluorescence values achieved at saturating calcium levels, improving inter-experiment comparisons [11, 13, 14].

3.5 Uncaging Glutamate

Glutamate uncaging is performed with a separate dedicated solid state pump laser with a Ti:Sapphire based crystal, tuned to 720 nm (*see* **Note** 7). This laser is directed to the desired focal volume using a separate set of galvanometers in the scan head, and the power is attenuated with a separate dedicated Pockels cell (Fig. 2). It is typically cost-prohibitive to dissolve MNI-glutamate in quantities of recording solution normally used for an experiment. For this reason, we either recirculate recording solution containing MNI-glutamate, or locally perfuse it over the striatal region of interest. We will focus on the local perfusion method here (*see* **Note 8**).

1. Load a 1 ml glass syringe with 5 mM MNI-glutamate dissolved in HEPES-buffered ACSF (*see* **Note 9**).
2. Place the syringe in a constant flow pump, set to 0.4 ml/h, and connect the needle of the syringe to a multi-barreled perfusion manifold with appropriate gauge tubing. Take great care to avoid air bubbles.
3. Place the perfusion manifold on the surface of the slice within 1–2 mm of the targeted SPN.
4. Once the SPN is patched and loaded as described above, turn the pump flow on. Allow the solution to equilibrate for 5–10 min before attempting to uncage glutamate.

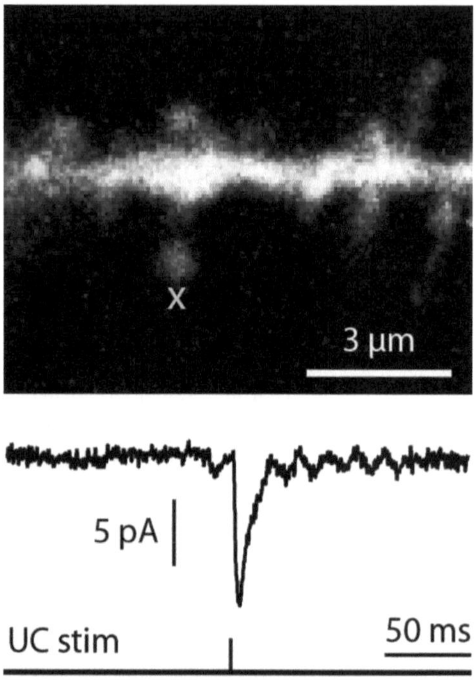

Fig. 2 Example of a somatic response to dendritic glutamate uncaging. High magnification 2PLSM reference image of a distal (~100 μm from soma) SPN dendrite. "X" indicates the location where MNI-glutamate was uncaged. The resulting somatic excitatory postsynaptic current is shown below, in register with the uncaging (UC) stimulus (1 ms duration)

5. Focus on the dendritic target (i.e., spine) of interest and take a reference image. Define the target coordinates and set the galvanometers to direct the uncaging laser to that point (*see* **Note 10**).
6. Define the timing and duration (typically 0.5–1 ms) of the uncaging pulse. This step is software and protocol specific.
7. Adjust the laser power with the dedicated Pockels cell. There are several approaches to choosing the appropriate laser power (*see* **Note 11**). The method we typically use is to incrementally adjust the power until a single uncaging pulse evokes an excitatory postsynaptic potential or current (EPSP or EPSC, respectively) that has a similar amplitude and kinetics to spontaneously occurring events [15]. Care should be taken to avoid photodamage induced by excessive exposure to the uncaging laser, particularly high power exposure (*see* **Note 12**).

4 Notes

1. Since the anatomical and calcium indicator dyes are introduced to the soma, loading time is directly related to dendritic distance (i.e., proximal dendrites fill with dye more quickly than

distal dendrites) and dye concentration. Insufficient loading time will therefore underestimate dendritic length. Allowing for sufficient loading time is especially crucial for measuring calcium dynamics. As the cell loads, calcium indicator concentration increases. If care is not taken to make calcium measurements at similar time points in the loading process, comparisons will be of dendrites containing different dye concentrations, which will show different calcium transient amplitudes and kinetics [16].

2. Two main parameters should be kept in mind when choosing which calcium indicator to use in an experiment. (1) The dye should have a linear emission curve within the calcium concentrations that will be induced by the protocol and (2) the dye should be used at the minimal concentration that produces sufficient signal-to-noise and avoids saturation. As an example, some standard protocol-dependent dye scenarios are: 100 μM of the high affinity indicator Fluo-4 to measure bAP-evoked calcium transients, 300 μM of the moderate affinity indicator Fluo-5F to measure synaptically evoked calcium transients, and 500 μM of the low affinity indicator Fluo-4FF to measure dendritic plateau potential-associated calcium transients [13, 15].

3. Even with commercially available 2PLSM systems and software, integrating physiology with imaging requires a customized solution. The major issue is that numerous imaging parameters (i.e., multiple Pockels cells and multiple galvanometer pairs) often must be manipulated during a given protocol, meaning simple trigger-synching is not a viable option. Our lab has dealt with this using WinFluor. We use WinFluor to deliver and acquire physiology protocols, but also to control (1) the imaging laser power directly through a second dedicated Pockels cell, (2) the galvanometer pair guiding the uncaging laser (using a graphical interface to target specific coordinates on a reference image), and (3) uncaging laser timing and gain via a third dedicated Pockels cell. A WinFluor protocol controlling all of these parameters can be engaged independently or by Prairie View software (if calcium imaging will be performed), and creates a single file containing both physiology and imaging data.

4. 2PLSM resolution is higher in the X/Y plane vs. the axial (Z) plane. For example, our system has a theoretical X/Y resolution of 0.38 μm, but an axial resolution of 1.43 μm. This axial "blurring" is often reduced using deconvolution software, which uses repetitive algorithms based on the point spread function of the imaging hardware to reduce Poisson-distributed noise. When deconvolution software will be used to process images, it is useful to over-sample in the Z-plane. For instance, although the theoretical resolution in this plane is near 1.5 μm,

sampling at less than 0.5 μm will improve the output generated by the deconvolution algorithm. If a maximum intensity projection (MIP) image (a collapsed two-dimensional rendering in the *X*/*Y* plane) is sufficient for the particular experiment, deconvolution is typically not necessary.

5. Several options exist for using a laser to excite calcium indicator dyes. One option is to "park" the imaging laser at a fixed position. This has the advantage of offering the fastest acquisition rate, but limits data collection to one point in space. Further, the risk of photodamage is considerable as the laser dose at this point will be high. Alternatively, a region of interest (ROI) may be used, which excites a defined and spatially connected number of pixels. The acquisition rate will depend upon the size of the ROI. Alternatively, line scans (or continuous scanning, such as curves, freehand drawn lines, etc.) allow for multiple regions to be imaged during the same protocol, given that all pixels are connected in the same plane. A useful application of line scans is measuring calcium transients in a dendritic spine and its parent shaft during the same protocol.
6. To reduce the photon dose delivered to the sample during a line scan protocol, we use a second dedicated Pockels cell effectively as a fast "shutter," so the SPN is only exposed to the laser during the period immediately flanking the physiological event of interest. This second Pockels cell is in series with the primary Pockels cell used to control the gain of the imaging laser, and is fully attenuated or fully opened by WinFluor at appropriate times, according to the protocol being used.
7. The theoretical optimal doubled-wavelength needed to uncage MNI-glutamate is 630 nm. However, most commercially available 2P lasers will not achieve such small wavelengths. Lasers working closer to 720 nm are readily accessible however, and are sufficient for this application.
8. An advantage of using a recirculating pump for MNI-glutamate application is that drug delivery will be uniform throughout the depth of the brain slice. This means a defined laser power at the sample plane can be correlated with a defined quantity of glutamate uncaging between experiments [17]. However, recirculating multiple solutions is challenging and expensive, thus limiting pharmacological analysis. Local perfusion will not deliver a uniform concentration of MNI-glutamate with depth, but if the uncaging laser power is tuned to a physiological parameter (i.e., epsp or epsc amplitude), this is not an issue. The benefit of local perfusion is that it allows for multiple pharmacological manipulations to be performed on the same cell, by using a multi-barreled perfusion pump. Typical concentrations of MNI-glutamate are between 2.5 mM (for bath application) and 5 mM (for local perfusion).

9. MNI-glutamate, anatomical dyes, and calcium indicators should be shielded from light. Shield tubing used to deliver MNI-glutamate to the recording stage from light.
10. The galvanometers used to guide the uncaging laser are controlled by WinFluor. These can be calibrated by determining the voltage required to center the laser in the *X* and *Y* directions, and the voltage required to move each galvanometer a known linear distance. The location of the uncaging laser can be visualized by photobleaching a stationary fluorescent target slide.
11. An alternative method of tuning the uncaging laser power has been described by Bloodgood and Sabatini [17]. Using bath application of MNI-glutamate, the uncaging laser is targeted directly to a spine and the laser power is tuned to produce a predetermined amount of photobleaching of the red anatomical dye. This method assumes MNI-glutamate concentration is identical throughout the slice and between experiments, and is therefore not suitable when MNI-glutamate is locally perfused.
12. The most common sign of photodamage in 2PLSM experiments is blebbing of the dendrites. If a high to moderate affinity calcium indicator is included in the recording electrode, elevated basal calcium signal is also an indication of diminished dendrite health.

Acknowledgements

We thank Drs. David Wokosin and Michelle Day for their invaluable contributions to establishing this technology in our laboratory. This work was funded by CHDI.

References

1. Smith AD, Bolam JP (1990) The neural network of the basal ganglia as revealed by the study of synaptic connections of identified neurones. Trends Neurosci 13:259–265
2. Gerfen CR, Surmeier DJ (2011) Modulation of striatal projection systems by dopamine. Annu Rev Neurosci 34:441–466. doi:10.1146/annurev-neuro-061010-113641
3. Kress GJ, Yamawaki N, Wokosin DL et al (2013) Convergent cortical innervation of striatal projection neurons. Nat Neurosci 16:665–667. doi:10.1038/nn.3397
4. Wilson CJ (1992) Dendritic morphology, inward rectification and the functional properties of neostriatal neurons. In: McKenna TM, Davis JL, Zornetzer SF (eds) Single neuron computation (neural networks: foundations to applications), 1st edn. Academic Press Professional Inc, San Diego, CA, pp 141–171
5. Gertler TS, Chan CS, Surmeier DJ (2008) Dichotomous anatomical properties of adult striatal medium spiny neurons. J Neurosci 28:10814–10824. doi:10.1523/JNEUROSCI.2660-08.2008
6. Svoboda K, Yasuda R (2006) Principles of two-photon excitation microscopy and its applications to neuroscience. Neuron 50:823–839. doi:10.1016/j.neuron.2006.05.019

7. Denk W, Strickler JH, Webb WW (1990) Two-photon laser scanning fluorescence microscopy. Science 248:73–76
8. Zipfel WR, Williams RM, Webb WW (2003) Nonlinear magic: multiphoton microscopy in the biosciences. Nat Biotechnol 21:1369–1377. doi:10.1038/nbt899
9. Plotkin JL, Guzman JN, Schwarz N et al (2011) Optical approaches to studying the basal ganglia. In: Lane EL, Dunnett SB (eds) Animal models of movement disorders: Volume I, vol 61, Neuromethods. Springer Science+Business Media, New York, pp 191–220. doi:10.1007/978-1-61779-298-4_10
10. Wokosin DL, Squirrell JM, Eliceiri KW, White JG (2003) Optical workstation with concurrent, independent multiphoton imaging and experimental laser microbeam capabilities. Rev Sci Instrum 74:193–201. doi:10.1063/1.1524716
11. Day M, Wokosin D, Plotkin JL et al (2008) Differential excitability and modulation of striatal medium spiny neuron dendrites. J Neurosci 28:11603–11614. doi:10.1523/JNEUROSCI.1840-08.2008
12. Day M, Wang Z, Ding J et al (2006) Selective elimination of glutamatergic synapses on striatopallidal neurons in Parkinson disease models. Nat Neurosci 9:251–259. doi:10.1038/nn1632
13. Carter AG, Sabatini BL (2004) State-dependent calcium signaling in dendritic spines of striatal medium spiny neurons. Neuron 44:483–493. doi:10.1016/j.neuron.2004.10.013
14. Higley MJ, Sabatini BL (2010) Competitive regulation of synaptic Ca^{2+} influx by D2 dopamine and A2A adenosine receptors. Nat Neurosci 13:958–966. doi:10.1038/nn.2592
15. Plotkin JL, Day M, Surmeier DJ (2011) Synaptically driven state transitions in distal dendrites of striatal spiny neurons. Nat Neurosci 14:881–888. doi:10.1038/nn.2848
16. Maravall M, Mainen ZF, Sabatini BL, Svoboda K (2000) Estimating intracellular calcium concentrations and buffering without wavelength ratioing. Biophys J 78:2655–2667. doi:10.1016/S0006-3495(00)76809-3
17. Bloodgood BL, Sabatini BL (2007) Nonlinear regulation of unitary synaptic signals by CaV(2.3) voltage-sensitive calcium channels located in dendritic spines. Neuron 53:249–260. doi:10.1016/j.neuron.2006.12.017

Chapter 11

Investigation of Synaptic Microcircuits Using Patch-Clamp Paired Recordings in Acute Brain Slices

Giovanni Russo and Stefano Taverna

Abstract

Multiple patch clamp recordings represent a powerful tool to investigate functional synaptic connectivity among individual cells. This technical approach is particularly useful to study the synaptic organization of microcircuits in certain brain areas, e.g., the striatum, which are characterized by heterogeneous cell populations and an apparent lack of an anatomically ordered cytoarchitecture. Fast-spiking interneurons (FSIs) represent less than 1 % of striatal neurons, but despite their rareness they exert a strong influence on signal processing by striatal microcircuits and principal cells output. Little is known about the functional properties of chemical and electrical synapses connecting striatal FSIs to each other. Here we describe a simple dual patch clamp approach to investigate electrical and GABAergic synapses in mouse neostriatal slices.

Key words Striatum, Dual patch-clamp, Gap junctions, Synaptic connectivity, IPSC, GABA

1 Introduction

Brain networks are extremely complex structures due to the enormous amount of neurons involved and the intricate nature of their dendritic and axonal projections. Many cell types may coexist in any given area and the elucidation of their specific role in local signal processing and output generation represents a great challenge in neurobiology. While the advent of optogenetic techniques has provided a major breakthrough in the functional dissection of neuronal circuits [1], a detailed investigation of synaptic transmission between specific cell types requires simultaneous electrophysiological recordings using two or more microelectrodes in order to (1) quantify connectivity rates and (2) induce specific patterns of firing activity and measure voltage changes and ionic currents from presynaptic and postsynaptic cells at the same time. Dual patch clamp recordings have been widely adopted to study properties of electrically connected cells [2–13] and dynamics of chemical synapses [14–20] in several brain regions. The striatum—the principal input area of the basal ganglia—is characterized by a heterogeneous cell

Marzia Martina and Stefano Taverna (eds.), *Patch-Clamp Methods and Protocols*, Methods in Molecular Biology, vol. 1183, DOI 10.1007/978-1-4939-1096-0_11, © Springer Science+Business Media New York 2014

population where different neuronal types use GABA as a primary neurotransmitter [21, 22]. During the past decade, dual patch clamp recordings have been highly useful in dissecting the puzzling synaptic organization of the striatum. A particularly interesting class of cells is represented by fast-spiking interneurons (FSIs), which directly control the output of principal cells and are themselves interconnected through both gap junctions and GABAergic synapses [23]. Here we describe the use of dual patch clamp recordings to investigate electrophysiological properties of synaptic and electrical connections between pairs of FSIs or medium spiny neurons (MSNs) in mouse neostriatal slices.

2 Materials

2.1 Slice Preparation

1. Young adult mice of both sexes (age p20 to p40).
2. Injectable anesthesia (a mixture of 100 mg/kg ketamine and 10 mg/kg xylazine).
3. Peristaltic pump with tubing connected to a 25 G (0.5 mm) needle for intra-cardiac perfusion.
4. Surgical tools (scissors, tweezers, spatula, vannas scissors, scalpel blade).
5. Two glass beakers (50 ml) filled with ice-cold extracellular solution (*see* **item 1** in Subheading 2.3).
6. Gas dispersion tube for bubbling a mixture of 95 % O_2–5 % CO_2 in extracellular solution.
7. A glass petri dish (diameter: 75 mm) filled with ice and with cap topped with a disk of filter paper.
8. Cyanoacrylate glue.
9. Filter paper cuts (2 × 2 cm).
10. Vibratome with blade mounted on vibrating arm.
11. Thermostatic bath.
12. Slice storage chamber (*see* **Note 1**).

2.2 Experimental Setup

1. Vibration control table mounted with a Faraday cage.
2. Upright microscope equipped with infrared differential interference contrast (IR-DIC) or similar optics, a fluorescence illumination system, and a 40× objective.
3. Dual three-axis micromanipulator and microscope movement control system.
4. Slice recording chamber mounted with flowing solution inlet and a vacuum suction outlet.
5. U-shaped platinum wire with two to three parallel nylon threads connecting the two arms, to be used to hold the slice in place inside the recording chamber.

6. Two three-way stopcocks with flexible tubing and manometer to control pressure applied to micropipette tips.
7. Dual-channel patch clamp operational amplifier, or two single-channel amplifiers.
8. Analog/digital converter.
9. Video camera mounted on the microscope adapter and connected with an analog or digital monitor.
10. Personal computer with data acquisition and camera control software.
11. Borosilicate glass capillaries.
12. Micropipette puller.

2.3 Solutions and Chemicals

1. Extracellular solution (artificial cerebrospinal fluid, or ACSF): 125 mM NaCl, 25 mM $NaHCO_3$, 2.5 mM KCl, 1.25 mM NaH_2PO_4, 2 mM $CaCl_2$, 1 mM $MgCl_2$, 11 mM glucose, saturated with 95 % O_2 and 5 % CO_2, pH 7.3, prepared from a 10× stock solution diluted in bi-distilled water (*see* **Note 2**).
2. Internal solution (high intracellular [Cl^-]): 30 mM KH_2PO_4, 100 mM KCl, 10 mM NaCl, 2 mM $MgCl_2$, 10 mM HEPES, 0.5 mM EGTA, 2 mM Na_2-ATP, 0.02 mM Na-GTP, pH 7.2 (adjusted with 1 N KOH), prepared in 50 ml stocks diluted in bi-distilled water and frozen at −20 °C in 1 ml aliquots.
3. Internal solution (low intracellular [Cl^-]): 124 mM KH_2PO_4, 10 mM NaCl, 2 mM $MgCl_2$, 10 mM HEPES, 0.5 mM EGTA, 2 mM Na_2-ATP, 0.02 mM Na-GTP, pH 7.2, prepared in 50 ml stocks diluted in bi-distilled water and frozen at −20 °C in 1 ml aliquots (*see* **Note 3**).

3 Methods

3.1 Slice Preparation

1. Fill the two beakers with ice-cold ACSF, bubble the solution with the O_2/CO_2 mixture, and keep the beakers in ice. Use one of them as a reservoir for the intra-cardiac perfusion solution (*see* **Note 4**) and the other for temporarily storing the explanted brain prior to slicing.
2. Anesthetize the mouse with an intraperitoneal injection of the ketamine–xylazine mixture and wait 3–4 min until limb and eye reflexes have disappeared.
3. Cut the ventral skin and the diaphragm in order to expose the abdominal and the thoracic cavities to allow intra-cardiac perfusion.
4. Hold the heart with flat-tip tweezers and gently insert the perfusion needle in the left ventricle in order to inject ice-cold ACSF using the peristaltic pump at a rate of 10 ml/min (*see* **Note 5**).

5. As soon as the ACSF starts being pumped into the left ventricle, cut the heart right atrium using a vannas scissor to induce blood efflux for ~1–2 min until the liver turns from a dark-red to a pale-whitish color.
6. Remove the needle and quickly decapitate the animal using scissors. Place the head on top of the ice-filled petri dish and cut the skin longitudinally along the skull median line using small scissors. Cut the skin away laterally as well, in order to fully expose the skull.
7. Cut the skull bone transversally and posteriorly at the level of the cerebellum, then gently cut each side laterally, advancing from back to front and taking care of not damaging the brain with the scissor tip. Make one last cut anteriorly and transversally, then remove the skull top by holding its anterior end using tweezers.
8. Gently remove the brain using a spatula and put it in a beaker filled with ice-cold ACSF (remove the gas bubbler first) for 15–20 s, then place it on the filter paper on top of the ice-filled petri dish. Separate the hemispheres by cutting the midline using a scalpel blade.
9. Apply a small drop of glue on the surface of the vibratome stage and spread it using a micropipette tip in order to obtain a thin film.
10. Gently put one hemisphere (with the median surface facing down) onto a spatula. Dry out the ACSF in excess by carefully touching the lower side of the brain using two to three pieces of filter paper sequentially, then place the brain over the glue film on the vibratome stage (*see* **Note 6**). If necessary, the second hemisphere may be temporarily placed in ice-cold ACSF for cutting extra-slices later.
11. Fill the stage chamber with ice-cold ACSF (bubbled with a 95 % O_2 and 5 % CO_2 gas mixture) and cut 300-μm-thick sagittal slices using a vibration frequency of 80–90 Hz and an advancing speed of 0.1–0.15 mm/s. Stop the vibrating arm just before the blade has cut a slice completely and make the final cut manually using a scalpel blade. Make sure that the slice stays adherent to the blade and does not move around or sink to the bottom of the chamber (in which case it should be discarded).
12. Carefully pick up any individual slice using a painter brush (make sure that the slice smoothly lies parallel to the major axis of the bristles). Place the slices inside the incubation chamber at 32 °C for 15 min, then put the chamber at room temperature for the rest of the day.

3.2 Dual Patch Clamp Recordings

This protocol is designed to perform dual patch clamp recordings in mouse neostriatal slices in order to study properties of synaptic connectivity between pairs of medium spiny neurons (MSNs) or striatal fast spiking interneurons (FSIs). These cells are connected by inhibitory GABAergic synapses and, in the case of FSI pairs, electrical synapses mediated by gap-junctions [23].

1. Prior to starting experiments, pull recording pipettes from borosilicate capillaries (1.65 mm OD/1.1 mm ID; *see* **Note 7**). Adjust the puller controls to obtain a tip resistance of 2–6 MΩ depending on the size of the soma to be patched.
2. Submerge an individual slice in the microscope recording chamber in which the ACSF is continuously flowing (2–3 ml/min) at 32 °C (*see* **Note 8**). Place the U-shaped platinum wire in order to hold the slice steady at the bottom of the chamber.
3. Identify two cells of interest using a 40× objective (*see* **Note 9**).
4. Fill half volume of two pipettes with intracellular solution (*see* **Note 10**) using a 1 ml syringe attached to a flexible capillary (e.g., a 28-gauge Microfil, WPI). Remove air bubbles by gently tapping the pipette in the vicinity of the taper.
5. Insert an electrode into each of the two holders connected to the amplifier head stages. Drive the tip of one electrode into the chamber containing the extracellular solution (just below the liquid surface) leaving the other electrode momentarily out. Apply a positive pressure (25–100 mbar; *see* **Note 11**) to the submerged tip using a 10 ml syringe and then focus the tip under the objective.
6. Apply a continuous voltage step (−10 mV, 300 ms) to check the electrode resistance in voltage-clamp mode. If the resistance is larger than expected, remove air bubbles or small debris by briefly applying negative pressure using the syringe (restoring the positive pressure immediately thereafter). If debris stick onto the outside of the tip, apply extra positive pressure to push them away. Make sure that the pipette tip is steadily clean and the resistance value corresponds to the expected one before approaching the cell. Use the amplifier control circuit to cancel any baseline linear offset.
7. Lower the electrode to approach the cell. Gently touch the slice surface with the tip and continue lowering until getting close to the cell. Focus alternately on the tip and the cell to have a better visual control of the tip movement.
8. When the electrode touches the cell, a small indentation of the membrane becomes visible at the level of the tip, while the current amplitude decreases as the tip resistance increases. At this point, reverse the pressure from positive to negative to create a high-resistance seal (in the order of GΩ) between the pipette tip and the cell membrane.

9. Use the amplifier compensation circuit to remove fast capacitive transients. Clamp the membrane potential at a voltage level similar to the one known as the resting membrane potential (rmp) for the cell of interest (e.g., –70 mV for striatal FSIs, –85 mV for MSNs).
10. Gently apply negative pressure to rupture the membrane patch beneath the pipette tip and obtain a whole-cell configuration (indicated by a sudden increase in the amplitude of capacitive transients).
11. Switch to current-clamp mode and inject hyperpolarizing and depolarizing current steps of various amplitudes (e.g., from –300 to +500 pA in 100 pA steps, 500 ms) to measure the input resistance and elicit action potential firing.
12. If the cell's electrophysiological response satisfies desired parameters (i.e., a steady rmp at a relatively hyperpolarized level <–60 mV, a sizeable input resistance, and repetitive firing with overshooting action potentials), repeat **steps 5–11** in order to patch the other cell using the second micromanipulator (*see* **Note 12**).
13. To test whether the two patched cells are synaptically connected, inject a suprathreshold current step (e.g., +400 pA, 500 ms) in cell 1 ("presynaptic") in current-clamp and check whether cell 2 ("postsynaptic") responds with excitatory or inhibitory postsynaptic potentials (EPSPs or IPSPs) depending on the phenotype of the presynaptic cell (i.e., glutamatergic or GABAergic, respectively; Fig. 1a).

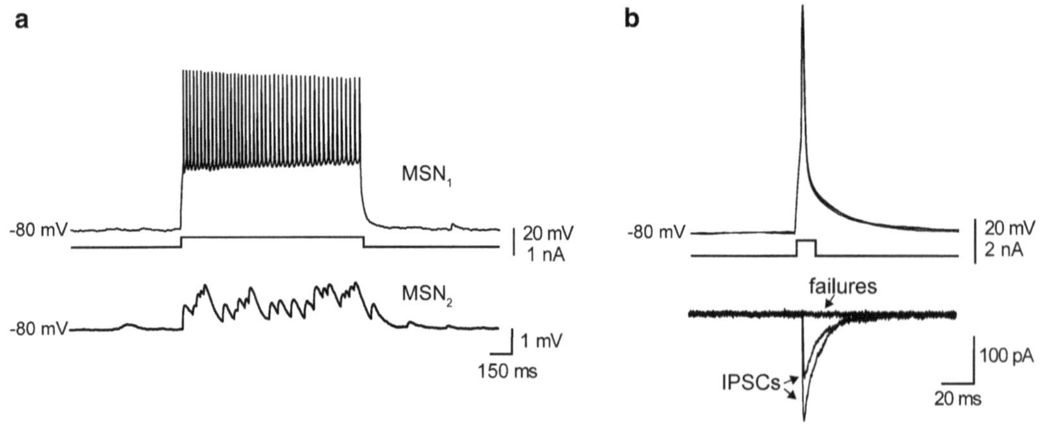

Fig. 1 Synaptic GABAergic transmission revealed by dual patch clamp recordings on striatal MSNs using a high [Cl⁻] intracellular solution. (**a**) A presynaptic train of action potentials elicited by suprathreshold current injection in MSN_1 evoked an array of GABAergic depolarizing PSPs in MSN_2. (**b**) Superimposed individual presynaptic spikes (*top*) evoked unitary IPSCs of different amplitudes or failures (*bottom*)

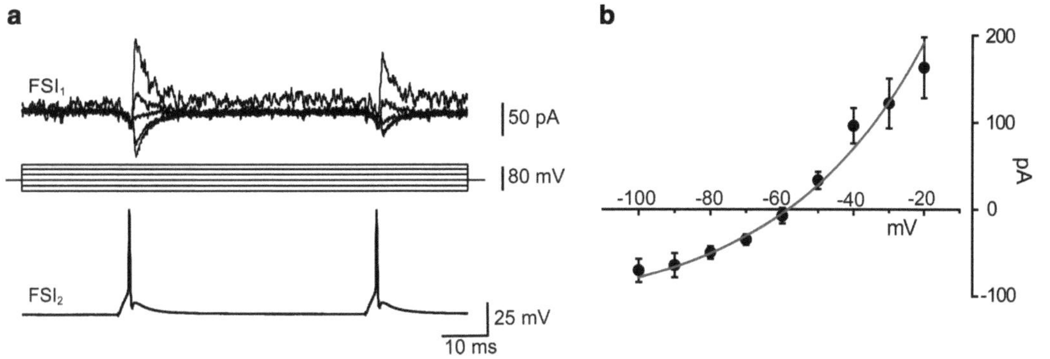

Fig. 2 Paired-pulse stimulation and current–voltage relationship in paired striatal FSIs. (**a**) GABAergic IPSCs recorded in voltage-clamp in FSI_1 in response to two action potentials evoked in current-clamp in FSI_2 at a time distance of 50 ms. Note paired-pulse depression of IPSCs. The protocol was repeated at different holding potentials, from −100 to −20 mV in 10 mV steps (starting from $V_m = -70$ mV; for clarity, only a few current and potential traces are represented). (**b**) Current–voltage (*I*–*V*) curve obtained by plotting peak amplitude values of GABAergic IPSCs against holding potentials. The reversal potential is ~−60 mV

14. Use a stimulating protocol designed to repetitively elicit one or two individual presynaptic action potential (Figs. 1b and 2) in order to obtain a series of unitary postsynaptic currents (PSCs) recorded in voltage-clamp (*see* **Note 13**).
15. Apply different voltage steps (e.g., from −100 mV to +40 mV in 10 mV steps, 500 ms) to the postsynaptic cell in order to construct a current–voltage curve (*I*–*V*) and measure the reversal potential for the synaptic current (Fig. 2).
16. Repeat **step 13** in reverse order to test whether cell 2 sends a synaptic projection to cell 1.
17. In case of absence of any connection, remove one electrode from a given cell and try to patch a different one using a new pipette (*see* **Note 14**).
18. When pairs are connected through electrical synapses (e.g., striatal fast-spiking interneurons, Fig. 3), injection of positive and negative current steps in one cell elicits membrane responses in both cells due to an electrical coupling effect mediated by gap-junctions. In addition, presynaptic action potentials elicit postsynaptic "spikelets" at small latencies (>1 ms).
19. Inject a negative current step (−100 pA, 500 ms) in cell 1 to induce a negative deflection in the membrane potential of cell 1 and cell 2. Calculate the coupling coefficient (CC) by dividing the steady state V_m2 value by the steady state V_m1 value (Fig. 3).

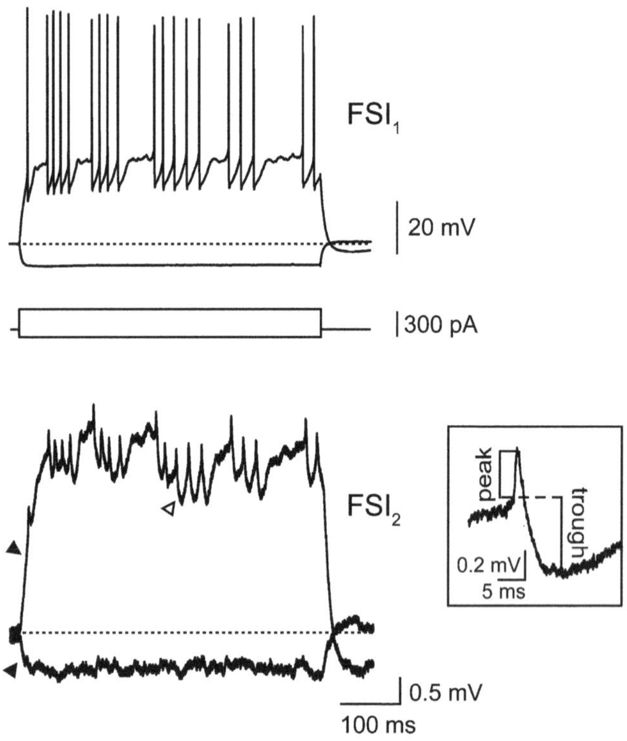

Fig. 3 A pair of striatal FSIs connected through electrical synaptic transmission. Injection of hyperpolarizing and depolarizing current pulses in FSI_1 caused membrane potential changes in FSI_1 but also in FSI_2 (*black arrowheads*). Presynaptic action potentials evoked spikelets in FSI_2. *Inset*: enlarged view of an individual spikelet marked by the *white arrowhead*

4 Notes

1. Slice incubation chambers are commercially available or can be custom-made by substituting the bottom of a plastic petri dish (35 mm diameter) with a soft nylon mesh (use cyanoacrylate glue to attach the mesh to the plastic rim). Fit the dish at half-height inside a 50 ml glass beaker, fill with ACSF, and bubble with a 95 % O_2, 5 % CO_2 gas mixture. Use a small tubing and moderate gas pressure to avoid the formation of excessively large bubbles which rupture may damage the slices. Cover the beaker top with a plastic cap.
2. We store the 10× stock solution at 4 °C and use it for 1 week. The stock is filtered using a vacuum filter unit (0.22 μm). After 1:10 dilution, it is used both as a slice-cutting solution and as an extracellular solution during recordings.
3. A high [Cl^-] intracellular solution is useful to increase the amplitude of GABAergic IPSCs by shifting the reversal potential for Cl^- ions to relatively depolarized values and consequently

increase the driving force. Such conditions are suitable for quantitatively assess connectivity rates. Conversely, a low intracellular [Cl^-] solution is the best choice to maintain physiological conditions to study the impact of GABAergic synapses on postsynaptic cells properties, e.g., when recording pairs connected by both electrical *and* GABAergic synapses. In this case, a strong driving force for Cl^- is not recommended because it would artificially mask the contribution of gap junctions (*see* **step 18** in Subheading 3.2).

4. Intra-cardiac perfusion of ice-cold, oxygenated ACSF minimizes the hypoxic damage induced by the preparation and consequently improves the slice quality. The striatum is particularly susceptible to hypoxia, thus intra-cardiac perfusion is recommended especially in animals older than p30–p35.
5. The needle must be inserted at the bottom tip of the heart for about 1 cm parallel to the longer axis of the heart, so that the needle tip does not perforate the interventricular septum and the ACSF is pumped directly into the aorta.
6. Drying away the liquid in excess is very important since it ensures that the brain will perfectly glue onto the vibratome stage. Failing to dry out may cause the brain to detach from the stage during slicing.
7. In order to keep glass capillaries clean and dust-free, we put them in 70 % ethanol for 30 min, then dry them with tissue paper and put them in oven for 2 h at 100 °C. Pipettes are freshly pulled before experiments and are not used at subsequent days.
8. Always use extreme care when handling the slices. The ability to record synaptic connections using dual recordings relies heavily on slice viability. While individual whole-cell recordings may often be obtained in less-than-optimal preparations, a sizeable cell-to-cell synaptic transmission is much more susceptible of the mechanical, chemical, or thermal stress imposed to the slice. Consequently, using non-ideal preparations may result in underestimating connectivity rates.
9. Striatal cells with a smooth, round- or oval-shaped soma of 12–15 μm diameter are usually healthy and patchable. Unhealthy or dead cells have an excessively large and swollen (balloon-shaped) or, conversely, shrunken soma. Moreover, the highest probability of finding synaptic connections in the striatum is found when the somatic distance of pairs does not exceed 80–100 μm.
10. The intracellular solution should be filtered using a syringe-driven-filter unit (0.22 μm). A fresh aliquot should be used at any day of experiment.

11. The amount of pressure depends on the depth of the desired cell (the deeper the cell, the larger the pressure). Always be sure that there is no pressure loss in the pipette. Positive pressure is used to avoid tissue debris to stick to the tip and to improve the visibility of the soma of the cell to patch. If pressure is not applied, the formation of a giga-seal will be much more difficult if not impossible.
12. In alternative to approaching the two cells sequentially (i.e., patching one first and then repeating the procedure for the other), one may choose to lower both electrodes "in parallel" in order to reduce the time the first cell has to wait in whole-cell conditions before the other one is also patched. However, we find it easier to leave one electrode out while patching the first cell since this minimizes the risk of clogging or soiling the tips with debris, etc., or accidentally touching one pipette while moving the other one.
13. Unitary PSCs are essential to measure fundamental synaptic parameters such as 10–90 % rise-time, decay time constant, peak amplitude, failure rate (i.e., the number of events in which a presynaptic action potential does not elicit a PSC; *see* Fig. 1b) and to perform quantal [24] and non-stationary noise analysis [16]. A protocol designed to evoke two presynaptic action potentials separated by different time windows (e.g., 20, 50, 100, 200 ms) is useful to analyze properties of PSC short-term plasticity (e.g., paired-pulse facilitation or depression; Fig. 2). In addition, dual recordings allow pairing of presynaptic and postsynaptic action potentials at different time intervals to study spike-timing dependent plasticity [25].
14. Care should be taken when trying to withdraw an electrode from one cell without damaging the other one being patched. We found that a quick withdrawal (best achieved with motorized manipulators set at their highest speed) minimizes the probability of damaging the second cell, especially when the pair is located in the deeper layers of the slice.

References

1. Zhang F, Aravanis AM, Adamantidis A et al (2007) Circuit-breakers: optical technologies for probing neural signals and systems. Nat Rev Neurosci 8:577–581
2. Spray DC, Harris AL, Bennett MVL (1979) Voltage dependence of junctional conductance in early amphibian embryos. Science 204: 432–434
3. Gibson JR, Beierlein M, Connors BW (1999) Two networks of electrically coupled inhibitory neurons in neocortex. Nature 402:75–79
4. Mann-Metzer P, Yarom Y (1999) Electrotonic coupling interacts with intrinsic properties to generate synchronized activity in cerebellar networks of inhibitory interneurons. J Neurosci 19:3298–3306
5. Galarreta M, Hestrin S (1999) A network of fast-spiking cells in the neocortex connected by electrical synapses. Nature 402:72–75
6. Landisman CE, Long MA, Beierlein M et al (2002) Electrical synapses in the thalamic reticular nucleus. J Neurosci 22:1002–1009

7. Bennett MV, Zukin RS (2004) Electrical coupling and neuronal synchronization in the mammalian brain. Neuron 41:495–511
8. Connors BW, Long MA (2004) Electrical synapses in the mammalian brain. Annu Rev Neurosci 27:393–418
9. Christie JM, Bark C, Hormuzdi SG et al (2005) Connexin36 mediates spike synchrony in olfactory bulb glomeruli. Neuron 46: 761–772
10. Long MA, Jutras MJ, Connors BW, Burwell RD (2005) Electrical synapses coordinate activity in the suprachiasmatic nucleus. Nat Neurosci 8:61–66
11. Vervaeke K, Lorincz A, Gleeson P et al (2010) Rapid desynchronization of an electrically coupled interneuron network with sparse excitatory synaptic input. Neuron 67: 435–451
12. Curti S, Hoge G, Nagy JI, Pereda AE (2012) Synergy between electrical coupling and membrane properties promotes strong synchronization of neurons of the mesencephalic trigeminal nucleus. J Neurosci 32:4341–4359
13. Otsuka K, Kawaguchi Y (2013) Common excitatory synaptic inputs to electrically connected cortical fast-spiking cell networks. J Neurophysiol 110:795–806
14. Koos T, Tepper JM (1999) Inhibitory control of neostriatal projection neurons by GABAergic interneurons. Nat Neurosci 2:467–472
15. Beierlein M, Gibson JR, Connors BW (2003) Two dynamically distinct inhibitory networks in layer 4 of the neocortex. J Neurophysiol 90:2987–3000
16. Taverna S, Ilijic E, Surmeier DJ (2008) Recurrent collateral connections of striatal medium spiny neurons are disrupted in models of Parkinson's disease. J Neurosci 28: 5504–5512
17. Gittis AH, Nelson AB, Thwin MT et al (2010) Distinct roles of GABAergic interneurons in the regulation of striatal output pathways. J Neurosci 30:2223–2234
18. Avermann M, Tomm C, Mateo C et al (2012) Microcircuits of excitatory and inhibitory neurons in layer 2/3 of mouse barrel cortex. J Neurophysiol 107:3116–3134
19. Szydlowski SN, Pollak Dorocic I, Planert H et al (2013) Target selectivity of feedforward inhibition by striatal fast-spiking interneurons. J Neurosci 33:1678–1683
20. Bugaysen J, Bar-Gad I, Korngreen A (2013) Continuous modulation of action potential firing by a unitary GABAergic connection in the globus pallidus in vitro. J Neurosci 33: 12805–12809
21. Gerfen CR (1988) Synaptic organization of the striatum. J Electron Microsc Tech 10: 265–281
22. Kreitzer AC (2009) Physiology and pharmacology of striatal neurons. Annu Rev Neurosci 32:127–147
23. Russo G, Nieus T, Maggi S, Taverna S (2013) Dynamics of action potential firing in electrically connected striatal fast-spiking interneurons. Front Cell Neurosci. doi:10.3389/fncel.2013.00209
24. Bekkers JM, Stevens CF (1995) Quantal analysis of EPSCs recorded from small numbers of synapses in hippocampal cultures. J Neurophysiol 73:1145–1156
25. Egger V, Feldmeyer D, Sakmann B (1999) Coincidence detection and changes of synaptic efficacy in spiny stellate neurons in rat barrel cortex. Nat Neurosci 2:1098–1105

Chapter 12

Outside-Out "Sniffer-Patch" Clamp Technique for In Situ Measures of Neurotransmitter Release

Émilie Muller-Chrétien

Abstract

The mechanism underlying neurotransmitter release is a critical research domain for the understanding of neuronal network function; however, few techniques are available for the direct detection and measurement of neurotransmitter release. To date, the sniffer-patch clamp technique is mainly used to investigate these mechanisms from individual cultured cells. In this study, we propose to adapt the sniffer-patch clamp technique to in situ detection of neurosecretion. Using outside-out patches from donor cells as specific biosensors plunged in acute cerebral slices, this technique allows for proper detection and quantification of neurotransmitter release at the level of the neuronal network.

Key words Sniffer patch clamp, Outside-out patch, In situ neurotransmitter detection

1 Introduction

A wide variety of techniques have been employed for the detection and measurement of neurotransmitter release from biological preparations, most of them offering impressive levels of sensitivity. One of these techniques, the "sniffer-patch," is seeing a revival in interest, as it combines sensitivity with the necessary temporal and spatial resolution required to study quantal neurotransmitter release from single cells. This technique consists in exploiting the natural high affinity of ligand-gated ion channels for their native neurotransmitter to provide a highly sensitive biological detection system. The basis is to use the outside-out configuration of the patch-clamp recording method [1] to excise small patches of receptor-rich membrane from a donor cell, and to use these patches to probe for sites of transmitter release [2]. Therefore, membrane patches can be used as biosensors to detect the release of any transmitter, provided an appropriate source of receptors is present. To date, the list of transmitters detected in this way includes acetylcholine [3–5], adenosine 5′-triphosphate [6], glutamate [7, 8], GABA [9], and glycine [10].

Marzia Martina and Stefano Taverna (eds.), *Patch-Clamp Methods and Protocols*, Methods in Molecular Biology, vol. 1183,
DOI 10.1007/978-1-4939-1096-0_12, © Springer Science+Business Media New York 2014

The first stage in the detection process is to bring donor cells into association with the cells under study. While donor cells were first expected or known to express receptors sensitive to the neurotransmitter of interest, cells transfected with chosen receptors avoid contamination of the analyses with other receptors also sensitive to this neurotransmitter, and provide a high density of specific receptors in the sniffer-patch to allow for both qualitative and quantitative detection.

To achieve the greatest sensitivity for transmitter detection, donor cells that express the highest possible density of the desired receptors are preferred. However, careful consideration should also be given to other characteristics and receptor subtypes expressed by the donor cell which could limit or otherwise complicate the detection process. One potential way of minimizing this type of problem is to use a cell line chosen as a *tabula rasa* to which the expression of the desired receptors can be added at the requisite density. Furthermore, choosing receptors displaying an unambiguously identifiable elementary conductance allows for the visualization of all channel openings, making them good sensors for the neurotransmitter of interest. Finally, as the number of channel openings observed reflects the number of channels expressed in the patch, the amount of neurotransmitter released in an experimental preparation can be estimated by first calibrating the sniffer-patch with a quick application of a known quantity of the neurotransmitter of interest [11].

So far, neurotransmitter release was mainly detected from separately cultured cells in cell-attached or whole-cell configuration, using brief current pulses to elicit action potentials. As an adjunct to their use as extracellular detectors, excised membrane patches can also be employed as intracellular detectors using a technique known as patch cramming [12]. Insertion through the cell membrane into the cytoplasm of an inside-out membrane patch containing second-messenger-sensitive ion channels allow for the measurement of the concentrations of intracellular messengers such as Ca^{2+}, cAMP, or inositol (1,4,5)-trisphosphate [12, 13]. While allowing for investigation of the mechanisms underlying neurotransmitter release at the cellular level, these techniques do not take into account interactions that exist at the level of the neuronal network. Interestingly, the sniffer-patch technique can be further adapted to in situ detection by plunging the sniffer-patch into an acute slice. Transmitter release can then be elicited through bath-application of high-potassium solution mimicking neuronal depolarization, or through electrical activation of presynaptic fibers afferent to the recording area using a stimulating electrode. In this chapter, we describe how to adapt the sniffer-patch clamp technique to in situ detection of neurotransmitter release, an approach recently used to demonstrate non-synaptic release of glycine during embryonic development of mouse spinal cord [10]. As an example, we have focused our attention on the demonstration of

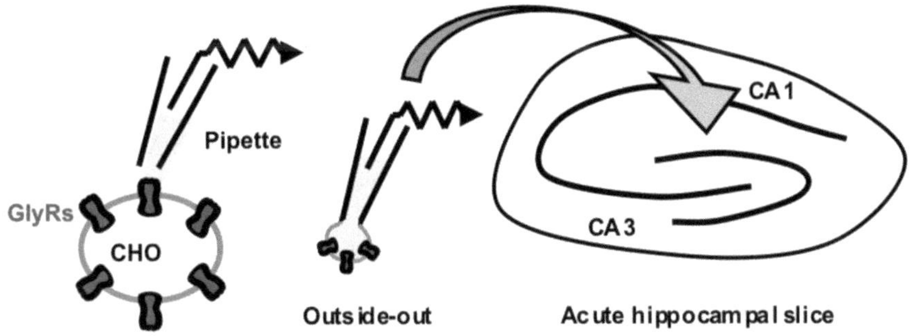

Fig. 1 An outside-out patch excised from CHO cells exclusively expressing α2-GlyRs sensitive to glycine is transferred and plunged into the CA1 *stratum radiatum* to probe for glycine release

glycine release in adult CA1 hippocampus under neuronal network activity (Fig. 1). Indeed, while modulating NMDA receptors function, the source of glycine required to co-activate these receptors in this cerebral area remains to be investigated.

2 Materials

2.1 Cloning and Transfection of Glycine Receptors in Donor Cells

Complete procedure is described in Mangin et al. [11].

1. CHO-K1; ATCC CCL61 cell line.
2. Superscript and oligoDT.
3. PCR reaction with 2.5 μg of High Fidelity Enzyme, 1.5 mM $MgCl_2$, rat GlyR α2 subunit (access no. X61159) forward (ATCACGGAAACAGGAATGAAC) and reverse primer (CATCTATTTCTTGTGGACATC).
4. Agarose gel and pCRII-TOPO vector.
5. pTRACER-CMV vector.
6. DAC30 lipofection reagent selected with Zeocin 500 μg/ml 48 h later.
7. 35 mm-diameter corning petri dish.

2.2 Sniffer Patch Clamping

Compared to classical outside-out patch-clamp protocols, the sniffer-patch technique does not require additional materials than those appropriate for donor cells.

1. 8–10-weeks-old wild-type Swiss mice.
2. Isoflurane vaporizer.
3. Carbogen (95 % O_2–5 % CO_2).
4. The Artificial Cerebrospinal Fluid (ACSF) solution contains: 126 mM NaCl, 2.5 mM KCl, 2 mM $CaCl_2$, 1 mM $MgCl_2$, 26 mM $NaHCO_3$, and 10 mM glucose. The osmolarity of the ACSF is adjusted to 300 mOsm and the pH to 7.2 using 1 M HCl.

5. Vibrating microtome.
6. Thick-wall borosilicate glass and horizontal puller.
7. The pipette intracellular solution contains: 140 mM CsCl, 4 mM $MgCl_2$, 10 mM HEPES, 10 mM EGTA and 2 mM Na^+-ATP. An alternative pipette solution contains: 130 mM CsCl, 4 mM $MgCl_2$, 4 mM Na_2ATP, 10 mM EGTA, and 10 mM HEPES. The pH is adjusted to 7.2 and the osmolarity to 300 mOsm using sucrose (*see* **Note 1**).
8. The stimulating solution contains: 10 μM glycine in ACSF, ACSF supplemented with 30 mM KCl.

3 Methods

3.1 Use of Stably Transfected CHO Expressing Glycine Receptors

For an appropriate detection of glycine release, we chose to transfect a Chinese Hamster Ovarian cell line (CHO-K1; ATCC CCL61) with α2 subunit-containing homomeric glycine receptors (α2-GlyRs) as described by Mangin et al. [11] (*see* **Note 2**).

1. Samples of 1 μg of total RNA extracted from adult rat spinal cord are reversely transcribed.
2. The cDNA obtained is used as a template for a PCR reaction according to the following protocol: after a 2 min denaturation initial step, 10 cycles at 94 °C for 15 s, at 50 °C for 30 s, and at 72 °C for 2 min are followed by 20 cycles at 94 °C for 15 s, at 60 °C for 30 s, and at 72 °C for 2 min with five additional seconds per cycle and by final extension at 72 °C for 2 min.
3. Following PCR product purification, the sequence is checked, the GlyR α2 subunit subcloned in BstXI sites and the various constructs and their orientations are verified by sequencing.
4. The constructs are transfected into CHO cells, selected recombinants are cloned by dilution, and several clones are tested and selected for the expression of the GlyR α2.

3.2 Preparation of Acute Hippocampal Slices

Hippocampal slices are prepared on the day of the experiment as previously described [14–16].

1. Acute coronal hippocampal brain slices (300 μm thick) are obtained from wild-type Swiss mice (8–10 weeks old).
2. Prior to decapitation, animals are anesthetized using an isoflurane vaporizer.
3. The brain is removed and placed in oxygenated (95 % O_2–5 % CO_2) ACSF at 4 °C. Slices were cut with a vibrating microtome and incubated for 1 h in an oxygenated chamber at room temperature (18–20 °C) before they were used for experiments.

3.3 Sniffer Patch-Clamp Experiment

1. Outside-out patch clamp electrodes (5–10 MΩ) were fire-polished to a final resistance of 3–5 MΩ (*see* **Note 3**).
2. The hippocampal slice was transferred to the recording chamber at room temperature, perfused with ACSF, and stabilized under a steel anchor to provide a full access to the *stratum radiatum* (*see* **Note 4**).
3. A 35 mm Corning petri dish containing a coverslip with plated cells expressing α2-GlyRs was taken from the incubator, and the coverslip broken into small pieces with the tip of a cutter blade. Using forceps, a piece of this coverslip is placed in the recording chamber close to the hippocampal slice. The petri dish was returned to the incubator with the remaining pieces of the coverslip.
4. Under the microscope, the coverslip was placed adjacent to the hippocampal slice and oriented so that the highest density of CHO cells was placed as close as possible to the *stratum radiatum* to decrease the sniffer-patch transfer distance.
5. An "outside-out" patch of one CHO cell membrane was excised (*see* **Note 5**). In this patch configuration, the extracellular portion of the ion channel is facing the bath, allowing for channel activation by extracellular agonists (i.e., glycine). Ideally, membrane seals should be as tight as possible (>10 GΩ).
6. The excised patch was voltage-clamped at a potential providing an appropriate driving force for chloride ion movement through the detector channels (i.e., −50 mV) (*see* **Note 6**).
7. The patch sensitivity (*see* **Note 7**) was then assessed by quickly and exogenously applying a low concentration of 10 μM glycine, allowing for the measure of the peak detector current (Fig. 2).

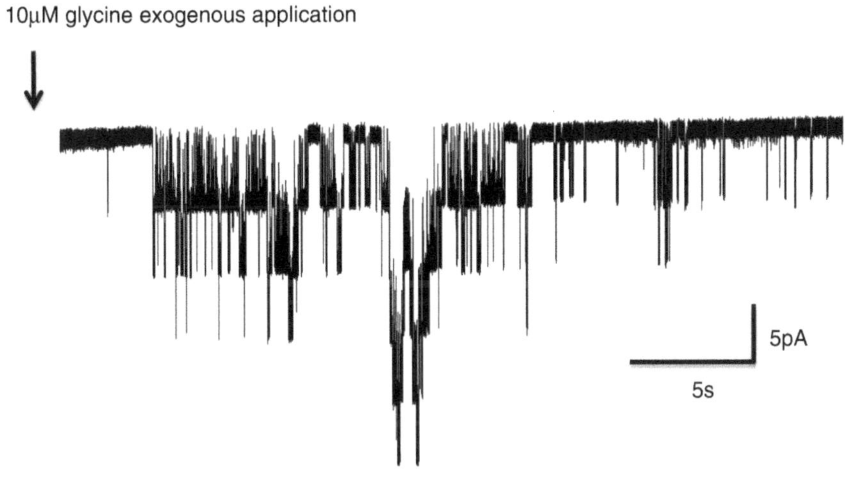

Fig. 2 Exogenous application of 10 μM glycine-containing solution on the outside-out patch pulled from a CHO expressing α2-GlyRs. Recording of channel openings reflects α2-GlyRs activation by glycine

8. Following patch excision, the electrode was gently moved horizontally above the CA1 *stratum radiatum*, lowered, and pushed inside the slice using a micromanipulator until the pipette tip completely disappeared within the tissue. Seal resistance was continuously monitored (*see* **Note 8**) to assure that it remained above 1 GΩ as an indicator of outside-out patch stability. In this configuration, the release of glycine into the extracellular space could activate α2-GlyRs contained in the sniffer-patch membrane.
9. Prior to recording, the sniffer-patch was left in the slice for 5 min to assure patch stability (i.e., seal resistance >1 GΩ).
10. α2-GlyRs-activity reflecting glycine release under basal conditions was then recorded.
11. At the end of this 5-min recording period, an ACSF solution containing 30 mM K^+ was bath-applied to mimic a strong depolarization of the neuronal network and induce neurotransmitter release from presynaptic terminals (*see* **Notes 9** and **10**).
12. Compared to normal ACSF, bath application of high-K^+ (30 mM) ACSF for 10–30 s induced strong glycine release, as shown by the numerous openings of α2-GlyRs contained in the patch (Fig. 3).
13. As a control for exclusive α2-GlyRs activation, strychnine added to the solution should not provide any channel opening (Fig. 4) (*see* **Note 11**).

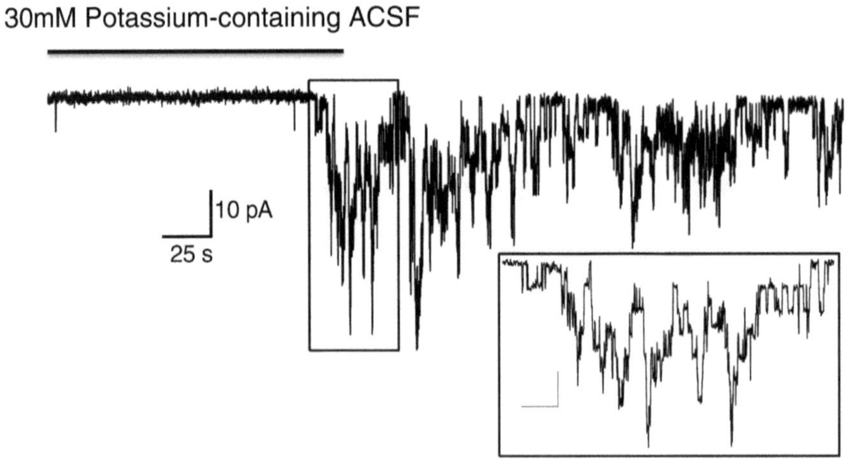

Fig. 3 Bath-application of ACSF containing 30 mM potassium induced strong glycine release, as detected by α2-GlyRs contained in the patch

Fig. 4 Sniffer-patch recording of glycine release in the CA1 hippocampus under glycine or glycine +1 μM of strychnine, a specific antagonist of all GlyRs

3.4 Analysis (See Note 12)

1. Single-channel currents were recorded using a patch-clamp amplifier. Recordings were sampled at 50 kHz, filtered at 10 kHz, and stored on a computer for further analysis.
2. Analysis of the sniffer currents was performed using appropriate single-channel recording software.
3. Decay rates of sniffer currents were estimated by measuring the duration of the sniffer current at 20 % of the peak amplitude.
4. One thousand consecutive points were averaged to measure changes in current amplitude with time (sampling rate, 50 kHz).

4 Notes

1. The inclusion of membrane-impermeable channel blockers or fluoride ions in the pipette-filling solution can be used to block unwanted channel activity and improve the overall signal-to-noise ratio of the patch. However, caution should be exercised, as some internal ions can produce unexpected reductions in patch sensitivity.
2. HEK 293 cells stably transfected with the non-desensitizing mutation L497Y were similarly used as donor cells for glutamate release detection in hippocampal slices.
3. The small size of membrane patches (typically 2–3 μm diameter) makes them particularly well suited for detecting release from discrete loci, such as varicosities or synaptic terminals. However, it is not possible to distinguish between several synaptic terminals in an in situ context.
4. As the concentration of neurotransmitter reaching the detector falls off very steeply with distance, very small fluctuations in the position of the patch will significantly alter the detected neurotransmitter concentration, which means that any changes in concentration due to mechanical drift will be amplified to produce a proportionally greater change in the output of the detector. Therefore, experiments require a very stable recording setup.

5. It is important to ensure that patches can be excised easily from the chosen donor cell, and that once excised they remain stable for at least 15–30 min with minimal loss of sensitivity. Exogenous application of neurotransmitter immediately following patch excision should provide similar channel activation as application 15 and 30 min later.
6. Patches with significant levels of spontaneous channel activity should be avoided, as they will reduce the effective detection threshold.
7. The sensitivity of a detector patch is determined by the number of receptors contained in the patch and their EC_{50}, with the detection threshold being the concentration of agonist that produces channel openings that can be distinguished from any background or spontaneous channel activity. The response or output of the patch is $I = iNp_o$ (where I and i are the detected and the unitary currents, respectively; N is the number of channels; and p_o is the channel open probability). From this, it follows that sensitivity can be enhanced by using patches containing large numbers (N) of receptors, thereby ensuring that even at agonist concentrations where the probability of individual channel opening (p_o) is low, Np_o remains relatively high.
8. It is possible to move the patch from one place to another, for example to get close to a suspected postsynaptic site of release. However, membrane stretch due to several pipette insertions into the slice generally caused loss of the patch following two attempts.
9. We hereby provide an example of glycine release detection following strong neuronal depolarization induced by bath-application of ACSF supplemented with 30 mM of K^+, mimicking neuronal depolarization. However, such a solution can also induce neurotransmitter release from glial cells surrounding neurons, and more specifically astrocytes. When investigating neurotransmitter release from presynaptic terminals, the following synaptic-like stimulation can be used:
   (a) Bath-application of ACSF supplemented with 10 mM of K^+, mimicking neuronal presynaptic depolarization rather than a global neuronal and glial activation under the stimulation of the whole network.
   (b) Release of neurotransmitter from presynaptic terminals elicited by electrical stimulation. In the case of glycine detection in the CA1 hippocampus, a bipolar stimulating electrode was positioned in the *stratum radiatum* of the CA1 region of the hippocampus, a region where the axons from CA3 neurons project and form Schaffer collateral synapses with CA1 neurons, before bringing plated cells expressing α2-GlyRs into the recording chamber.

The single stimulation protocol consisted of 100 μs current pulses (10–200 μA) evoked every 10 s.

In both cases, it has to be noted that such synaptic-like stimulation allows for transmitter detection provided that the sniffer-patch is plunged close to a corresponding postsynaptic site.

10. Bath application of high-potassium-containing ACSF solution had to be stopped and replaced with normal ACSF at the onset of channel opening to avoid losing the patch.
11. Using the sniffer-patch technique to probe for glycine release in the hippocampus under different conditions of stimulation, notably electrical stimulation of Schaffer collaterals and bath-application of 10 mM potassium-containing ACSF solution, we were able to confirm that glycine release originates from presynaptic terminals, as suggested by recent immunohistochemical studies [17]. It has been suggested that GlyT1 glycine transporters could contribute to this release; however, we observe that blocking these transporters did not impact α2-GlyRs activation following bath-application of ACSF containing 10 mM of potassium and a specific GlyT1 (data not shown).
12. A good knowledge of single channel electrophysiological properties, most notably conductance and desensitization kinetics, is essential for a proper quantification of neurotransmitter release. Using α2-GlyRs, displaying an unambiguously identifiable elementary conductance (100–120 pS), a long mean open time, and little desensitization, allowed for the visualization of all channel openings, and correlation between channel openings and the amount of neurotransmitter released. α2-GlyRs are therefore a good sensor for glycine [11].

References

1. Horn R, Patlak J (1980) Single channel currents from excised patches of muscle membrane. Proc Natl Acad Sci U S A 77:6930–6934
2. Young SH, Poo MM (1992) Measurement of transmitter release: moving patch method. In: Kettenmann H, Grantyn R (eds) Practical electrophysiological methods. Wisley-Liss, New York, pp 354–357
3. Hume RI, Role LW, Fischbach GD (1983) Acetylcholine release from growth cones detected with patches of acetylcholine receptor-rich membranes. Nature 305:632–634
4. Young SH, Poo MM (1983) Spontaneous release of transmitter from growth cones of embryonic neurones. Nature 305:634–637
5. Grinnell AD, Gundersen CB, Meriney SD et al (1989) Direct measurement of Ach release from exposed frog nerve terminals: constraints on interpretation of non-quantal release. J Physiol 419:225–251
6. Silinsky EM, Redman RS (1996) Synchronous release of ATP and neurotransmitter within milliseconds of a motor nerve impulse in the frog. J Physiol 492:815–822
7. Maeda T, Shimoshige Y, Mizukami K et al (1995) Patch sensor detection of glutamate release evoked by a single electrical shock. Neuron 15:253–257
8. Billups B, Attwell D (1996) Modulation of non-vesicular glutamate release by pH. Nature 379:171–174
9. Shimane M, Miyagawa K, Sugawara M (2006) Detection of gamma-aminobutyric acid-induced glutamate release in acute mouse

hippocampal slices with a patch sensor. Anal Biochem 353:83–92
10. Scain AL, Le Corronc H, Allain AE et al (2010) Glycine release from radial cells modulates the spontaneous activity and its propagation during early spinal cord development. J Neurosci 30:390–403
11. Mangin JM, Baloul M, Prado De Carvalho L et al (2003) Kinetic properties of the alpha2 homo-oligomeric glycine receptor impairs a proper synaptic functioning. J Physiol 553:369–586
12. Kramer RH (1990) Patch cramming: monitoring intracellular messengers in intact cells with membrane patches containing detector ion channels. Neuron 4:335–341
13. Van-Hooft JA, Vijverberg HP (1995) Phosphorylation controls conductance of 5-HT3 receptor ligand-gated ion channels. Receptors Channels 3:7–12
14. Martina M, Krasteniakov NV, Bergeron R (2003) D-Serine differently modulates NMDA receptor function in rat CA1 hippocampal pyramidal cells and interneurons. J Physiol 548:411–423
15. Martina M, Gorfinkel Y, Halman S et al (2004) Glycine transporter type 1 blockade changes NMDA receptor-mediated responses and LTP in hippocampal CA1 pyramidal cells by altering extracellular glycine levels. J Physiol 557: 489–500
16. Martina M, Turcotte ME, Halman S et al (2007) The sigma-1 receptor modulates NMDA receptor synaptic transmission and plasticity via SK channels in rat hippocampus. J Physiol 578:143–157
17. Muller E, Bakkar W, Martina M et al (2013) Vesicular storage of glycine in glutamatergic terminals in mouse hippocampus. Neuroscience 242:110–127

Chapter 13

A Cost-Effective Method for Preparing, Maintaining, and Transfecting Neurons in Organotypic Slices

Cary Soares, Kevin F.H. Lee, Denise Cook, and Jean-Claude Béïque

Abstract

The cellular and molecular mechanisms that underlie brain function are challenging to study in the living brain. The development of organotypic slices has provided a welcomed addition to our arsenal of experimental brain preparations by allowing both genetic and prolonged pharmacological manipulations in a system that, much like the acute slice preparation, retains several core features of the cellular and network architecture found in situ. Neurons in organotypic slices can survive in culture for several weeks, can be molecularly manipulated by transfection procedures and their function can be interrogated by traditional cellular electrophysiological or imaging techniques. Here, we describe a cost-effective protocol for the preparation and maintenance of organotypic slices and also describe a protocol for biolistic transfection that can be used to introduce plasmids in a small subset of neurons living in an otherwise molecularly unperturbed network. The implementation of these techniques offers a flexible experimental paradigm that can be used to study a multitude of neuronal mechanisms.

Key words Organotypic slice cultures, Hippocampus, Biolistic transfection, Electrophysiology, Patch-clamp

1 Introduction

For over half a century, the acute brain slice preparation has offered neurophysiologists a means to study cellular processes that underlie nervous system function [1, 2]. This in vitro experimental preparation retains much of the three-dimensional architecture and local connectivity of neural circuitries in situ, and has thus gained widespread popularity in modern neuroscience laboratories. Despite the development of new technologies that permit the interrogation of circuit function and cellular physiology in vivo, there still remain overarching barriers that restrict the implementation of many useful methodologies in the intact brain.

Notwithstanding their widespread acceptance and practicality, the limited life span of acute slices restricts their use in experiments

Marzia Martina and Stefano Taverna (eds.), *Patch-Clamp Methods and Protocols*, Methods in Molecular Biology, vol. 1183, DOI 10.1007/978-1-4939-1096-0_13, © Springer Science+Business Media New York 2014

involving prolonged pharmacological and/or genetic manipulations [3]. The development of organotypic slices has opened an important avenue to circumvent these limitations [4, 5]. This technique allows for thick sections of brain tissue to be maintained in culture for several weeks to months, enabling their use in experiments involving molecular and/or chronic pharmacological intervention [3, 6–12]. Importantly, unlike dissociated neurons, organotypic slices retain several of the key features of network and cellular physiology found in situ [3, 13–16].

Combining organotypic slices with a reliable transfection method, such as biolistic-mediated gene transfer [17], provides a powerful method for investigating the effects of molecular interventions on cellular functions. The biolistic method involves the bombardment of brain slices with DNA-coated micron-sized gold particles, leading to a sparse transfection of neurons located in an otherwise undisturbed network [18]. Consequently, this transfection method allows traditional molecular manipulations to be applied in a visually identified (i.e., by inclusion of a fluorescent reporter gene) subset of neurons, which can then be studied using electrophysiological or optical approaches. In addition, organotypic slices provide a highly controlled environment for evaluating the prolonged effect of pharmacological treatments on various neural functions [3, 6, 11].

Here, we describe a cost-effective method for preparing organotypic slices and provide several tips based on our own experience that facilitate their use in electrophysiological experiments. Although described here for hippocampal tissue, this approach can be modified to culture tissue from other parts of the brain. We also describe a biolistic approach that has been optimized for transfection of neurons in hippocampal organotypic slices. Collectively, these techniques enable electrophysiological and/or optical recordings of genetically modified neurons and, as such, provide a cost-effective paradigm to capitalize on the ever expanding repertoire of available genetically encoded biosensors, reporters, and molecular tools that manipulate the expression of proteins in living cells.

2 Materials

2.1 Solutions

1. Culture Media.

   Two culture media recipes have been used successfully in this protocol. We tend to prefer a Neurobasal based solution composed of (per 50 mL): 42.5 mL of 1× Neurobasal-A (without phenol red or glutamine; with 26 mM sodium bicarbonate), 0.25 mL of 200 mM L-glutamine, 1 mL of B27 supplement (50×), 0.5 mL of penicillin–streptomycin (100×), 0.75 mL of

45 % D-Glucose, and 5 mL of heat-inactivated horse serum. An alternative culture media (50 mL) is composed of: 25 mL of Minimal Essential Media (1×), 12 mL of Hank's Balanced Salt Solution (1×), 0.25 mL of 200 mM L-glutamine, 0.5 mL of penicillin–streptomycin (100×), 11.75 mL of heat-inactivated horse serum, and 0.5 mL of 45 % D-glucose in solution. The pH of the culture media is adjusted to 7.15 using a 1 M HCl solution, filtered with a sterile 0.22 μm filter and kept at 4 °C prior to use. Culture media should be prepared routinely and should not be stored for longer than 2 weeks.

2. Dissection solution.
   For brain dissection and slice preparation, we use a choline-based dissection solution, although other dissection solution recipes can be used. This solution is made fresh on the day of dissection and is composed of (in mM): 119 choline chloride, 2.5 KCl, 4.3 $MgSO_4$-$7H_2O$, 1 $CaCl_2$-$2H_2O$, 1 NaH_2PO_4, 26.2 $NaHCO_3$, 11 Glucose, and 1.3 Na-Ascorbate. To adjust pH, the solution is bubbled with 95/5 % carbogen for 20–30 min (equilibrates to pH: 7.3–7.4) and then frozen at −80 °C until a slush-like consistency is achieved.

2.2 Materials and Equipment for Tissue Culture

Slices for organotypic cultures can be made using any conventional vibratome or tissue chopper. For hippocampal slice cultures, we use a tissue slicer. Slice cultures are maintained on small rectangular islands of 0.4 μm Biopore membrane (Millipore) placed on top of membrane culture inserts (0.4 μm pore) in a 6-well culture plate (*see* Fig. 1a). An incubator that can be equilibrated to 34 °C with 5 % CO_2 is required, and to avoid contamination, part of this procedure should be performed in a sterile biosafety cabinet. For biolistic transfection of organotypic slices, we use a Gene Gun System.

3 Methods

3.1 Preparation of Organotypic Slices

This is a modified version of an organotypic slice protocol originally described by Stoppini et al. [5], which has been optimized to reduce cost and improve the transfer of slices to an electrophysiological or imaging chamber. One important difference is that individual brain slices are not directly cultured on the surface of the porous membrane inserts. Rather, slices are maintained on small rectangular pieces of Biopore membrane that are placed on the surface of the membrane inserts prior to dissection (*see* Fig. 1a). This additional piece of membrane at the slice–media interface provides practical advantages by facilitating the transfer of brain slices from culture conditions to experimental (electrophysiological or imaging) chambers, and an economical advantage by enabling recycling and repeated use of costly membrane inserts.

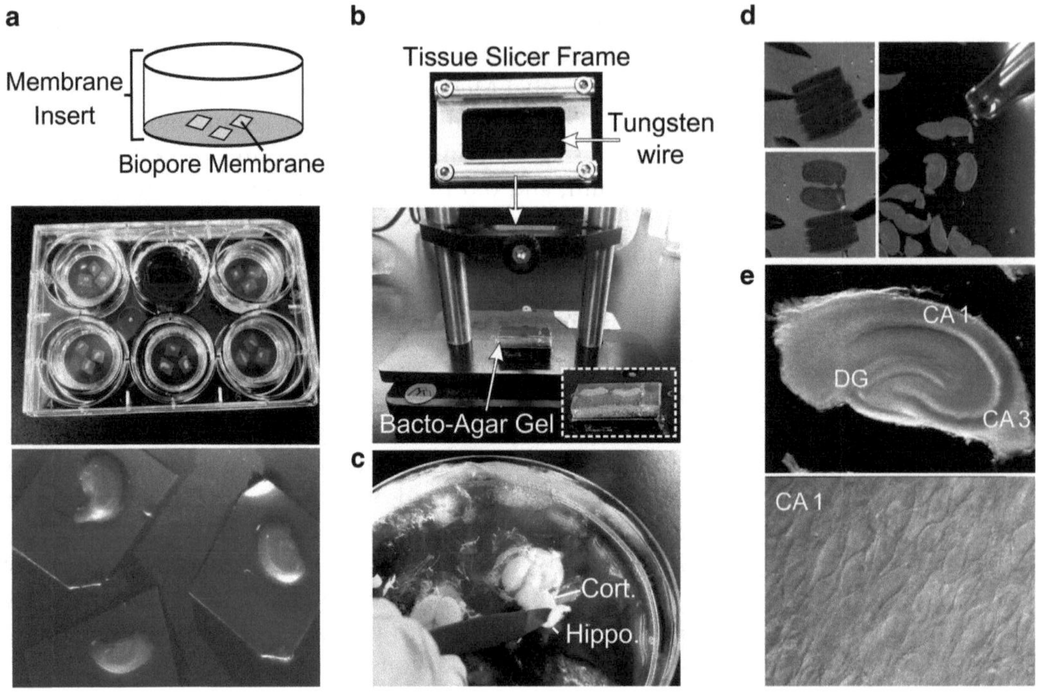

Fig. 1 Protocol for generating organotypic hippocampal slices. (**a**) Small pieces of Biopore membrane are placed on membrane inserts, each of which will accommodate a single hippocampal slice, which allows for easy transfer of slices to the experimental chamber. (**b**) The tissue slicer frame for the MX-TX tissue chopper is threaded with tungsten wire to cut hippocampal tissue into 400 μm slices. *Inset*, hippocampi are positioned on the Bacto Agar base such that the longitudinal axis is perpendicular to the threads of the tissue slicer. (**c**) Once the septal and temporal most regions of the hippocampus are cut, the hippocampus (Hippo.) should easily fold out from the overlying cortex (Cort.). (**d**) Hippocampal slices under a dissection microscope. Slices can remain adhered to one another following dissection, resembling a loaf of sliced bread. Using a pair of forceps, the slices can be carefully separated from one another. A plastic Pasteur pipette with a cut tip is used to transfer slices to membrane inserts. (**e**) A single hippocampal slice under the dissection microscope. All three major hippocampal regions (*CA1* Cornu Ammonis 1, *CA3* Cornu Ammonis 3, *DG* Dentate Gyrus) should be intact and readily recognized in each slice

1. Biopore membrane is acquired as a large sheet and is cut into small pieces using scissors prior to the dissection day. The dimensions of the cut membrane will depend on size of the slice intended for culturing (*see* **Note 1**).
2. On the morning of dissection, transfer the pieces of Biopore membrane into a 10 cm petri dish containing 3 mL of 100 % ethanol. Leave membranes to dry in a biosafety cabinet.
3. Prepare culture media (for recipe and directions, *see* Subheading 2.1) and add 1 mL of media into five wells of a 6-well plate. In the last remaining well, add 3 mL of culture media.
4. Using round forceps, place a sterile membrane insert in each of the five wells containing 1 mL of culture media. Avoid air bubbles from forming underneath the membrane insert by

slowly lowering the insert into the media diagonally from one edge. If membrane inserts are being recycled (*see* **Note 2**), be sure to remove them from ethanol storage in advance so they are dry when used.

5. Using rounded forceps, dip the dry rectangles of Biopore membrane in culture media (use the well with 3 mL of media) and place them on top of the membrane inserts. Three to four pieces of Biopore membrane will fit on a single membrane insert, and each small piece of membrane will accommodate a single hippocampal slice (*see* Fig. 1a).
6. Once the plates are prepared place them in a 5 % CO_2 incubator set at 34 °C for a minimum of 1 h to allow for proper temperature and pH equilibration of the culture media.
7. Prepare dissection solution (for recipe, *see* Subheading 2.1) and bubble with 95/5 % carbogen gas for a minimum of 30 min to equilibrate pH (should reach 7.3–7.4). Store at −80 °C until partially frozen. Slices should be prepared in an icy slush of dissection solution.

Steps 7–17 describe our dissection protocol for generating organotypic hippocampal slices using a tissue slicer. Organotypic slices can be prepared from other brain regions using a vibratome and following standard acute slice preparation protocols (*see* **Note 3**). If slices are prepared using a vibratome, skip **steps 8–18**.

8. Wind tissue chopper frames with teflon-coated tungsten wire according to the manufacturer's protocol. We use a thread spacing of 400 μm. Rinse with 100 % ethanol and let dry in a sterile biosafety cabinet. Once dry, attach the frame to a tissue slicer (*see* Fig. 1b).
9. A batch of 1 % Bacto Agar plates are made in advance and are required for the hippocampal dissection using a tissue slicer. Make a solution of 1 % Bacto Agar in water (100 mL, autoclaved), microwave on high for 2 min, and let the agar solution cool at room temperature until warm. Next, aliquot 20–25 mL of the warm Bacto Agar solution into 10 cm petri dishes and let the gel solidify at room temperature. The gel should be 1–2 cm thick. Seal edges with Parafilm and store agar plates at 4 °C until dissection (can be stored for several months).
10. Remove one Bacto Agar plate from the refrigerator and cut the gel with a razor blade to fit the base of the tissue slicer. This provides a raised base for the hippocampi to rest which will allow the threads of the tissue slicer to fully traverse the tissue (*see* Fig. 1b).
11. Remove the dissection solution from the freezer and let it thaw partially. Mix at room temperature until slushy. Meanwhile, clean dissection tools with 70 % ethanol.

12. Rats or mice (postnatal day 6–12) should be sacrificed according to local institutional animal care guidelines. We anesthetize animals by isoflurane inhalation prior to decapitation.
13. Remove the whole brain from the cranial cavity as quickly as possible and let it cool in a 100 mL beaker filled with ice-cold dissection solution for a few minutes. Multiple whole brains can be extracted simultaneously (*see* **Note 4**).
14. Use cyanoacrylate glue to adhere the ventral side of the brain(s) to the base of a dry 10 cm petri dish, and fill the dish with ice-cold dissection solution to completely immerse the brain(s).
15. Use a scalpel to make a shallow cut down the midline of the brain, severing the corpus callosum to separate the two cerebral hemispheres.
16. Expose the hippocampi by everting each hemisphere laterally with a spatula. Transect both the septal and temporal most regions along the longitudinal axis of the hippocampus with fine forceps. Once cut, use a spatula to flip the hippocampi out from the overlying cerebral cortex (*see* Fig. 1c). Remove the hippocampi, which should have a "banana-like" shape, using a scalpel.
17. Use a spatula to transfer the hippocampi to the Bacto Agar gel at the base of the tissue slicer (*see* Fig. 1b). It is critical that the longitudinal axis of the hippocampus be perpendicular to the threads of the tissue slicer (*see* **Note 5**). Release the tissue slicer and ensure that the threads have sufficiently penetrated into the gel (*see* **Note 6**). Transfer slices into a new 10 cm dish containing fresh ice-cold dissection solution.
18. Under a sterile dissection microscope, separate slices in cold dissection solution using two pairs of sterile Dumont forceps, or other fine dissection tools according to preference (*see* **Note 7** and Fig. 1d). All three major regions of the hippocampal circuit should be readily visible in each slice (*see* **Note 8** and Fig. 1e). Transfer the best slices to a new petri dish with ice-cold dissection solution using a plastic Pasteur pipette (*see* Fig. 1d).
19. Remove one 6-well plate at a time from the incubator. Using a fine tip plastic Pasteur pipette, transfer the slices first to the well containing extra culture media (to minimize transfer of dissection solution to culture inserts), and then onto the Biopore membrane (*see* **Note 9**). These steps require patience and finesse, as it is critical that the slices are carefully positioned on the small rectangular islands of membrane (*see* Fig. 1a). Aspirate excess media surrounding each slice with a 100 μL pipette (*see* **Note 10**) and return finished plates to the incubator.
20. Culture media should be changed the day after dissection (~24 h) and then every 2–3 days thereafter. To change culture

media, raise each membrane insert using rounded forceps and replace the entire volume in the culture well with fresh, equilibrated, culture media (*see* **Note 11**).

3.2 Preparation of Gene Gun Cartridges

Gene gun cartridges are prepared largely according to the manufacturer's protocol and previously published protocols [18, 19], although several steps have been modified to improve particle transfer and transfection efficiency of neurons in organotypic slices. This section contains the steps required to precipitate DNA onto gold microcarriers and coat these particles uniformly along tubing, which is then cut into separate cartridges for the gene gun.

1. Cut Tefzel tubing (outer diameter: 0.125″; inner diameter: 0.093″) to fit the tubing apparatus, allowing a short overhang at the open end. Dry the tubing by commencing flow of 100 % nitrogen gas through the tubing at >0.5 L/min for 20–30 min.
2. Weigh 8–10 mg of 1 μm gold microcarriers and transfer to a 1.5 mL test tube. Add 100 μL of 0.05 M spermidine solution (*see* **Note 12**).
3. Briefly sonicate and vortex the gold–spermidine solution to suspend the microcarriers, then immediately add DNA plasmids (*see* **Note 13**). The total amount of DNA required will vary between constructs and will also depend on the number of plasmids added (*see* **Note 14**). For expression of a cytosolic marker (e.g., mCherry or EGFP) 10–15 μg of DNA is often sufficient. Although we routinely keep total DNA to about 50 μg, we have used on occasion up to 80 μg of total DNA for expression of multiple plasmids (*see* Fig. 2b).
4. Vortex the tube at a slow rate and add 100 μL of 1 M $CaCl_2$ drop wise to the gold–DNA mixture. This will precipitate the DNA onto the gold microcarriers.
5. Allow the gold to settle in the tube at room temperature for 10 min. Meanwhile, transfer 3.3 mL of fresh 100 % ethanol to a 15 mL conical tube, and set aside until **step 8**. In the manufacturer's protocol, polyvinylpyrrolidone (PVP) is added to this ethanol solution; however, we avoid this step (*see* **Note 15**).
6. Once the gold has settled, place the 1.5 mL tube in a tabletop centrifuge and spin at 8,000–10,000 rcf for 10 s to form a pellet.
7. Remove and discard the supernatant, then add 1 mL of 100 % fresh ethanol to the gold pellet. Resuspend the microcarriers in the ethanol solution (*see* **Note 16**) and repeat the centrifuge step to reform the pellet. Repeat this washing step three times.
8. After the final wash, remove the ethanol and add 500 μL of 100 % ethanol from the 3.3 mL set aside in Subheading 3.2, **step 5**. Resuspend the gold in the ethanol solution and transfer

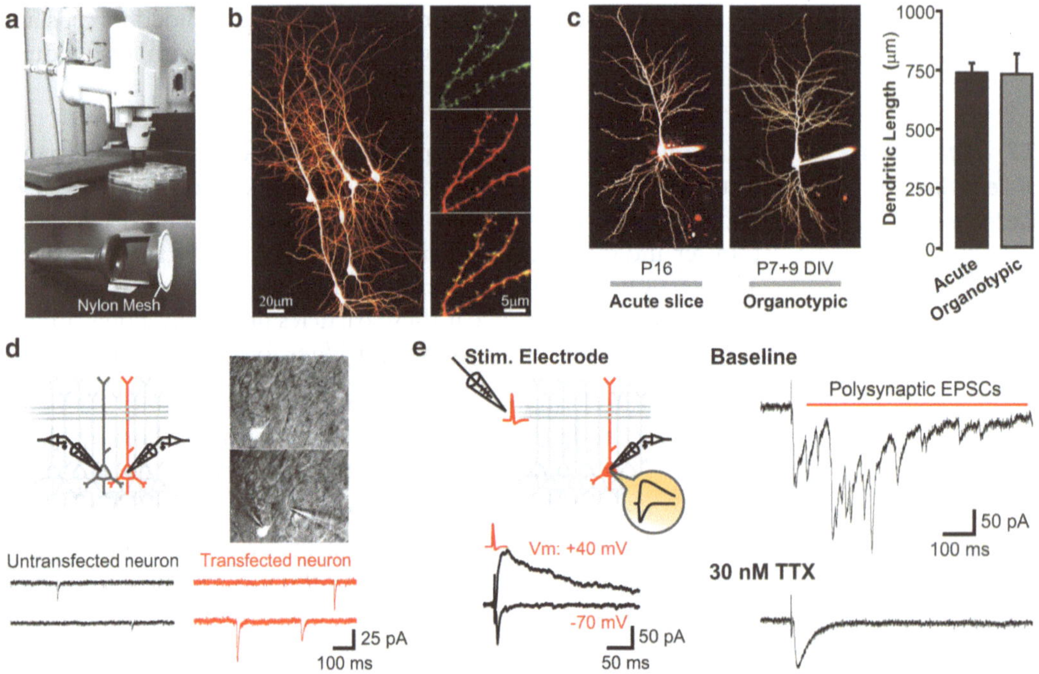

Fig. 2 Biolistic transfection for single cell expression of DNA plasmids in hippocampal slice culture. (**a**) Mounting the gene gun to the lab bench has enabled the optimization of biolistic conditions. The addition of a nylon mesh (90 μm pore) to the end of the manufacturer's barrel shunts the blast of helium coming through the chamber and allows for shooting the gene gun at higher pressures. (**b**) A confocal image stack of hippocampal CA1 pyramidal neurons biolistically transfected with mCherry and a GFP-tagged protein of interest. (**c**) Two-photon images of age-matched CA1 pyramidal neurons in an acute and organotypic slice filled with Alexa 594 through the recording electrode. Measurement of dendrite development provides evidence that neurons in slice culture develop as they would in situ. (**d**) Paired recordings from a transfected and untransfected "control" neuron is an effective strategy to examine the effect of single cell genetic manipulation. In this example, both the transfected and untransfected neuron are voltage clamped at −70 mV and miniature excitatory postsynaptic currents (EPSCs) are recorded simultaneously in both cells. (**e**) Studies of evoked synaptic transmission in organotypic slices can be compromised by the occurrence of polysynaptic events. We suggest using a monopolar glass stimulating electrode to stimulate axons in organotypic slices. Several methods have been developed to dampen polysynaptic activity [7, 9]. A low concentration of tetrodotoxin (10–30 nM) in the extracellular solution is one approach we have taken to reduce the polysynaptic activity in organotypic slice recordings

this mixture to a new 15 mL conical tube. Repeat this step until all 3.3 mL of 100 % ethanol, including the gold microcarriers, are transferred to the new 15 mL conical tube.

9. Stop nitrogen flow at the tubing station and connect a 10 mL syringe to one end of the dry Tefzel tubing.
10. Briefly vortex the gold–ethanol solution to maximize suspension. Immediately after, insert dry tubing to the bottom of the gold–ethanol suspension. Apply negative pressure to the 10 mL syringe to draw the microcarrier solution up into the tubing. The gold will settle in the 15 mL conical tube quickly,

so speed will ensure an even transfer of gold along the length of tubing.

11. Insert Tefzel tubing back into the tubing apparatus and allow the gold to settle for approximately 5 min.
12. Slowly draw the ethanol solution from the tubing by applying negative pressure to the 10 mL syringe until all the ethanol has been removed (*see* **Note 17**). Disconnect the syringe and commence rotation of the tubing apparatus without nitrogen flow for approximately 30 s, which should promote an even distribution of microcarriers around the inner circumference of the tubing (*see* **Note 18**).
13. Start flow of 100 % nitrogen at a rate of 0.3–0.4 L/min to dry the tubing while maintaining rotation for an additional 5 min.
14. Terminate the flow of nitrogen, stop the rotation, and remove tubing from the apparatus.
15. Cut tubing into individual cartridges using the Bio-Rad tubing cutter. Cartridges are stored in small plastic vials with desiccation pellets at 4 °C, and can be kept for several months.

3.3 Biolistic Transfection of Organotypic Brain Slices

The biolistic method described below is an adapted version of the manufacturer's protocol that has been modified to preserve slice vitality and improve transfection efficiency in organotypic brain slices. Mounting the hand held gene gun to a vertical support (*see* Fig. 2a) has enabled us to systematically optimize the transfection conditions (i.e., shooting distance and pressures, see below) and is strongly recommended for the implementation of this technique.

1. Slices can be transfected as early as the day of dissection, and as late as several weeks in culture. Typically, we wait at least 1 week to allow the slices to recover and adapt to culture conditions (*see* **Note 19**).
2. Remove the vial containing gene gun cartridges from the 4 °C refrigerator and let stand at room temperature for at least 10 min prior to opening.
3. Insert cartridges into the revolver cylinder using rounded forceps.
4. Attach the revolver cylinder and barrel to the mounted gene gun. It is important to use separate barrels if more than one DNA construct is being transferred (*see* **Note 20**).
5. Connect the gene gun to a compressed tank containing 100 % helium. Raise the pressure at the regulator to 180–200 psi.
6. Fire a blank shot to remove debris from the chamber.
7. Remove plates from the incubator one at a time. Under a biosafety cabinet, carefully move slices into the center of each membrane insert using rounded forceps.

8. The gene gun barrel should be level with respect to the table and positioned above each well. The height is ideal when the tip of the barrel is in line with the height of the culture plate, ~1.5–2.5 cm from the table (*see* Fig. 2a).
9. Fire the gene gun. Label each well of the culture plate appropriately after shooting and return slices to the incubator.
10. The amount of time required for maximal protein expression will vary between constructs (*see* **Note 21**).

3.4 Electrophysiological Recordings from Single Neurons in Organotypic Slices

In this last section, we provide a set of experimental tips for obtaining whole-cell patch-clamp recordings from neurons maintained in organotypic slices. Recording techniques are largely the same as those from acute slices; however, there are some differences, which may need to be taken into account to facilitate recordings.

1. On the experiment day, remove an individual slice with its respective membrane from the culture insert and transfer it to a drop of equilibrated culture media in a small petri dish. The slice is then taken immediately to an electrophysiological and/or imaging chamber for experimentation.
2. A prerequisite for obtaining reliable electrophysiological recordings is to consistently have healthy tissue to work with. If tissue health degrades rapidly after the dissection protocol, several approaches can be taken to improve slice viability (*see* **Note 22**). If tissue degeneration tends to be a problem only after transfection, the biolistic conditions may need to be optimized (*see* **Note 23**).
3. It is often appropriate to run experiments within a defined age range (i.e., days in vitro) to reduce potential artifacts from studying slices at different developmental stages (*see* **Note 24**).
4. For whole-cell recordings, we prefer tissue that has been cultured for 7–14 days, as over time, the tissue flattens, leading to improved visualization of cell morphology (*see* Fig. 1d) [13, 20].
5. As the slice matures in culture, glial coverage will make it increasingly difficult to access neuronal cell bodies with glass microelectrodes. Approach the tissue with sufficient positive pressure to penetrate the glia while descending into the slice. Once the electrode is in proximity to the targeted neuron, reduce but maintain positive pressure to navigate toward the cell body of the neuron of interest prior to gigaohm seal formation.
6. For paired recordings (*see* Fig. 2d), it is best to obtain a gigaohm seal in both cells before gaining whole cell access (*see* **Note 25**).
7. For studies of evoked excitatory transmission (i.e., with pharmacological blockade of inhibitory $GABA_A$ transmission),

polysynaptic activity is often more prevalent and problematic in organotypic slices than in acute slices (*see* Fig. 2e). In these conditions, electrical stimulation of axons is best achieved by positioning a monopolar glass patch electrode (rather than a bipolar electrode) filled with ACSF in proximity to the recorded neuron. This will not only minimize the number of axon fibers stimulated, but will also enable the experimenter to easily maneuver the position of the stimulating electrode to reduce the occurrence of evoked polysynaptic events. Several methods have been developed to minimize polysynaptic activity during electrical synaptic stimulation (*see* **Note 26**).

4 Notes

1. To accommodate hippocampal slices, Biopore membrane is cut into ~0.5 cm rectangles (*see* Fig. 1a).
2. Used membrane inserts can be stored in a large beaker of distilled water until enough are gathered for cleaning. To clean, transfer the membrane inserts into a beaker containing a 10 % acetic acid solution and let stand at room temperature for 10–15 min. Rinse the inserts by transferring them to a separate beaker containing 1–2 L of milliQ double distilled H_2O. Replace the water every 10 min for a total of three washes to dilute the residual acetic acid solution. After the third wash, store the membrane inserts in a beaker filled with 100 % ethanol until subsequent use. Use Parafilm to seal the beaker to minimize ethanol evaporation.
3. Although the details of this procedure are for culturing slices of hippocampal tissue sectioned with a MX-TS tissue slicer, we have also had success culturing coronal/horizontal cortical brain sections (300–400 μm) using a standard vibratome-based technique. Ensure that the vibratome chamber and blade are properly cleaned and sterilized prior to dissection.
4. To save time, sacrifice two to three animals simultaneously. This is not recommended, however, for individuals that are new to this protocol as best results are obtained when the duration between sacrificing the animal to completion of the culture procedure is minimized. Once adeptness in the dissection technique is established, sacrificing multiple animals at once provides an efficient way to prepare many slices without the need to perform repetitions of the protocol.
5. The longitudinal hippocampal artery, which runs along the longitudinal axis of the hippocampus [21], should be readily observable with the naked eye. Orient the hippocampi such that the longitudinal axes are perpendicular to the threads of the tissue slicer and position the longitudinal hippocampal artery face-up, towards the tissue slicer frame (*see* Fig. 1c).

6. In advance, adjust the final height of the tissue slicer frame such that the threads will pass into the Bacto Agar gel block once released. This will ensure complete sectioning of the hippocampal tissue.
7. In some cases, slices will remain partially adherent to one another following tissue-slicer sectioning. These blocks of tissue will resemble a loaf of bread. Use two pairs of Dumont forceps with a 45° tip to gently separate the slices from one another (*see* Fig. 1d).
8. In each slice, the main hippocampal sub regions should be clearly identifiable (CA1, CA3, dentate gyrus) (*see* Fig. 1e). If this is not the case, there may have been a problem with the orientation in which the slices were cut (*see* **Note 5**).
9. Sterile, individually wrapped, plastic Pasteur pipettes are used to transfer slices. Tips are cut to accommodate the size of the slice, but not too large, as larger pipette bores make it difficult to manipulate single slices when transferring them onto membrane inserts. Furthermore, smaller pipette bores reduce the amount of dissection media that accompanies the slice on to the Biopore membrane (*see* Subheading 3.1, **step 19**).
10. To remove excess media from the slices once they are transferred to the Biopore membrane, use a 100 μL pipette. Slices suspended in a drop of media will tend to follow the direction of solution flow while aspirating. To prevent the slice from entering the pipette, make physical contact between the pipette tip and the base of the Biopore membrane, and maintain this contact while aspirating excess media. This will help draw the slice downward toward the membrane.
11. The culture media needs to be warmed and equilibrated to the conditions of the incubator prior to media exchange. On a "feeding" day, add fresh culture media (5 mL) to the empty well of the 6-well plate and return the plate to the incubator for at least an hour before transferring to the slices.
12. Store spermidine aliquots (100 μL, 0.05 M) at −20 °C for no longer than 2–3 months. If aliquots are used past their shelf life, the transfection efficiency of the biolistic procedure can drop significantly.
13. Once the DNA has been added to the gold–spermidine mixture, sonication should be avoided, as this can lead to sheering of DNA [22].
14. We prefer using DNA with a concentration above 1 μg/μL, although this is not a requirement. Effective co-transfection of plasmids can be problematic, and needs to be optimized for each set of DNA constructs. As a typical starting point, a total of 50 μg of DNA is used, of which 30–40 μg are of the DNA

for the protein of interest and 10–15 μg are of that coding a cytosolic morphological reporter (e.g., mCherry). When possible, expressed constructs should be tagged with a distinguishable fluorescent reporter, to ensure that co-transfection has been achieved. Although co-transfection rates are typically high, they are usually not 100 %.

15. In the manufacturer's protocol, the ethanol solution that is used to transfer the gold microcarriers to the Tefzel tubing includes a low concentration of polyvinylpyrrolidone (PVP), which helps adhere the gold microcarriers to the tubing. PVP is, however, not used in our protocol. In our experience, when PVP is used, even at half the recommended concentration, a significant amount of gold microcarriers remain adhered to the tubing cartridges after shooting. We find that excluding PVP enhances microcarrier delivery into tissue, which in turn provides higher transfection efficiency.
16. Gently flicking the tube should be sufficient to resuspend the pellet. If this does not work, it is likely that the gold pellet is too compact and the speed/duration of the centrifuge step should be reduced.
17. Here, timing is important. Removal of ethanol should be done quickly, but slow enough to prevent resuspending the gold microcarriers. The ethanol removal should take about 30 s, and if performed well, no gold should be visible in the ethanol that is drawn into the syringe. However, if this step is done too slowly, the tubing will dry and the gold will non-uniformly adhere to the inside wall of the tubing. Timely removal of ethanol from the tubing and immediate rotation of the tubing apparatus is the best way to achieve uniform coating of the microcarriers.
18. If uneven coating of microcarriers persists, try knocking the apparatus against the surface of the bench during the first few rotations, prior to resuming nitrogen flow. The impact will promote a more uniform distribution of gold particles around the inner surface of the tubing.
19. During the first week of culturing, the slices will flatten from a thickness of 400 μm to less than 200 μm [13]. During this time, healthy cells will persist while those damaged during the cutting process will gradually be sloughed away. Between 1 and 2 weeks after dissection, the slices tend to appear the healthiest and most appropriate for biolistic transfection. We obtain excellent transfection efficiency when slices are shot between 8 and 10 days in culture.
20. We have optimized transfection conditions by modifying the original gene gun barrel to include a nylon mesh at the tip (*see* Fig. 2a). The mesh has a 90 μm pore diameter and is glued

to the end of the barrel using cyanoacrylate glue. We found that inclusion of the mesh prevents tissue damage when shooting at high pressures (180–200 psi), since it shunts the blast of helium coming through the chamber. At these high pressures, we found enhanced penetration of the microcarriers into the tissue resulting in higher transfection efficiency. A separate barrel/nylon mesh is used for each set of DNA–gold microcarriers to avoid cross-contamination.

21. Depending on the constructs being expressed, the amount of time required for maximal protein expression will vary. For example, when protein expression is suppressed using shRNA constructs, more time is typically required to accommodate for turnover of existing proteins.

22. In the event that slices are obviously unhealthy early in the culture period (i.e., first 2–3 days) it is likely that there was a problem with the dissection itself, or with culture media formulation. If this is consistently a problem, inclusion of a broad glutamate receptor antagonist (e.g., 0.5–1 mM kynurenic acid) in the dissection solution may help.

23. If problems with tissue health arise only after transfection, the biolistic approach can be modified to reduce slice damage. It is possible that the pressure blast coming from the gene gun while shooting is physically damaging the tissue. One way to circumvent this is to reduce the pressure or move the gene gun further away from the sample. It is also possible that the density of microcarriers in the tissue it too high which can be detrimental to slice health. To reduce microcarrier density, either reduce the amount of gold used in the protocol, or increase the shooting distance from the slice.

24. Neurons in slice culture continue to develop largely as they would in situ (e.g., *see* Fig. 2c, and [3, 13–16]). Since organotypic slices are typically prepared from young neonatal animals during a period of rapid neuronal and circuit development, it may be difficult to compare results obtained from neurons at 1 and 3 weeks in culture. We suggest that a defined experimental time course should be established during experimental design to avoid potential complications associated with ongoing development of neurons in slice culture.

25. Movement of the electrode in the slice will inevitably displace tissue in the local environment, especially when positive pressure is being applied. By first obtaining a seal on the transfected neuron, the experimenter can then decide which control neuron is most accessible during the second approach.

26. Several methods have been described to reduce polysynaptic responses when evoking excitatory synaptic transmission in organotypic brain slices. Increasing the divalent ion concentration (e.g., to 4 mM Ca^{2+}, 4 mM Mg^{2+}) and/or including

the adenosine A1 receptor agonist 2-chloro adenosine in the extracellular recording solution has been used in several studies to minimize polysynaptic activity [7, 9]. In our hands, we find that inclusion of a low concentration of tetrodotoxin (10–30 nM), a Na^+ channel blocker, dampens network excitability while maintaining the ability to evoke synaptic transmission, thereby effectively minimizing polysynaptic responses (*see* Fig. 2e). For hippocampal slice cultures, we also routinely cut the CA1 flanking regions with a scalpel prior to mounting the slice on the microscope (i.e., at both the CA3–CA1 and CA1–subiculum interface).

References

1. Li CL, McIlwain H (1957) Maintenance of resting membrane potentials in slices of mammalian cerebral cortex and other tissues in vitro. J Physiol 139(2):178–190
2. Yamamoto C, McIlwain H (1966) Potentials evoked in vitro in preparations from the mammalian brain. Nature 210(5040):1055–1056
3. Gahwiler BH, Capogna M, Debanne D et al (1997) Organotypic slice cultures: a technique has come of age. Trends Neurosci 20(10): 471–477
4. Gahwiler BH (1981) Organotypic monolayer cultures of nervous tissue. J Neurosci Methods 4(4):329–342
5. Stoppini L, Buchs PA, Muller D (1991) A simple method for organotypic cultures of nervous tissue. J Neurosci Methods 37(2):173–182
6. Noraberg J (2004) Organotypic brain slice cultures: an efficient and reliable method for neurotoxicological screening and mechanistic studies. Altern Lab Anim 32(4):329–337
7. Schnell E, Sizemore M, Karimzadegan S et al (2002) Direct interactions between PSD-95 and stargazin control synaptic AMPA receptor number. Proc Natl Acad Sci U S A 99(21): 13902–13907
8. Beique JC, Andrade R (2003) PSD-95 regulates synaptic transmission and plasticity in rat cerebral cortex. J Physiol 546:859–867
9. Barria A, Malinow R (2005) NMDA receptor subunit composition controls synaptic plasticity by regulating binding to CaMKII. Neuron 48(2):289–301
10. Beique JC, Imad M, Mladenovic L et al (2007) Mechanism of the 5-hydroxytryptamine 2A receptor-mediated facilitation of synaptic activity in prefrontal cortex. Proc Natl Acad Sci U S A 104(23):9870–9875
11. Soares C, Lee KF, Nassrallah W et al (2013) Differential subcellular targeting of glutamate receptor subtypes during homeostatic synaptic plasticity. J Neurosci 33(33):13547–13559
12. Hayashi Y, Shi SH, Esteban JA et al (2000) Driving AMPA receptors into synapses by LTP and CaMKII: requirement for GluR1 and PDZ domain interaction. Science 287(5461): 2262–2267
13. Buchs PA, Stoppini L, Muller D (1993) Structural modifications associated with synaptic development in area CA1 of rat hippocampal organotypic cultures. Brain Res Dev Brain Res 71(1):81–91
14. Muller D, Buchs PA, Stoppini L (1993) Time course of synaptic development in hippocampal organotypic cultures. Brain Res Dev Brain Res 71(1):93–100
15. De Simoni A, Griesinger CB, Edwards FA (2003) Development of rat CA1 neurones in acute versus organotypic slices: role of experience in synaptic morphology and activity. J Physiol 550(Pt 1):135–147
16. Mellentin C, Møller M, Jahnsen H (2006) Properties of long-term synaptic plasticity and metaplasticity in organotypic slice cultures of rat hippocampus. Exp Brain Res 170(4):522–531
17. Johnston SA (1990) Biolistic transformation: microbes to mice. Nature 346(6286):776–777
18. McAllister AK (2000) Biolistic transfection of neurons. Sci STKE 2000(51):pl1
19. Woods G, Zito K (2008) Preparation of gene gun bullets and biolistic transfection of neurons in slice culture. J Vis Exp (12). pii: 675. doi:10.3791/675
20. Miller LD, Petrozzino JJ, Mahanty NK et al (1993) Optical imaging of cytosolic calcium, electrophysiology, and ultrastructure in pyramidal neurons of organotypic slice cultures from rat hippocampus. Neuroimage 1(2):109–120
21. Coyle P (1976) Vascular patterns of the rat hippocampal formation. Exp Neurol 52(3):447–458
22. O'Brien J, Lummis SC (2002) An improved method of preparing microcarriers for biolistic transfection. Brain Res Brain Res Protoc 10(1): 12–15

Chapter 14

Acute Brain Slice Methods for Adult and Aging Animals: Application of Targeted Patch Clamp Analysis and Optogenetics

Jonathan T. Ting, Tanya L. Daigle, Qian Chen, and Guoping Feng

Abstract

The development of the living acute brain slice preparation for analyzing synaptic function roughly a half century ago was a pivotal achievement that greatly influenced the landscape of modern neuroscience. Indeed, many neuroscientists regard brain slices as the gold-standard model system for detailed cellular, molecular, and circuitry level analysis and perturbation of neuronal function. A critical limitation of this model system is the difficulty in preparing slices from adult and aging animals, and over the past several decades few substantial methodological improvements have emerged to facilitate patch clamp analysis in the mature adult stage. In this chapter we describe a robust and practical protocol for preparing brain slices from mature adult mice that are suitable for patch clamp analysis. This method reduces swelling and damage in superficial layers of the slices and improves the success rate for targeted patch clamp recordings, including recordings from fluorescently labeled populations in slices derived from transgenic mice. This adult brain slice method is suitable for diverse experimental applications, including both monitoring and manipulating neuronal activity with genetically encoded calcium indicators and optogenetic actuators, respectively. We describe the application of this adult brain slice platform and associated methods for screening kinetic properties of Channelrhodopsin (ChR) variants expressed in genetically defined neuronal subtypes.

Key words Acute brain slice, Adult animals, Patch clamp recording, Protective recovery method, NMDG aCSF, Optogenetics, GCaMP, Channelrhodopsin

1 Introduction

For several decades the vast majority of brain slice physiologists have relied upon a "protective cutting" method for preparing healthy brain slices from juvenile and adolescent animals. This method is based on the premise that passive sodium influx and subsequent water entry and cell swelling during the slice cutting step is the major insult that leads to poor survival of neurons, particularly for those neurons located in the superficial layers that are most likely to sustain direct trauma from the blade movement.

Marzia Martina and Stefano Taverna (eds.), *Patch-Clamp Methods and Protocols*, Methods in Molecular Biology, vol. 1183, DOI 10.1007/978-1-4939-1096-0_14,

Thus, the implementation of a protective cutting solution having equimolar replacement of sodium chloride (NaCl) with sucrose (e.g., low sodium aCSF, sucrose-substituted aCSF, or simply sucrose aCSF) provides a notable improvement in neuronal preservation, especially for difficult to preserve brain areas such as brainstem and other highly myelinated regions [1].

The sucrose aCSF protective cutting solution has been widely adopted for preparing acute brain slices from virtually all brain regions following the initial description over 20 years ago. Variations of the protective cutting method have subsequently been described, including modified sucrose cutting aCSF formulations with optimized osmolarity [2], mixed NaCl/sucrose [3] or alternative sodium ion substitutes such as choline [4], NMDG [5], glycerol [6], or K-Gluconate [7]. While each particular study reports improved neuronal viability in juvenile or young adult brain slices using different protective cutting solutions in distinct brain regions, no clear consensus has emerged in the field to support a most effective formulation (*see* also ref. 8, 9), and the protective cutting method still leaves much to be desired for brain slice preparation from mature adult animals regardless of the particular aCSF formulation implemented.

Through systematic investigation of the parameter space for both aCSF formulations and procedures we were able to identify and optimize several key determinants for preparing healthy acute brain slices specifically from mature adult animals:

1. Sodium ion replacement during the initial phases of slice recovery, but not during the slicing procedure per se, is both necessary and sufficient to drastically curb the majority of neuronal swelling and subsequent pyknosis. Thus, we call this simple but highly effective procedural modification the "protective recovery" method in order to draw a clear distinction from the commonly employed methods based solely on implementation of "protective cutting" solutions.
2. The choice of sodium ion replacement applied to protective recovery is crucial. Among those cations we have tested *N*-methyl-D-glucamine (NMDG) is the most versatile substitute for a wide range of adult ages and applications. In more general terms, the use of methylated organic cations (e.g., NMDG, choline, Tris) as a replacement for sodium ions leads to greatly diminished permeability of sodium channel pores on neuronal membranes [10]. This property may account for the enhanced neuroprotective benefits of NMDG aCSF relative to sucrose aCSF in the preparation of adult brain slices.
3. Adult brain slices undergo edema during the course of incubation in normal aCSF and this is ameliorated by inclusion of 20 mM HEPES.

Each of these aspects contribute positive additive effects, and collectively the NMDG protective recovery method we describe here enables routine preparation of healthy mature adult brain slices across diverse cell types, brain regions, species, and animal ages. Earlier versions of this method have appeared in our previously published work [11–13].

2 Materials

2.1 Solutions

1. NMDG aCSF: 92 mM NMDG, 2.5 mM KCl, 1.25 mM NaH_2PO_4, 30 mM $NaHCO_3$, 20 mM HEPES, 25 mM glucose, 2 mM thiourea, 5 mM Na-ascorbate, 3 mM Na-pyruvate, 0.5 mM $CaCl_2{\cdot}4H_2O$, and 10 mM $MgSO_4{\cdot}7H_2O$. Titrate pH to 7.3–7.4 with concentrated hydrochloric acid (*see* **Note 1**).
2. HEPES holding aCSF: 92 mM NaCl, 2.5 mM KCl, 1.25 mM NaH_2PO_4, 30 mM $NaHCO_3$, 20 mM HEPES, 25 mM glucose, 2 mM thiourea, 5 mM Na-ascorbate, 3 mM Na-pyruvate, 2 mM $CaCl_2{\cdot}4H_2O$, and 2 mM $MgSO_4{\cdot}7H_2O$.
3. Recording aCSF: 119 mM NaCl, 2.5 mM KCl, 1.25 mM NaH_2PO_4, 24 mM $NaHCO_3$, 12.5 mM glucose, 2 mM $CaCl_2{\cdot}4H_2O$, and 2 mM $MgSO_4{\cdot}7H_2O$.
4. All solutions should be made using water of high purity, such as from a Milli-Q water system. Trace metals in distilled or tap water can lead to suboptimal slice quality through various pro-oxidative effects. All glassware used to contain aCSF should be thoroughly cleaned either by autoclave cycle or by rinsing with dilute (0.1 M) nitric acid followed by copious amounts of Milli-Q water.
5. All aCSF solutions must be saturated with carbogen (95 % O_2/5 % CO_2) prior to use to ensure stable pH buffering and adequate oxygenation. The pH should be adjusted to 7.3–7.4 and osmolarity measured and adjusted to 300–310 mOsm. It is recommended that solutions be made fresh on the day of the experiment (*see* **Note 2**).
6. Intracellular pipette solution: 145 mM K-Gluconate, 10 mM HEPES, 1 mM EGTA, 2 mM Mg-ATP, 0.3 mM Na_2-GTP, and 2 mM $MgCl_2$ (pH 7.3, 290–300 mOsm).
7. 1.25 % Avertin stock solution (mix 2.5 g of 2,2,2-Tribromoethanol with 5 mL of 2-methyl-2-butanol and then gradually dissolve in 200 mL phosphate buffered saline, pH 7.0–7.3). Avertin stock solution should be 0.22 μm sterile filtered and warmed to body temperature before use to minimize potential adverse effects on the animal. The use of isoflurane with a vaporizer breathing system is a suitable alternative to injectable anesthetics.

2.2 Equipment and Reagents

1. Tissue slicer machine: Various models are commercially available, many of which can provide excellent performance when optimally calibrated. We prefer the Compresstome VF-200 from Precisionary Instruments for adult brain slice applications. The machine uses agarose embedding of the tissue (*see* **Note 3**) and slight compression as the sample contacts the blade edge in order to create highly uniform slices with minimal surface chatter marks. These features produce slices with very clean surfaces, which is essential for preserving neurons in the superficial layers that are easily accessible for patch clamp recording (*see* **Note 4**).
2. Blades: Endurium ceramic injector style blades EF-INZ10 (Cadence/Specialty Blades). These highly durable blades outperform stainless steel and carbon steel feather blades and are best for cutting through heavily myelinated regions in adult brain. The performance of the ceramic blades is on par with the far more expensive and fragile sapphire blades.
3. Dissection tools: curved blunt forceps, fine dissecting "super cut" scissors (for cutting through skull), large heavy duty scissors (for decapitation), fine spatula, heavy duty spatula, scalpel handle and #10 blades, plastic transfer pipette (cut off end to produce a wide mouth), super glue, filter paper, glass petri dish (for dissection surface), razor blade.
4. Transcardial perfusion tools: a large dish filled with Sylgard for pinning anesthetized animal, dissecting pins, 30 mL syringe with 25 5/8 gauge needle.
5. pH meter: standard model—must be dedicated use with no contact with fixatives.
6. Vapor pressure osmometer. The osmometer should be calibrated very frequently and the thermocouple should be cleaned as needed for optimal performance.
7. Heater water bath with thermometer: 2.5 L or other shallow bath design, large enough bath dimensions for holding the slice incubation chamber. The temperature should be maintained at 32–34 °C.
8. Thermomixer with 24×1.5 mL thermoblock. The temperature should be set to 42 °C and mixing speed at 600 rpm to maintain the 1.5 % low melting point agarose in the molten state prior to use.
9. Multi-slice incubation chambers: Brain Slice Keeper-4 (BSK-4, AutoMate Scientific), or similar custom design. Slice chamber must have a minimally submerged netting for the slices to rest on, a fine gas diffuser stone for infusion of carbogen into the aCSF, and ideally some gentle constant flow to circulate solution around and/or through the slices.

10. Carbogen supply: either provided by refillable compressed gas cylinders or an in-wall line served by the building facilities, depending on existing infrastructure.
11. Electrophysiology rig suitable for patch clamp recording in acute brain slices. This will generally be composed of an upright microscope equipped with infrared differential interference contrast (IR-DIC) optics and a fluorescence illumination system, a patch clamp amplifier and data digitizer, motorized micromanipulators and microscope platform, vibration isolation table, Faraday cage, and solution heating and perfusion system. The sample chamber and platform should be designed for submerged slice recording.
12. Data acquisition software package for data analysis.
13. Laser scanning confocal microscope equipped with a 488 nm laser line for live calcium imaging experiments with GCaMP. The sample chamber must be adapted for continuous solution flow.
14. Blue laser for ChR2 photostimulation experiments with 200 μm core diameter optical fiber patch cord. Green laser for VChR1 photostimulation experiments. The lasers should have a digital or analog power controller box and option for TTL trigger. High power LED systems may also be suitable.
15. Boroscilicate glass capillaries type 8250 or KG-33 and pipette puller for fabricating microelectrodes.

2.3 Mice

1. Many unique lines of transgenic mice for fluorescently labeling diverse genetically defined neuronal subtypes are commercially available. Such lines exhibit stable and heritable expression of fluorophores in the nervous system and thereby greatly facilitate patch clamp recording from rare or broadly distributed cell populations. We have developed several transgenic mouse lines for optogenetics-based investigations (both manipulation and monitoring of neuronal activity), and these lines have been previously deposited to the Jackson Laboratory (JAX). Please *see* ref. 14 for a comprehensive review of these lines and their demonstrated utility in diverse experimental applications. The stock numbers are provided below for many of the lines we used in developing our brain slice methods (*see* also ref. 12, 13, 15–20):

   Thy1-GCaMP3 (JAX stock #017893)

   VGAT-ChR2(H134R)-EYFP (*JAX stock* #014548)

   ChAT-ChR2(H134R)-EYFP line 6 (*JAX stock* #014546)

   TPH2-ChR2(H134R)-EYFP (*JAX stock* #014555)

   Pvalb-ChR2(H134R)-EYFP (*JAX stock* # 012355)

   Thy1-ChR2(H134R)-EYFP line 18 (JAX stock #007612)

Thy1-vChR1-EYFP line 8(JAX stock #012348)
Thy1-eNpHR2.0-EYFP line 2 (JAX stock #012332)
R26-2xChETA (JAX stock #017455)
Ai32 (JAX stock #012569)

3 Methods

3.1 Brain Slice Procedure

1. Prepare for the slicing procedure by melting the required 1.5 % agarose block in the microwave (in appropriate container) and then aliquot the molten agarose into 1.5 mL tubes. Maintain the agarose in the molten state using a thermomixer set to 42 °C at 600 rmp. Care must be taken to ensure that the molten agarose does not solidify prematurely. Place the chilling block accessory device on ice to precool.
2. Glue a ceramic injector blade onto the blade holder and optimally align the blade for slicing. If the blade edge is not physically damaged, it can be reused for many weeks or even months without replacement.
3. Deeply anesthetize mature adult mice aged 2–8 months by intraperitoneal injection of Avertin (250 mg/kg: 0.2 mL of 1.25 % Avertin working stock solution per 10 g body weight) and then perform transcardial perfusion with 25–30 mL of room temperature carbogenated NMDG aCSF (*see* **Note 5**). If the perfusion is successful the liver will change in color from deep red to pale yellow, and in some cases clear fluids can be observed exiting the nostrils towards the end of the procedure. The transcardial perfusion is an important step when working with adult animals. For best results it is not advisable to omit transcardial perfusion.
4. Following perfusion the mice are decapitated, and the brains gently extracted from the skull within 1 min and placed into the cutting solution for an additional 30 s. Block the brain for the desired brain region and slicing angle (coronal for corticostriatal slices or transverse for hippocampal slices) and glue the tissue onto the mounting cylinder, which looks similar to a lipstick tube.
5. The tissue block is withdrawn into the lipstick tube and rapidly embedded by pouring a 1.5 mL tube of molten agarose directly over the tissue until completely submerged, taking care to avoid air bubbles. The agarose is rapidly solidified by clamping the tube with the cold chilling block for 5–10 s, and then the lipstick tube with embedded brain block is inserted into the receiver of the slicing platform of the Compresstome VF-200 (*see* Fig. 1a, b).

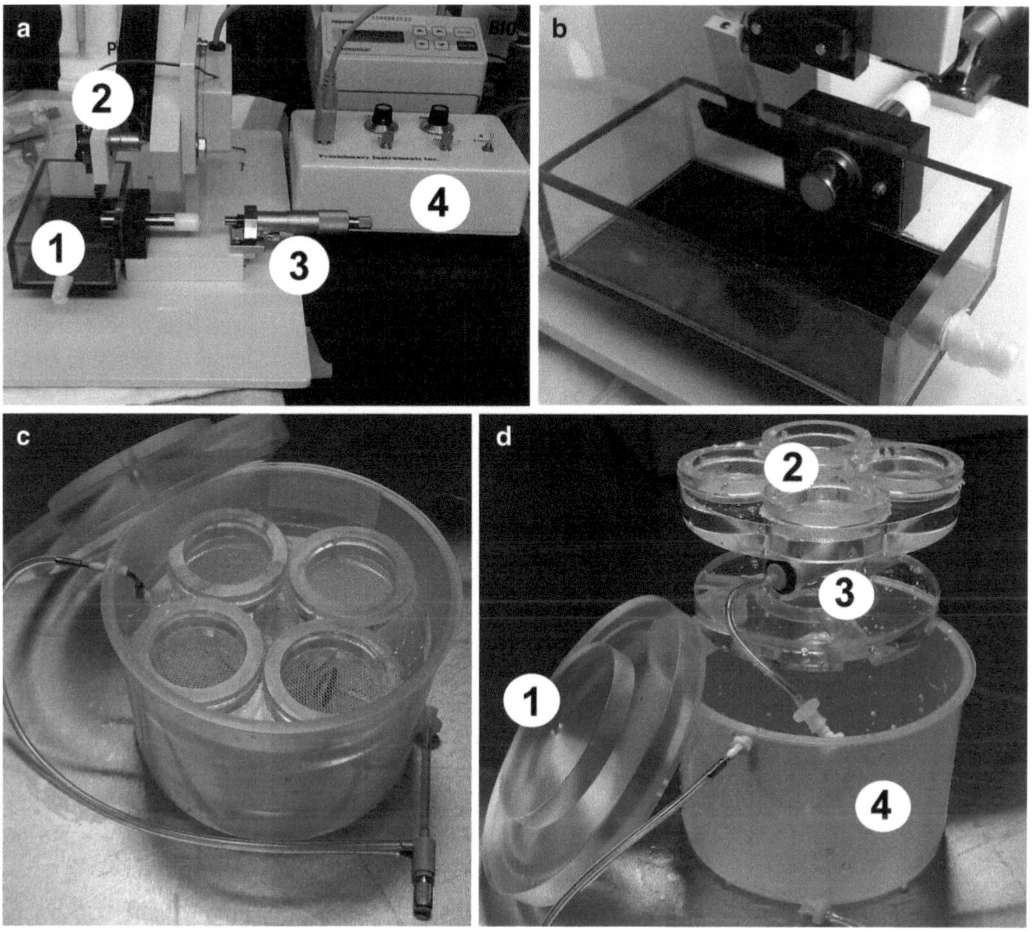

Fig. 1 Equipment for preparation of brain slices. (**a**) Compresstome VF-200 slicing machine with major components labeled as follows: 1—slicing chamber, 2—blade arm, 3—micrometer, 4—controller box. (**b**) Alternate view of the blade arm with lipstick inserted into the receiver of the slicing platform. (**c**) Brain Slice Keeper-4 apparatus. (**d**) Disassembled BSK-4 with major components labeled as follows: 1—lid, 2—four chambers with netting, 3—gas diffuser stone, 4—outer container

6. Fill the slicing chamber with NMDG aCSF solution and ensure continued carbogenation throughout the procedure. The micrometer should now be adjusted and locked in place. Manually advance the tissue for sectioning at 300 μm thickness. The optimal settings on each machine should be determined empirically. Set the advance to the slowest possible speed (~20 s per pass). Adjust the oscillation to a low-moderate setting where the blade is moving rapidly. The optimal blade movement produces a gentle humming sound but without any harsh buzzing noise. The total time for the slicing procedure should be less than 15 min.
7. Critical protective recovery step: transfer slices using a cut-off plastic Pasteur pipette into a pre-warmed BSK-4 (*see* Fig. 1c, d)

containing carbogenated NMDG aCSF. Allow the initial protective recovery to proceed for ≤12 min at 32–34 °C. Care must be taken to ensure that this step is precisely timed and that the temperature is properly maintained (*see* **Notes 6** and 7).

8. After the initial recovery period, transfer the slices into a new holding chamber (additional BSK-4 or similar design) containing room-temperature HEPES holding aCSF under constant carbogenation. Care should be taken to ensure minimal carry-over of the NMDG aCSF into the new holding chamber. At this point the slices can be stored for 1–5 h before transfer to the recording chamber for use. The presence of the 20 mM HEPES and ascorbate/thiourea combination reduces slice edema and slows deterioration.

3.2 Evaluation of Slice Quality: Morphological Integrity

1. The slices were transferred one at a time to the recording chamber of the microscope equipped with epifluorescence and IR-DIC optics (900 nm IR filter) aligned for Köhler illumination (*see* **Note 8**). The slices were constantly perfused with room-temperature (22–25 °C) carbogenated recording aCSF at a rate of 4 mL per minute.
2. A direct comparison of the traditional sucrose protective cutting method and the NMDG protective recovery method was performed using age-matched littermate mice at 5 months of age. Acute brain slices are prepared and IR-DIC images acquired at defined early time windows (0–5 min and 25–30 min) following the slicing procedure to evaluate the extent of morphological preservation. This approach revealed a dramatic improvement with the NMDG protective recovery method across brain regions, with much less initial neuronal swelling evident at 0–5 min and much reduced subsequent neuronal shrinkage and pyknosis at 25–30 min post slicing (*see* Fig. 2). Thus, the number of neurons that appear accessible for patch clamp recordings at later time points throughout the recording session is greatly increased, particularly for superficial layers of the slices (*see* **Note 9**).

3.3 Evaluation of Slice Quality: Functional Integrity

1. Additional testing was conducted to broadly assess functional responding of large populations of neurons in adult brain slices prepared with the protective recovery method. This was carried out using population calcium imaging and high-potassium depolarization in brain slices prepared from 5-month-old Thy1-GCaMP3 transgenic mice.
2. The slices were transferred one at a time to the sample chamber of the laser scanning confocal system. The slices were constantly perfused with room-temperature (22–25 °C) carbogenated recording aCSF at a rate of 4 mL per minute.

a

	'protective cutting' method	'protective recovery' method
Transcardial perfusion:	sucrose aCSF (0-2°C)	NMDG aCSF (RT)
Brain Slicing:	sucrose aCSF (0-2°C)	NMDG aCSF (RT)
Initial recovery:	standard aCSF (32-34°C) 30 min	NMDG aCSF (32-34°C) ≤12 min
Additional recovery:	standard aCSF (room temp) minimum 30 min	standard aCSF (room temp) minimum 60 min

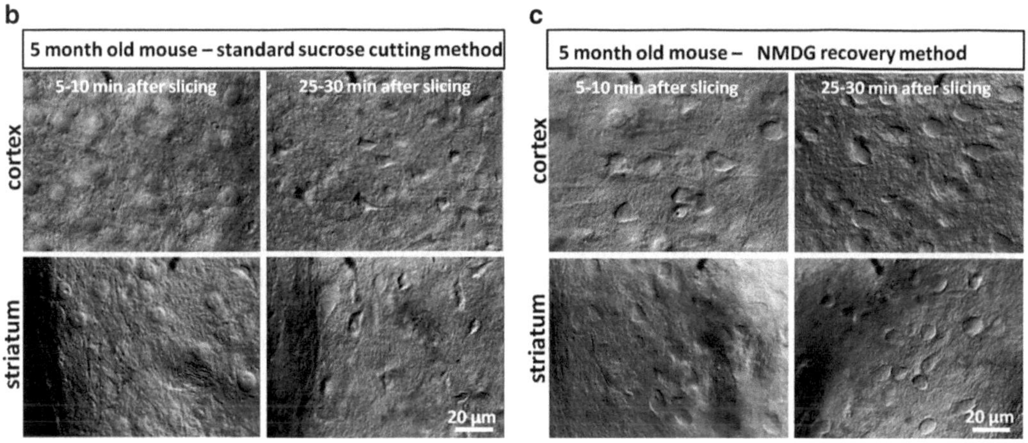

Fig. 2 The protective recovery method yields superior neuronal preservation for acute brain slice preparation from mature adult animals. (**a**) Comparison of procedural steps in the protective cutting versus protective recovery methods. (**b**) Rapid neuronal swelling and subsequent shriveling in acute brain slices prepared from 5-month-old adult mice with the sucrose aCSF protective cutting method. (**c**) Reduced swelling and improved neuronal preservation in acute brain slices prepared from 5-month-old adult littermate mice with the protective recovery method

3. GCaMP3 was excited with a 488 nm laser line at 5–10 % power and the green fluorescent signal was detected using a band-pass filter (505–525 nm). The selected region was imaged at three frames per second. We primarily focused on imaging calcium transients in hippocampal dentate gyrus granule cells because this population is uniformly labeled with high level GCaMP3 expression in the selected transgenic mouse line. We imaged a focal plane below the superficial layers (~20–50 µm deep) to avoid damaged neurons and autofluorescence from cellular debris.
4. To induce synchronous depolarization and neuronal firing we chose transient bath perfusion of high K^+ aCSF using a peristaltic pump for solution delivery (*see* **Note 10**). The time of stimulation was marked from the time that the high K^+ solution first entered the bath and was terminated after 60 s, followed by return to perfusion with normal aCSF.
5. Images were analyzed as previously reported [13]. For each image series we selected at least 20 regions of interest (ROI) for analysis. Each ROI corresponded to single granule cell bodies.

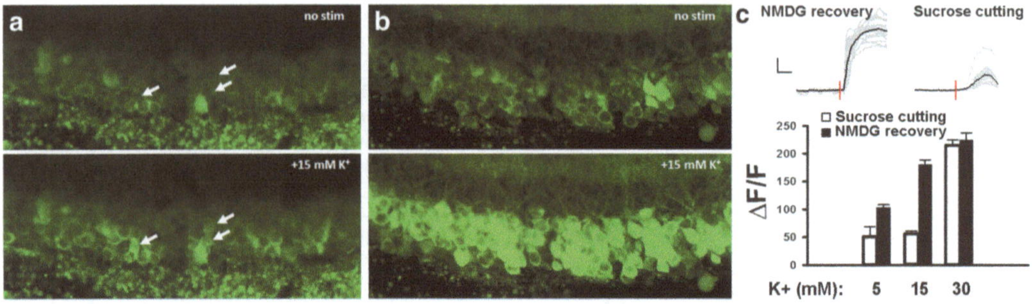

Fig. 3 GCaMP3 calcium imaging to assess functional integrity of mature adult brain slices. (**a**) Transient bath application of high K^+ solution (15 mM) evokes weak fluorescence increase in hippocampal dentate granule cells of brain slices prepared from a 5-month-old Thy1-GCaMP3 transgenic mouse using the standard sucrose aCSF protective cutting method. Arrows mark examples of weakly responding neurons. (**b**) The same high K^+ (15 mM) perfusion evokes a robust increase in fluorescence throughout the entire granule cell layer in brain slices prepared from an age matched littermate animal using the NMDG protective recovery method. (**c**) Example raw traces of fluorescence intensity measured over time for selected regions of interest (ROIs) demonstrating the effect of high K^+ bath perfusion. Scale bars: 50 %, 50 s. Summary data comparing ΔF/F for the two slice preparation methods over a range of extracellular K^+ concentrations

The fluorescence intensity increase upon high K^+ stimulation was calculated as peak fluorescence intensity divided by baseline fluorescence intensity (ΔF/F), corrected for background fluorescence.

6. The imaged neuronal populations in brain slices from the protective recovery group exhibited larger calcium transients at moderate stimulation levels (5 and 10 mM K^+ stimulation) and there were far more total numbers of functionally responding neurons with nearly every granule cell showing robust fluorescence increases (*see* Fig. 3). In addition, in several brain regions there were fewer neurons exhibiting elevated fluorescence in the basal condition, indicating fewer dead or damaged neurons as compared to slices prepared with the sucrose protective cutting method.

3.4 Application: Measuring Kinetic Properties of Channelrhodopsin Variants

1. The patch clamp electrophysiology rig was modified to enable focal delivery of laser light for precisely timed photostimulation of ChR-expressing neurons in adult brain slices. This was accomplished by adding a micromanipulator for positioning of a 200 μm core diameter optic fiber directly above patch clamped neurons in submerged slices (*see* Fig. 4). A glass capillary was mounted on the probe holder and the optic fiber was inserted through the glass capillary and secured in place with the flat cleaved end extending approximately 1 cm beyond the glass. The other end of the optic fiber was connected to a blue or green laser source under TTL control in the pClamp10 data acquisition software (*see* **Note 11**).

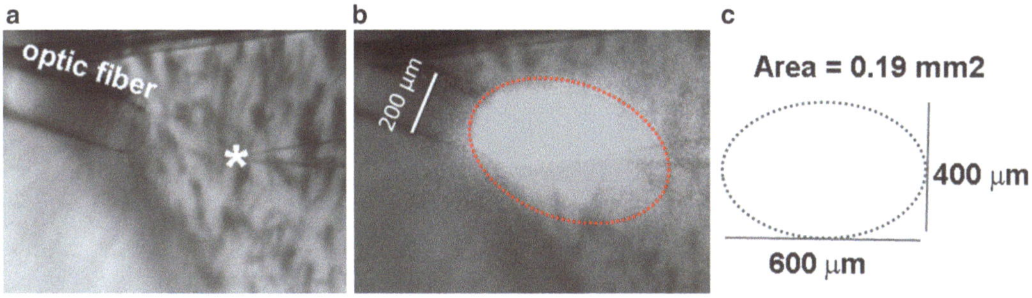

Fig. 4 Positioning of laser-coupled optical fibers for focalized laser stimulation in brain slices. (**a**) Example positioning of a 200 μm core optic fiber directly over a recorded striatal neuron (*asterisk*). (**b**) Each laser pulse illuminates an ellipse surrounding the target neuron. (**c**) Estimated area of illumination for determining power output as a function of area (irradiance)

2. A simple stimulation protocol was designed for triggering a brief 2-ms light pulse using the pClamp10 software. Each sweep was 1-s long and the stimulus was triggered at the 100 ms time point.
3. Adult brain slices were prepared using the NMDG recovery method and transgenic mouse lines with stable expression of ChR variants in cortical pyramidal neurons (Thy1-$ChR2_R$-EYFP line 18 and Thy1-VChR1-EYFP line 8).
4. Slices were transferred one at a time to the recording chamber of the patch clamp rig. Transgene-expressing neurons were targeted for whole-cell patch clamp recordings by combined IR-DIC imaging and fluorescence.
5. Recordings were obtained under identical conditions in order to standardize the light-evoked photocurrents mediated by transient activation of the ChR variants. The rate of channel closure at the termination of the light stimulus (Tau_{off}) was quantified using a mono-exponential fit of the decay of the photocurrent after light offset (*see* **Note 12**). To perform the decay fitting the data files were opened in Clampfit and cursor 1 aligned to the peak of the response. A second cursor was set to +300 ms after cursor 1 (to ensure complete decay of the response). Select analyze > fit > standard exponential ($n = 1$), then click OK. The tau value is now exported into the results sheet for the selected traces.
6. Tau_{off} was measured for cortical pyramidal neurons expressing $ChR2_R$ (mean = 26.3 ms) or VChR1 (mean = 93.6 ms) (*see* Fig. 5a).
7. A screen was conducted to determine Tau_{off} for cortical pyramidal neurons expressing the novel ChR variants $ChETA_{ARC}$ and $oChIEF_{AC}$. $ChETA_{ARC}$ denotes the ChR2 variant with E123A/H134R/T159C mutations, a novel fast kinetic variant

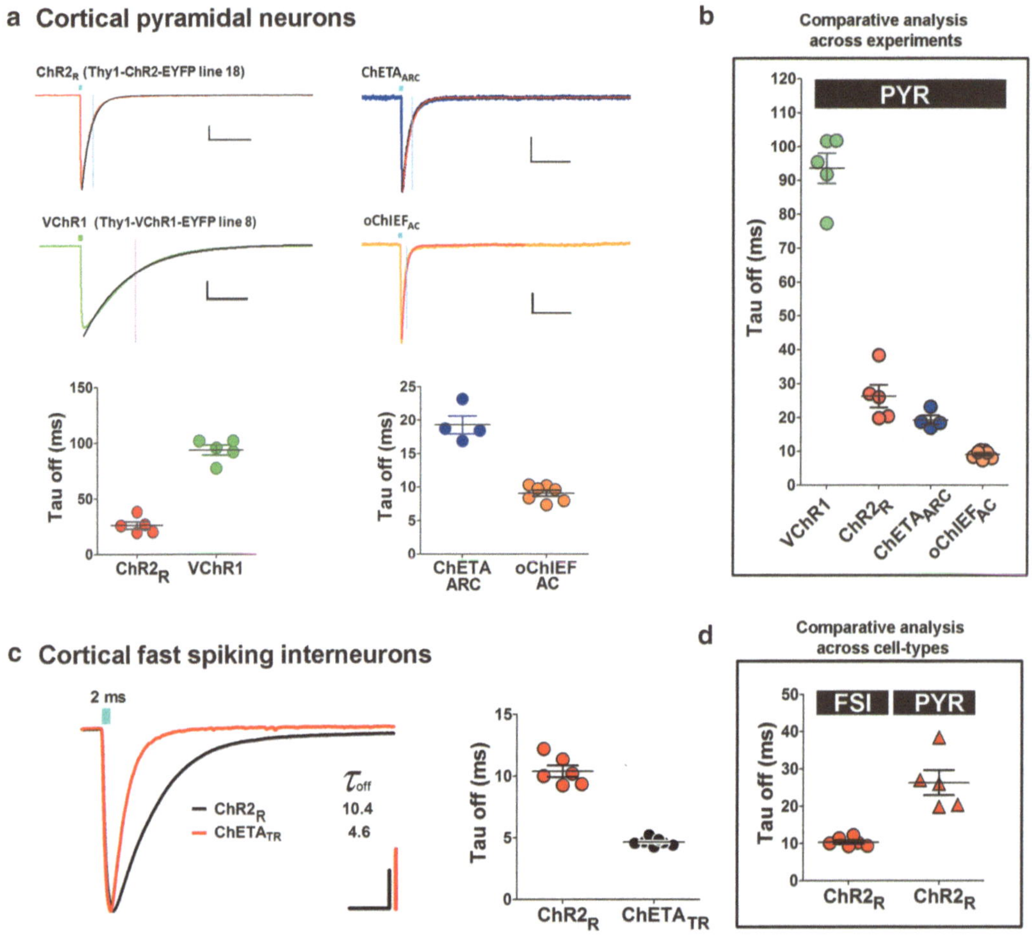

Fig. 5 Comparative analysis of ChR kinetic variants in distinct cell types. Measurement of the kinetics of channel closure (Tau_{off}) following a brief 2-ms light stimulation was performed using whole-cell voltage clamp (−70 mV) as a sensitive assay for screening novel ChR kinetic variants. (**a**) Experiment #1: comparison of Tau_{off} measured in cortical pyramidal neurons with transgenic expression of $ChR2_R$ or VChR1. Scale bars: 200 pA, 100 ms. Experiment #2: comparison of Tau_{off} measured in cortical pyramidal neurons expressing the novel variants $ChETA_{ARC}$ or $oChIEF_{AC}$. Scale bars: 50 pA, 100 ms. (**b**) Combined data for pyramidal neurons. (**c**) Comparison of Tau_{off} measured in cortical fast spiking interneurons with transgenic expression of $ChR2_R$ (VGAT-ChR2R-EYFP line 8) or $ChETA_{TR}$ (R26-2X$ChETA_{TR}$/Pvalb-IRES-Cre). Scale bars: 200 pA, 10 ms. (**d**) Summary data comparing Tau_{off} measured from either cortical fast-spiking interneurons or cortical pyramidal neurons both expressing $ChR2_R$. Intrinsic cell type differences influence measured kinetic properties and thus preclude comparative analysis across cell-types

designed for higher photocurrent amplitude than the previously reported ChETA variants. $oChIEF_{AC}$ denotes the oChIEF variant with the dual E163A/T199C mutations, a second novel fast kinetic variant designed for large photocurrents and fast kinetics. By comparing data sets across experiments but obtained from the identical cells type we found that Tau_{off} was reduced for both $ChETA_{ARC}$ (19.3 ms) and oChIEF$_{AC}$ (9.1 ms) relative to the ChR2$_R$ variant (26.3 ms) (*see* Fig. 5b). With a Tau$_{off}$ of nearly half that of ChETA$_{ARC}$, we propose that oChIEF$_{AC}$ is the most promising new fast kinetic variant for applications requiring ultrafast optogenetic stimulation. We also found that photocurrents measured from oChIEF$_{AC}$ expressing neurons were approximately five times larger than those measured from ChETA$_{ARC}$ expressing neurons.

8. Additional experiments were conducted to evaluate the effect of the ChETA$_{TR}$ fast kinetic variant compared to ChR2$_R$ in cortical fast-spiking interneurons. For these experiments the NMDG recovery method was used to prepare adult brain slices from transgenic mouse lines with stable expression of ChR variants in cortical interneurons (VGAT-ChR2$_R$-EYFP line 8 or R26-2xChETA$_{TR}$/Pvalb-IRES-Cre double transgenic mice). ChR2$_R$ denotes the ChR2 variant with the H134R mutation. ChETA$_{TR}$ denotes the ChR2 variant with dual E123T/H134R mutations.
9. Tau$_{off}$ was measured for cortical fast spiking interneurons expressing ChR2$_R$ (mean = 10.4 ms) or ChETA$_{TR}$ (mean = 4.6 ms) (*see* Fig. 5c). A direct comparison of Tau$_{off}$ values measured for ChR2$_R$-expressing cortical pyramidal neurons versus ChR2$_R$-expressing cortical fast spiking interneurons revealed a 2.5-fold difference, demonstrating that kinetic properties of ChR variants are highly dependent on cell type (*see* **Note 13**).

3.5 Improved Methods for Visually Guided Patch Clamp Recording of ChR2-Expressing Neurons in Slices

An important consideration for patch clamp analysis in adult brain slices is the fluorescent labeling of cellular populations under investigation. In many cases it is not practical to target defined neuronal subsets without the aid of a fluorescent label, particularly when the targeted population is rare, highly distributed, or intermingled with morphologically identical but functionally distinct subsets. In addition, for optogenetics-based investigations fluorescent reporters are important to identify neurons with functional expression of the optogenetic probe. Fortunately many transgenic animals and viral expression strategies have been developed to label diverse neuronal populations and allow access for patch clamp recordings and monitoring or manipulation of neuronal activity.

In the case of ChR expression strategies, the majority of the available viral vectors and transgenic lines were developed for expression of ChR variants as fusions with EYFP (ChR2-EYFP) with fewer studies using the ChR2-mCherry fusion design [21]. Due to the strong trafficking of ChR to the plasma membrane the physically linked fluorescent reporter is also confined to this domain and is largely excluded from the cell body. This scenario leads to difficulty in identifying transgene expressing cells for patch clamp recording in slices (or in vivo), especially when the labeled population is relatively dense. The resulting signal is often a dense, fluorescent neuropil with no clearly discernible cell bodies (*see* Fig. 6a, b).

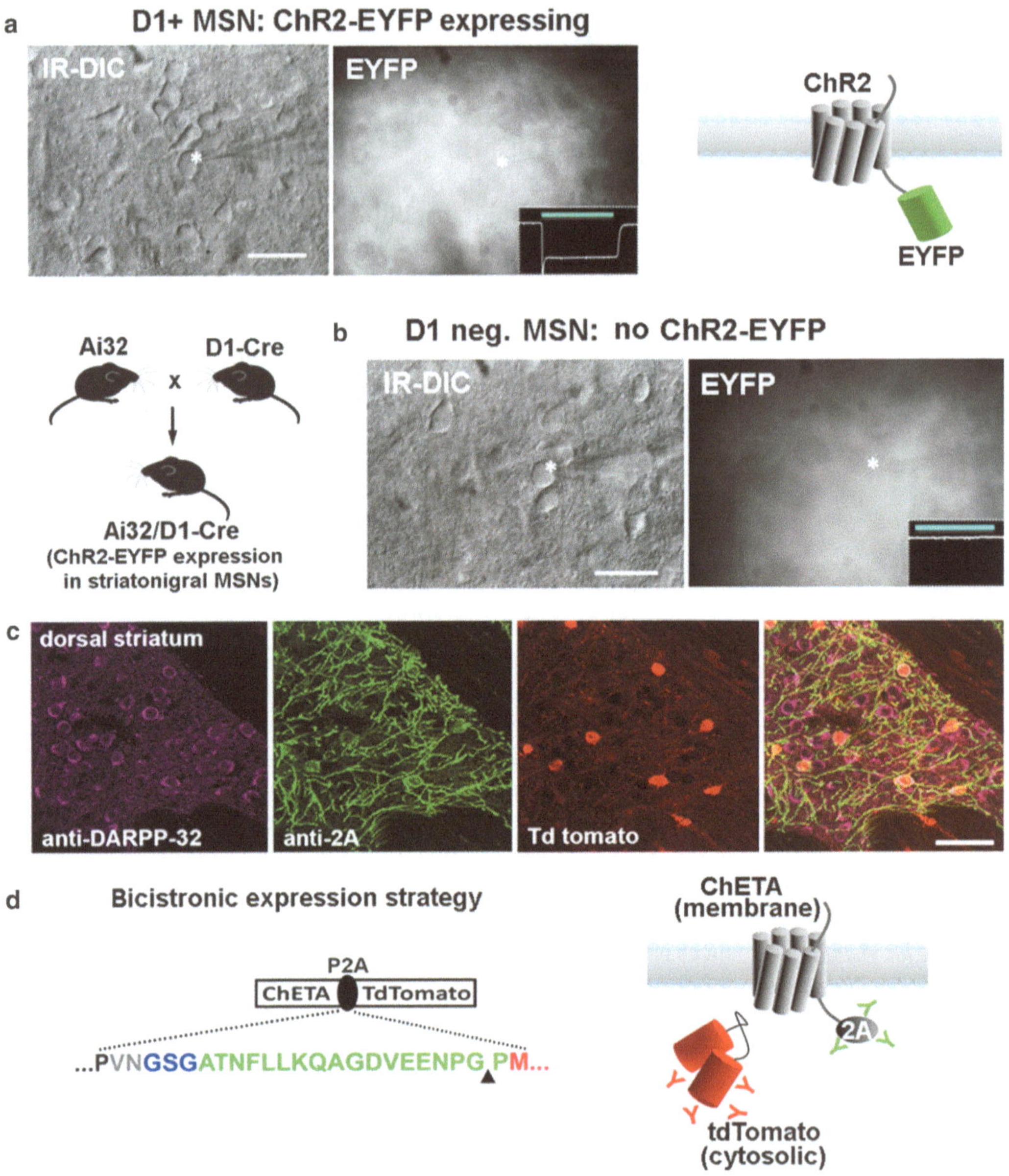

Fig. 6 Improved visualization of opsin-expressing neurons for targeted patch clamp recordings in brain slices using viral P2A linkers. (**a**) The challenge of identifying ChR2-EYFP expressing neurons is examined in acute striatal brain slices from Ai32/D1-Cre mice. This line has strong expression of the ChR2-EYFP transgene in roughly half of all striatal medium spiny neurons. The ChR2-EYFP gene fusion is localized to the cell membrane and produces a dense fluorescent neuropil with little signal from cell bodies. A recorded neuron is shown (*asterisk*) along with an inset of recorded photocurrent, thus confirming the identity as a D1+ MSN. (**b**) A recorded MSN in a nearby region had no photocurrent and was presumed D1 negative (D1−/D2+ MSNs account for the other half of the MSN population). The recorded neurons were indistinguishable on the basis of morphology or live EYFP fluorescence. (**c**) Analysis of native tdTomato fluorescence together with double immunostaining with anti-2A (indicating localization of ChR2) and DARPP32 (indicating all MSNs in the striatum region) demonstrates the unambiguous identification of ChR2 expressing neurons with the opsin-2A-XFP expression strategy. The example shown here is from our Cre-inducible ChETA-P2A-tdTomato reporter line crossed to the RGS9-Cre driver mice for labeling a subset of striatal MSNs. (**d**) The use of the viral P2A linker (*green sequence*) allows for physical uncoupling of opsin and fluorophore, and the 2A epitope tag can then be used to track the localization of the membrane-targeted opsin protein. The cytosolic fluorophore, in this case tdTomato, fills the entire cell body. Scale bars: 20 μm in panels **a** and **b**, 50 μm in panel **c**

A preferable expression strategy would be one in which physically separate but equal levels of opsin and fluorophore are produced in the labeled populations. This can be accomplished by incorporating viral 2A linkers, as was first demonstrated in the mouse acute slice preparation following in vivo injections of AAV-Synapsin-ChR2-2A-tDimer and AAV-Synapsin-NpHR-2A-Venus into the hippocampus [22]. Despite the demonstrated utility of this approach, the 2A expression strategy has been slow to be widely adopted, with only a few examples in recent studies [23–25]. We recently developed a transgenic mouse line for optogenetics-based research incorporating the 2A expression strategy for Cre-dependent expression of $ChETA_{TR}$ and a physically separate tdTomato fluorophore [19]. Using this line (when crossed to an appropriate Cre driver line) we were able to demonstrate improved visualization of ChR-expressing neurons by virtue of the cytosolic tdTomato fluorescence readily detected at the cell body (*see* Fig. 6c, d). In addition, because the 2A peptide linker segregates with the upstream encoded protein, it was still possible to visualize the membrane targeted ChR by immunostaining using a commercially available anti-2A antibody (*see* Fig. 6d). In essence, the 2A sequences are highly useful as epitope tags.

3.6 Avoiding Unintended Photoactivation During Cell Selection and Targeting

An additional important issue related to the widespread use of ChR-EYFP fusions is the unintended photoactivation of transgene expressing neurons while searching for neurons to record from in acute brain slices. This is due to the extensive overlap of the ChR activation spectrum with the excitation range for EYFP. Thus, visual identification of EYFP expression in a targeted neuron results in a transient but strong photoactivation of the targeted neuron and the surrounding transgene expressing neurons in the area of blue light illumination. This activation may induce alterations in cellular or synaptic function lasting variable durations, and perhaps in some cases leading to irreversible changes.

In order to circumvent this issue it is important to select judicious pairings of opsin and fluorophore. For example, our Cre-inducible $ChETA_{TR}$ knock-in mouse line allows for expression of $ChETA_{TR}$ and a physically separate tdTomato fluorophore [19]. By using a custom designed tdTomato filter set with only minimal overlap of the ChETA activation range we were able to titrate the excitation intensity to a point where visual identification of tdTomato positive neurons was easily accomplished without inducing neuronal firing. In contrast, a typical Texas red filter set was not suitable for this purpose since even low intensity light still induced robust neuronal firing (*see* Fig. 7). It is important to note that direct fusion of ChR2 to tdTomato was reported to impair ChR2 function to some extent in the Ai27 mouse line [20]. The 2A expression strategy overcomes this issue; however, it may still be advantageous to explore additional opsin–fluorophore combinations.

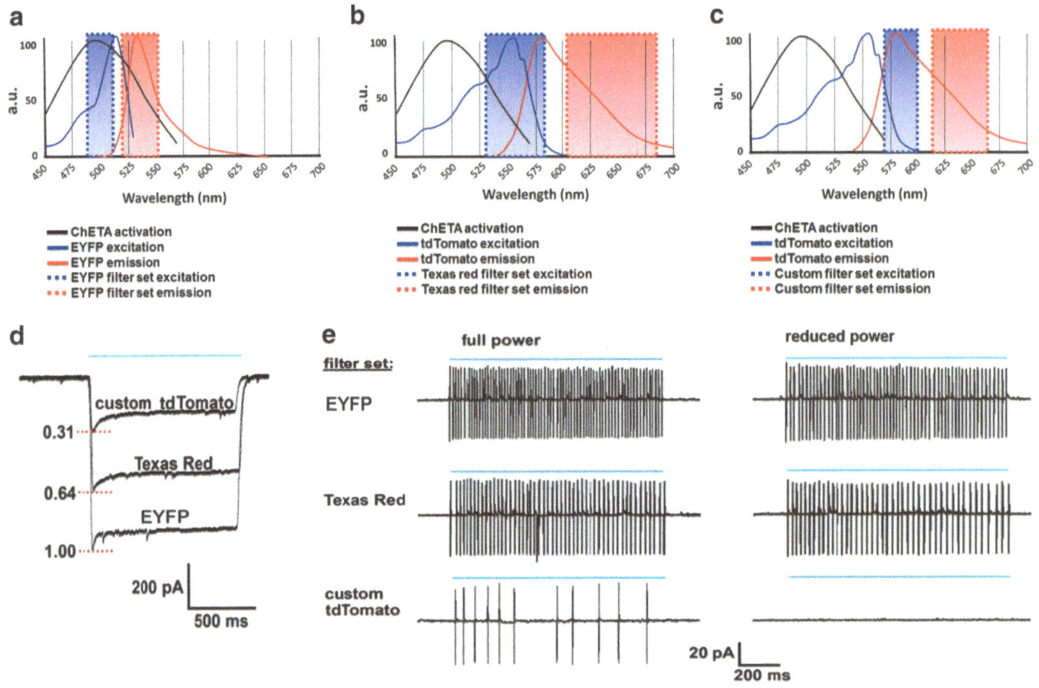

Fig. 7 Avoiding photoactivation while searching for ChR2-expressing cells. (**a–c**) The excitation/emission spectra of various optical filter sets are plotted together with the activation spectrum for ChETA. Considerable overlap with ChETA activation is observed for the excitation range using EYFP (**a**) and Texas red (**b**) filter sets but not with a custom tdTomato filter set (**c**). (**d**) Raw traces of maximal photocurrents evoked by 1-s epifluorescent illumination (blue line) using the various filter sets for a ChETA-P2A-tdTomato expressing neuron recorded in whole-cell voltage clamp. (**e**) Cell-attached recordings demonstrating light-evoked spiking with the various filter sets at full and reduced light intensity. Complete elimination of spiking was only achieved with the custom tdTomato filter set at reduced light intensity

Although tdTomato was selected in our work due to its bright fluorescence, we have found that the more red-shifted mKate2 is ideal for most applications in combination with ChR2 variants. With a suitable filter set the ChR2-P2A-mKate2 or ChR2-mKate2 pairing allows for virtual elimination of photocurrents during fluorophore visualization.

3.7 Strategies for Further Improving the Quality of Adult Brain Slices

The adult brain slice method we have described has been successfully implemented in a variety of experimental contexts for analysis of diverse brain regions and cell types. However, we would encourage adopters to view this method as a work in progress, and we believe there is still substantial room for systematic improvement. As a case in point, we have observed that mature adult brain slices experience high levels of oxidative stress due in large part to rapid depletion of cellular antioxidants including ascorbate and reduced glutathione (GSH). This can lead to lipid peroxidation, neuronal membrane rigidity, and tissue deterioration. There appears to be a non-uniform susceptibility to this form of oxidative damage, for example, CA1 and CA3 pyramidal neurons are particularly vulnerable, making patch clamp recording of these cells difficult in brain slices from adult and aging animals in spite of the protective recovery method.

The specific restoration of intracellular pools of neuronal GSH (e.g., supplementation with the cell-permeable GSH-ethyl ester) is highly effective at curbing deterioration and prolonging slice viability under these circumstances. Thus, we have been able to further improve the NMDG recovery method by devising strategies for stimulating de novo synthesis of glutathione during acute brain slice preparation and incubation. This is most readily accomplished by adding the inexpensive GSH precursor N-acetyl-L-cysteine (NAC, 5–12 mM) to the NMDG aCSF and HEPES holding aCSF formulas, but not the recording aCSF (*see* **Note 14**). NAC is cell-permeable and has been shown to specifically increase neuronal glutathione levels in brain tissue [26]. Within 1–2 h of slice preparation we are able to observe notable improvements in the general appearance of neurons and in the ease of patch clamp recording, and the slices are able to be maintained in a healthy state for extended time periods.

Although these more advanced methods are not absolutely required for successful adult brain slice patch clamp recordings (as demonstrated by the specific application we have described in this chapter), we include this information in hopes of providing more options to extend the versatility of our method for particularly challenging applications. Glutathione restoration is highly effective at maintaining healthy brain slices but may not be appropriate in every experimental context, e.g., investigations focusing on oxidative stress in the aging brain. On the other hand, without implementing the NMDG protective recovery method together with glutathione restoration strategy, targeted patch clamp analysis in brain slices from very old animals is prohibitively challenging.

4 Notes

1. Addition of the NMDG powder will make the solution very alkaline. When making NMDG aCSF it is important to carefully titrate the pH to 7.3–7.4 with concentrated hydrochloric acid. This step should ideally be performed prior to addition of divalent cations to avoid precipitation. In addition, the HEPES and thiourea plus ascorbate were included as critical components to reduce edema and oxidative damage during slicing, recovery and extended slice incubation [27, 28]. Due to the presence of the HEPES it is necessary to elevate $NaHCO_3$ in order to maintain the proper pH with constant carbogenation.

2. Solutions can be stored at 4 °C for 1 week without causing any detrimental impact on experiments. However, the pH and solution color should be carefully monitored to ensure stability. If Na-ascorbate is included the color will gradually change to a yellowish-orange hue over time, and this is a clear indication that the solution needs to be discarded. Many people prefer to use concentrated stock solutions for brain slice work. Given the difficulty in preserving healthy slices derived from adult animals and the numerous potential sources for failure, it is strongly recommended to prepare solutions fresh each day and avoid using stock solutions.
3. Freshly cut brain slices will detach from the surrounding agarose more easily when the low melt agarose type I-B is dissolved in phosphate buffered saline rather than water. Gently collecting and then expelling the slices with a wide mouth cut-off plastic Pasteur pipette is generally sufficient to dislodge the agarose. Repeat as necessary to free the slices.
4. The top of the line slicers will have a "zero-z" feature, meaning that the displacement of the cutting blade in the z-axis during blade vibration can be tuned to nominal zero. Some machines do not offer the ability to tune z-axis deflection, but nonetheless, empirically exhibit suitably low values. We have used the Leica VT1000, the Leica VT1200S with Vibrocheck, and the Precisionary Instruments Compresstome VF-200 all with excellent results. There are several features of the Compresstome VF-200 that make this slicer particularly well-suited for acute brain slice from adult and aging animals: (1) The speed of slicing is about two times faster than other models due to the stability provided by agarose embedding. (2) This machine also allows uniform thin sections (~100–150 μm) to be prepared from live tissue, which may be advantageous when examining densely myelinated regions where visualization is more difficult. (3) The Compresstome VF-200 has a very small footprint and can be easily moved or transferred between rooms or lab sites. (4) The cost of the Compresstome VF-200 is about half the cost of the top-of-the-line machines.
5. The temperature of the transcardial perfusion and slicing solution is not as critical when using NMDG aCSF in this protocol. The NMDG aCSF formulation is particularly effective at shutting down metabolic activity of the brain tissue. When using less protective formulations (e.g., sucrose aCSF or standard aCSF) it is much more crucial to chill the solutions to ~2–4 °C prior to use. A recent study demonstrated that warm slicing solution is advantageous for functional preservation of some cell types in brain slices derived from adult animals [29]. In our experience room temperature NMDG aCSF is highly effective

and obviates prolonged preparation times for cooling. If chilled solution is preferred, we advise cooling to 2–4 °C. Partially frozen slushy aCSF (as is commonly used for juvenile brain slice) was observed to be detrimental for adult brain tissue.

6. The exact duration of the recovery period was critical for obtaining the optimal balance between morphological and functional preservation of the brain slices, and the timing of this recovery step exhibited clear temperature dependence, as determined by extensive empirical testing. Proper implementation of this brief protective recovery step using our NMDG-based aCSF formula greatly reduced initial neuronal swelling during rewarming and enabled routine preparation of healthy acute brain slices for targeted whole-cell recordings from mature adult and aging mice.
7. A protective recovery step of ≤12 min is suitable for diverse brain regions and adult ages. However, the use of the NMDG protective recovery method is not optimized for animals younger than 5–6 weeks of age. We observe that it is difficult to completely wash out the NMDG aCSF in juvenile slices following the short protective recovery step, which often leads to difficulty in forming giga-ohm seals. Thus, the application of the NMDG protective recovery method for brain slices from animals younger than 5–6 weeks old requires either shorter recovery duration (e.g., ≤5 min) or using a 50/50 mix of NMDG aCSF to standard aCSF solution. The procedures we developed are intended to facilitate patch clamp analysis in brain slices from adult and aging animals and we are not advocating the application towards juvenile animals. The standard sucrose cutting method is perfectly suitable for work with juvenile animal brain slices.
8. In our experience it is highly desirable to have a 900 nm IR filter rather than the more common 775 nm IR filter. In adult brain the myelination is more developed than for juvenile brain, which poses a challenge for clear visualization of neurons located deep in adult brain slices. In some instances it is desirable to be able to see clear neuronal profiles at depths up to 50–100 μm into the slice. This is most easily accomplished with 900 nm IR-DIC optics. Proper alignment for Köhler illumination is also important for clear visualization. For a step-by-step guide for IR-DIC optimization please *see* ref. 30.
9. In our experience the morphological appearance of neuronal membranes under IR-DIC optics is well correlated with ease of patch clamping and physiological state of the neurons. This is in excellent agreement with other patch clamp studies in adult brain slices where the morphology of neurons was tracked by time lapse imaging prior to patch recordings [2].

With high-quality optics and repeated effort it becomes very easy to target healthy neurons based on visual criteria alone. However, additional objective criteria must still be applied to ensure adequate recovery of normal electrophysiological properties for each recorded neuron.

10. High K^+ bath stimulation was selected to evaluate time-locked functional responding of the largest population of neurons possible. For more refined analysis of functional responding we have also used bath application of low concentrations of glutamate receptor agonists such as NMDA and glutamate, which induce more stochastic patterns of firing with slower onset and lasting for sustained periods.
11. Other more elegant solutions for light delivery are possible such as directing the optic fiber through the microscope objective or through the patch electrode holder (OptoPatcher). It is also possible to achieve photostimulation of particular regions of single neurons using focal light spots or grids with a 2-photon confocal system. In our screening application a light delivery system with crude spatial resolution was both practical and adequate.
12. Previous work has established that fast kinetic properties (shorter duration Tau_{off} values) of ChR variants are correlated with smaller plateau potential size, reduced light sensitivity, and smaller photocurrent amplitude. This evidence supports a theoretical tradeoff between the precision of spiking and the overall efficacy of photoactivation [31]. Thus, this protocol we describe for measuring Tau_{off} can be quite robust for inferring several very important parameters of ChR function. For the sake of brevity we have chosen to present a simple and straightforward optogenetic stimulation protocol. More detailed protocols are required to thoroughly address distinct properties of novel variants [31].
13. Based on the 2.5-fold difference in Tau_{off} for $ChR2_R$ in cortical pyramidal neurons versus fast spiking interneurons we project a Tau_{off} of 7.7 ms for $ChETA_{ARC}$ and 3.6 ms for $oChIEF_{AC}$ in cortical fast spiking interneurons. This prediction awaits direct experimental validation.
14. Addition of 12 mM NAC to the NMDG aCSF and HEPES holding aCSF formulas will substantially reduce pH and increase osmolarity. Thus, it is important to readjust the pH to 7.3–7.4 and osmolarity to 300–310 mOsm. Alternatively, it is feasible to revise the aCSF formulas to account for NAC addition. We contend that adding NAC fresh immediately before use is preferable to ensure that the NAC is not degraded and retains full potency.

Acknowledgements

This work was supported in part by a National Alliance for Research on Schizophrenia and Depression: The Brain and Behavior Research Foundation Young Investigator Award to J.T.T., and US National Institutes of Health Ruth L. Kirschstein National Research Service Awards to J.T.T. (F32-MH084460).

References

1. Aghajanian GK, Rasmussen K (1989) Intracellular studies in the facial nucleus illustrating a simple new method for obtaining viable motoneurons in adult rat brain slices. Synapse 3:331–338. doi:10.1002/syn.890030406
2. Moyer JR Jr, Brown TH (1998) Methods for whole-cell recording from visually preselected neurons of perirhinal cortex in brain slices from young and aging rats. J Neurosci Methods 86:35–54
3. Bischofberger J, Engel D, Li L, Geiger JR, Jonas P (2006) Patch-clamp recording from mossy fiber terminals in hippocampal slices. Nat Protoc 1:2075–2081. doi:10.1038/nprot.2006.312
4. Mainen ZF, Maletic-Savatic M, Shi SH et al (1999) Two-photon imaging in living brain slices. Methods 18:231–239. doi:10.1006/meth.1999.0776
5. Tanaka Y, Furuta T, Yanagawa Y, Kaneko T (2008) The effects of cutting solutions on the viability of GABAergic interneurons in cerebral cortical slices of adult mice. J Neurosci Methods 171:118–125. doi:10.1016/j.jneumeth.2008.02.021
6. Ye JH, Zhang J, Xiao C, Kong JQ (2006) Patch-clamp studies in the CNS illustrate a simple new method for obtaining viable neurons in rat brain slices: glycerol replacement of NaCl protects CNS neurons. J Neurosci Methods 158:251–259. doi:10.1016/j.jneumeth.2006.06.006
7. Dugue GP, Dumoulin A, Triller A, Dieudonne S (2005) Target-dependent use of co-released inhibitory transmitters at central synapses. J Neurosci 2:6490–6498. doi:10.1523/JNEUROSCI.1500-05.2005
8. Aitken PG, Breese GR, Dudek FF et al (1995) Preparative methods for brain slices: a discussion. J Neurosci Methods 59:139–149
9. Lipton P, Aitken PG, Dudek FE et al (1995) Making the best of brain slices: comparing preparative methods. J Neurosci Methods 59: 151–156
10. Hille B (1971) The permeability of the sodium channel to organic cations in myelinated nerve. J Gen Physiol 58:599–619
11. Peca J, Feliciano C, Ting JT et al (2011) Shank3 mutant mice display autistic-like behaviours and striatal dysfunction. Nature 472:437–442. doi:10.1038/nature09965
12. Zhao S, Ting JT, Atallah HE et al (2011) Cell type-specific channelrhodopsin-2 transgenic mice for optogenetic dissection of neural circuitry function. Nat Methods 8:745–752
13. Chen Q, Cichon J, Wang W et al (2012) Imaging neural activity using Thy1-GCaMP transgenic mice. Neuron 76:297–308. doi:10.1016/j.neuron.2012.07.011
14. Ting JT, Feng G (2013) Development of transgenic animals for optogenetic manipulation of mammalian nervous system function: progress and prospects for behavioral neuroscience. Behav Brain Res 255:3–18. doi:10.1016/j.bbr.2013.02.037
15. Arenkiel BR, Peca J, Davison IG et al (2007) In vivo light-induced activation of neural circuitry in transgenic mice expressing channelrhodopsin-2. Neuron 54:205–218. doi:10.1016/j.neuron.2007.03.005
16. Asrican B, Augustine GJ, Berglund K et al (2013) Next-generation transgenic mice for optogenetic analysis of neural circuits. Front Neural Circuits 7:160. doi:10.3389/fncir.2013.00160
17. Ren J, Qin C, Hu F et al (2011) Habenula "cholinergic" neurons co-release glutamate and acetylcholine and activate postsynaptic neurons via distinct transmission modes. Neuron 69:445–452. doi:10.1016/j.neuron.2010.12.038
18. Wang H, Peca J, Matsuzaki M et al (2007) High-speed mapping of synaptic connectivity using photostimulation in Channelrhodopsin-2 transgenic mice. Proc Natl Acad Sci U S A 104:8143–8148. doi:10.1073/pnas.0700384104
19. Ting JT, Peca J, Daigle TL et al (2012) Ultrafast optogenetic control of diverse neuronal populations with cre-inducible ChETA knock-in mice. Soc Neurosci Abs 208:11
20. Madisen L, Mao T, Koch H et al (2012) A toolbox of Cre-dependent optogenetic transgenic mice for light-induced activation

and silencing. Nat Neurosci 15:793–802. doi:10.1038/nn.3078

21. Yizhar O, Fenno LE, Davidson TJ et al (2011) Optogenetics in neural systems. Neuron 71:9–34. doi:10.1016/j.neuron.2011.06.004
22. Tang W, Ehrlich I, Wolff SB et al (2009) Faithful expression of multiple proteins via 2A-peptide self-processing: a versatile and reliable method for manipulating brain circuits. J Neurosci 29:8621–8629. doi:10.1523/JNEUROSCI.0359-09.2009
23. Prakash R, Yizhar O, Grewe B et al (2012) Two-photon optogenetic toolbox for fast inhibition, excitation and bistable modulation. Nat Methods 9:1171–1179. doi:10.1038/nmeth.2215
24. Yonehara K, Balint K, Noda M et al (2011) Spatially asymmetric reorganization of inhibition establishes a motion-sensitive circuit. Nature 469:407–410. doi:10.1038/nature09711
25. Li Y, Tsien RW (2012) pHTomato, a red, genetically encoded indicator that enables multiplex interrogation of synaptic activity. Nat Neurosci 15:1047–1053. doi:10.1038/nn.3126
26. Aoyama K, Suh SW, Hamby AM et al (2006) Neuronal glutathione deficiency and age-dependent neurodegeneration in the EAAC1 deficient mouse. Nat Neurosci 9:119–126. doi:10.1038/nn1609
27. MacGregor DG, Chesler M, Rice ME (2001) HEPES prevents edema in rat brain slices. Neurosci Lett 303:141–144
28. Brahma B, Forman RE, Stewart EE et al (2000) Ascorbate inhibits edema in brain slices. J Neurochem 74:1263–1270
29. Huang S, Uusisaari MY (2013) Physiological temperature during brain slicing enhances the quality of acute slice preparations. Front Cell Neurosci 7:48. doi:10.3389/fncel.2013.00048
30. Davie JT, Kole MH, Letzkus JJ et al (2006) Dendritic patch-clamp recording. Nat Protoc 1:1235–1247. doi:10.1038/nprot.2006.164
31. Mattis J, Tye KM, Ferenczi EA et al (2012) Principles for applying optogenetic tools derived from direct comparative analysis of microbial opsins. Nat Methods 9:159–172. doi:10.1038/nmeth.1808

Chapter 15

Synaptic Connectivity in Engineered Neuronal Networks

Peter Molnar, Jung-Fong Kang, Neelima Bhargava, Mainak Das, and James J. Hickman

Abstract

We have developed a method to organize cells in dissociated cultures using engineered chemical clues on a culture surface and determined their connectivity patterns. Although almost all elements of the synaptic transmission machinery can be studied separately in single cell models in dissociated cultures, the complex physiological interactions between these elements are usually lost. Thus, factors affecting synaptic transmission are generally studied in organotypic cultures, brain slices, or in vivo where the cellular architecture generally remains intact. However, by utilizing engineered neuronal networks complex phenomenon such as synaptic transmission or synaptic plasticity can be studied in a simple, functional, cell culture-based system. We have utilized self-assembled monolayers and photolithography to create the surface templates. Embryonic hippocampal cells, plated on the resultant patterns in serum-free medium, followed the surface clues and formed the engineered neuronal networks. Basic whole-cell patch-clamp electrophysiology was applied to characterize the synaptic connectivity in these engineered two-cell networks. The same technology has been used to pattern other cell types such as cardiomyocytes or skeletal muscle fibers.

Key words Engineered networks, Hippocampal cultures, Synaptic connectivity, EPSC, Action potential, SAM, Photolithography, Serum free

1 Introduction

The behavior of the cells (attachment, proliferation, differentiation, and function) is determined by their internal programs and by extracellular clues. The origin of these extracellular signals can be the culture medium, other cells, or the culture surface. Recent developments in cell biology and fabrication technology have enabled the engineering of active surfaces to present these signaling molecules to the cells in a controlled manner [1–6]. Using functionalized self-assembled monolayers in combination with advanced surface patterning methods, the inherent differentiation and self-organizing programs utilized by the neurons can be controlled and guided to form directed networks [7–10]. These directed networks can be used to study synaptic transmission and synaptic plasticity in a controlled and systematic manner [11].

Marzia Martina and Stefano Taverna (eds.), *Patch-Clamp Methods and Protocols*, Methods in Molecular Biology, vol. 1183,
DOI 10.1007/978-1-4939-1096-0_15,

Synaptic transmission is a complex process, even between similar cell types [12, 13], regulated by several presynaptic and postsynaptic mechanisms [14]. It plays an important role in memory formation and several neurodegenerative diseases [15–18]. An in vitro functional system which would enable high-throughput testing of drugs on synaptic transmission and plasticity in a controlled environment would be high significance for drug development and physiology.

2 Materials

2.1 Surface Modification

1. Glass coverslips (22 × 22 mm).
2. Plasma cleaner.
3. Trimethoxysilylpropyl diethylenetriamine.
4. Toluene.
5. Molecular sieves, 4A.
6. Goniometer.
7. X-ray photoelectron spectroscope equipped with Al Kα X-ray source.

2.2 Photolithography

1. Layout editor.
2. Chrome/Quartz mask.
3. Ar/F laser combined with a beam homogenizer.
4. Tridecafluoro-1,1,2,2-tetrahydrooctyl-1-trichlorosilane (DETA).
5. Chloroform.
6. $PdCl_4$ solution: 10 mg $PdCl_4$ and 1.75 g NaCl/50 ml of water pH adjusted to 1 with concentrated HCl.
7. Dimethylamine borane (DMAB) solution: 1.7 g DMAB/50 ml of water.
8. Copper bath solution: 3 g copper sulfate, 14 g sodium potassium tartrate, 2 g of NaOH, and 1 ml of formaldehyde (37.2 %) in 100 ml distilled water.

2.3 Embryonic Hippocampal Cultures

1. Dissecting medium: 500 ml Hibernate E, 10 ml B27 supplement, 5 ml GlutaMAX, 5 ml Antibiotic/antimycotic.
2. Culture medium 2: 500 ml Neurobasal E, 10 ml B27, 5 ml GlutaMAX, 5 ml Antibiotic/Antimycotic.
3. Culture medium 1: Culture medium 2 plus 25 μM Glutamate.
4. Stereo microscope.
5. Laminar flow hood.
6. CO_2 incubator.
7. Hemocytometer.

8. Inverted cell culture microscope.
9. Trypan blue.
10. Hemocytometer.

2.4 Dual Patch-Clamp Recordings

1. Upright microscope.
2. Two micromanipulators.
3. Microscope platform.
4. Vibration isolation table.
5. Patch-clamp amplifier with two channels.
6. A/D converter.
7. Data acquisition software.
8. Pipette puller.
9. Borosilicate glass pipettes.
10. pH meter, balance, osmometer.
11. Extracellular solution: Neurobasal E medium, pH is adjusted to 7.3 with HEPES.
12. Intracellular solution: 140 mM K-gluconate, 1 mM EGTA, 2 mM $MgCl_2$, 2 mM Na_2ATP, 10 mM Hepes; pH = 7.2; 276 mOsm. In some experiments 1 % Alexa Fluor 488 dye was added.

3 Methods

3.1 Surface Modification

The glass coverslip surface is modified with DETA self-assembled monolayers [19].

1. Glass coverslips are cleaned using O_2 plasma cleaner for 30 min at 400 mTorr (*see* **Note 1**).
2. The self-assembled monolayer (SAM) is formed by the reaction of the cleaned surface with a 0.1 % (v/v) mixture of the organosilane (DETA) in freshly distilled toluene (*see* **Note 2**).
3. The coverslips are heated with the mixture to just below the boiling point of the toluene, rinsed with toluene, reheated to just below the boiling temperature, and then oven dried.
4. Surfaces are characterized by contact angle measurements using an optical contact angle goniometer and by X-ray photoelectron spectroscopy by monitoring the N 1 s peak (*see* **Note 3**) [20].
5. For contact angle measurements a static, sessile drop (5 μl) of deionized water is applied to the surface. Three measurements are taken and averaged. The stable contact angles (40.64 ± 2.9/ mean ± SD/) throughout the study period indicate a high reproducibility and quality of the DETA coatings.

6. For the XPS analysis the pass energy of the analyzer and the takeoff angle of photoelectron are set at 40 eV and 90°. Survey scans as well as high-resolution scans for fluorine (1 s), oxygen (1 s), nitrogen (1 s), carbon (1 s) and silicon (2p) are performed on each sample. Based on the high-resolution scans, elemental composition percentages are calculated for each element [19].

3.2 Photolithography (See Note 4)

DETA patterns are created on the surface by ablation using a 193 nm excimer laser and a quartz photomask. The ablated region is then backfilled with a second self-assembled monolayer (13F) [9]. The patterns (amino groups on the surface) are visualized by electroless copper deposition.

1. Masks are designed using the Clewin program (*see* **Note 5**) and written by a commercial vendor (Bandwidth Foundry).
2. DETA is ablated by exposure to UV light through the photomask (193 nm, 50 mJ/mm^2) (*see* **Note 6**, Fig. 1).
3. Patterns are backfilled by dipping the ablated coverslips in a 13F solution in chloroform (0.1 % v/v) (*see* **Note 7**). 13F is not permissive for cellular growth.
4. For the copper deposition, the coverslips are immersed in the $PdCl_4^{-2}$ solution for 15 min. They are rinsed thoroughly with water and immersed for 15 min in DMAB solution. The samples are again rinsed thoroughly with water and then immersed in the copper bath solution for 5 min (*see* **Note 8**).

3.3 Embryonic Hippocampal Cultures

The hippocampal cultures are prepared from E17 rat embryos.

1. Rats are euthanized in a pre-charged CO_2 chamber.
2. The hippocampal tissue is dissected from the embryos and placed in ice-cold dissociating medium.
3. The tissue is triturated using a sterile Pasteur pipette (*see* **Note 9**).
4. The dissociated tissue is centrifuged at 300 × *g* for 2 min at 4 °C.
5. The pellet is resuspended in 2 ml of Culture medium 1.
6. Cells are counted in a hemocytometer and plated on the patterns at a density of 100 cells/mm^2.
7. Half of the medium is changed every 4 days using Culture medium 2.

3.4 Dual Patch-Clamp Recordings

Dual patch-clamp recordings are performed on the neurons plated on the patterns. Synaptic connectivity is verified by evoking action potentials in one of the cells in current-clamp mode, whereas recording of the Excitatory Postsynaptic Currents (EPSCs) was done in voltage-clamp mode (Fig. 2).

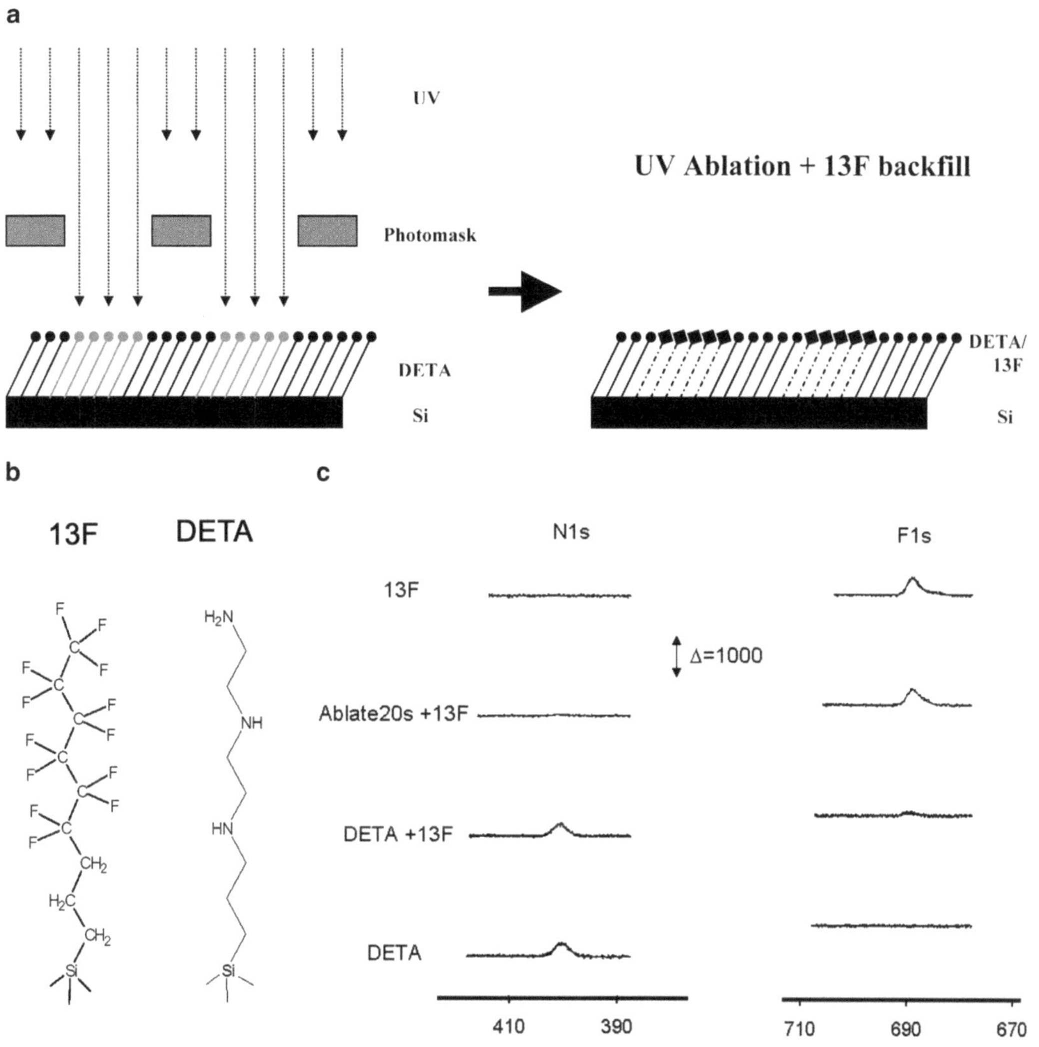

Fig. 1 Photolithographic patterning of SAMs. (**a**) DETA coverslips ablated through a photomask followed by backfilling the ablated areas by 13 F. (**b**) Chemical structure of DETA and 13 F. (**c**) High resolution XPS spectra of the surfaces at different stages of the patterning process. The area under the peaks should be proportional of the amount of the given element on the surface. DETA coverslips (N 1 s peak, no F 1 s peak) are ablated and backfilled with 13 F. Ablate 20s +13 F shows that the N 1 s peak disappeared and is replaced by the F 1 s peak of 13 F. DETA + 13 F shows some incorporation of 13 F into the DETA monolayer. *X* axis: binding energy, *Y* axis: arbitrary intensity unit. Reprinted from [23] with permission

1. On the 14th day of the culture, the patterned coverslips are transferred to a recording chamber on the stage of an upright microscope. Experiments are performed at room temperature.
2. Electrodes are pulled from borosilicate glass and filled with the intracellular solution. The resistance of the electrodes should be 6–10 MOhm.

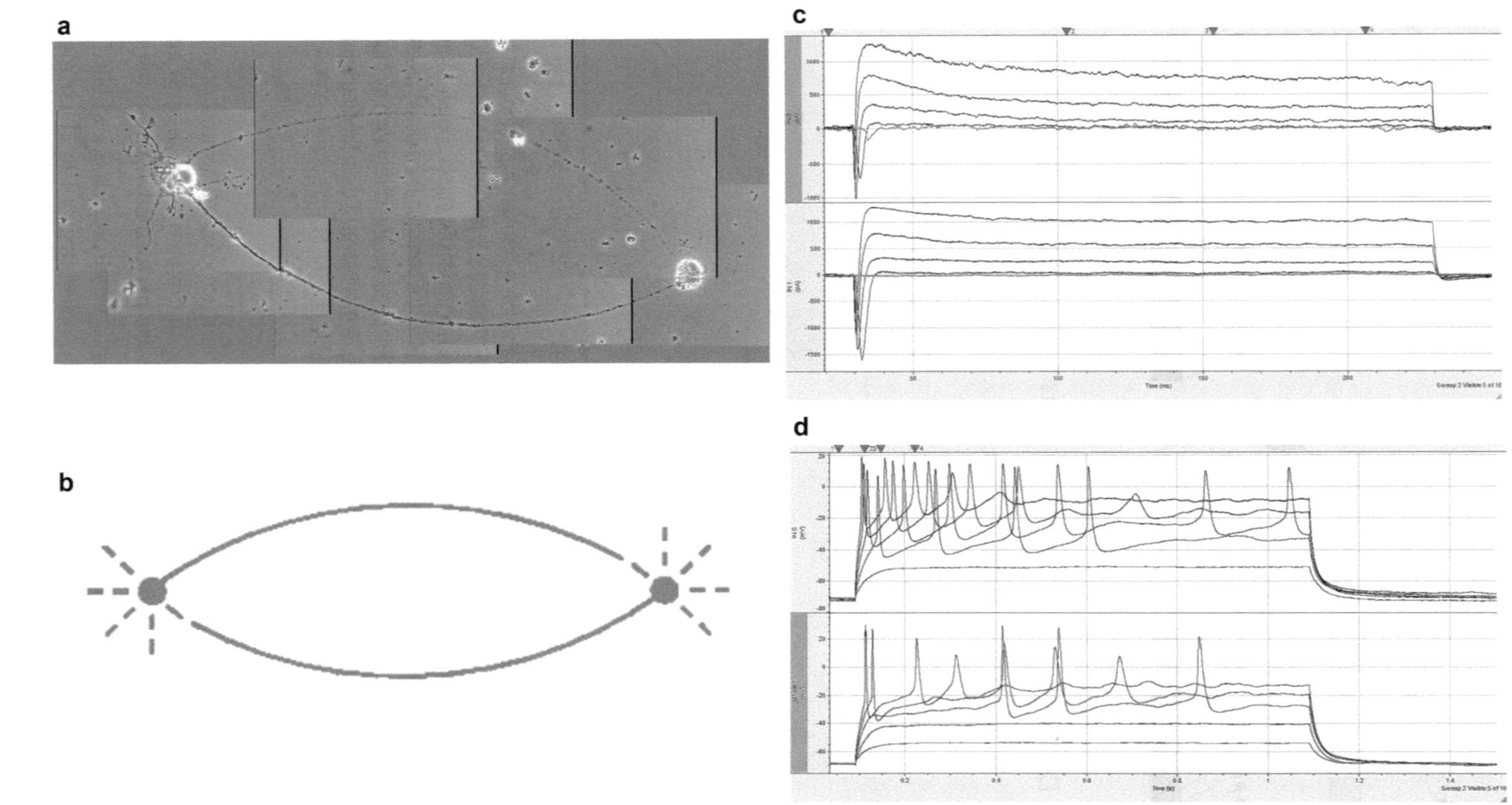

Fig. 2 Functional engineered networks. (**a**) Embryonic hippocampal cells complied with the surface patterns. (**b**) Mask design for two-cell networks. The diameter of the soma-attachment area is 25 μm. (**c**) Ionic currents recorded from the patterned cells in voltage-clamp mode. (**d**) Action potentials were evoked by current injections. Reprinted from [23] with permission

3. Dual current-clamp voltage-clamp experiments are performed on two apparently connected neurons on the patterns.
4. Two filled electrodes are placed into the two head stages; positive pressure (about 2 cc from a 10 cc syringe) is applied to both pipettes before touching the extracellular solution; the electrodes are brought close to the target cells; pipette offset are compensated at both channels; cells are touched with the tip of the electrodes (seal position) under visual control; −5 mV seal test is applied; pipette capacitances are compensated; −70 mV holding is applied on both channels (voltage-clamp mode); gigaseal is formed; the cell membrane is ruptured by the application of short suction pulses; resting membrane potential is measured in $I=0$ mode; whole-cell capacitance and resistance are compensated.
5. Voltage step protocols are applied in parallel to both cells. Signals are filtered at 2 kHz and digitized at 20 kHz. Passive membrane properties (membrane resistance, capacitance) and ionic currents (voltage-dependent sodium, potassium) are measured.
6. The action potential threshold is measured in current-clamp mode with increasing current injections.
7. Synaptic transmission is measured by evoking action potentials in one cell and recording EPSCs from the other cell in voltage-clamp mode. EPSCs are recorded at −70, −30, 0, and 20 mV holding potential to discriminate between glutamatergic and GABAergic synapses. The reversal potential for chloride currents is about −50 mV in our experimental conditions, whereas reversal of AMPA receptor mediated currents is about 0 mV (Fig. 3).
8. The cells could be filled with a fluorescent dye through the patch pipette for consequent visualization with a confocal microscope.

4 Notes

1. Instead of plasma cleaning of the coverslips, acid cleaning (30 min. in cc H_2SO_4) could also be used.
2. It is very important (and difficult) to control the amount of water in the solvent. The water content of the toluene and the surface determine the speed of the reaction. At high water content the organosilane can polymerize. We distilled the toluene and used molecular sieves to control the water content.
3. In our experience DETA forms a monolayer on the surface when the ratio of the areas below the N 1 s and Si 2p peaks is about $1{,}500 \pm 500$.

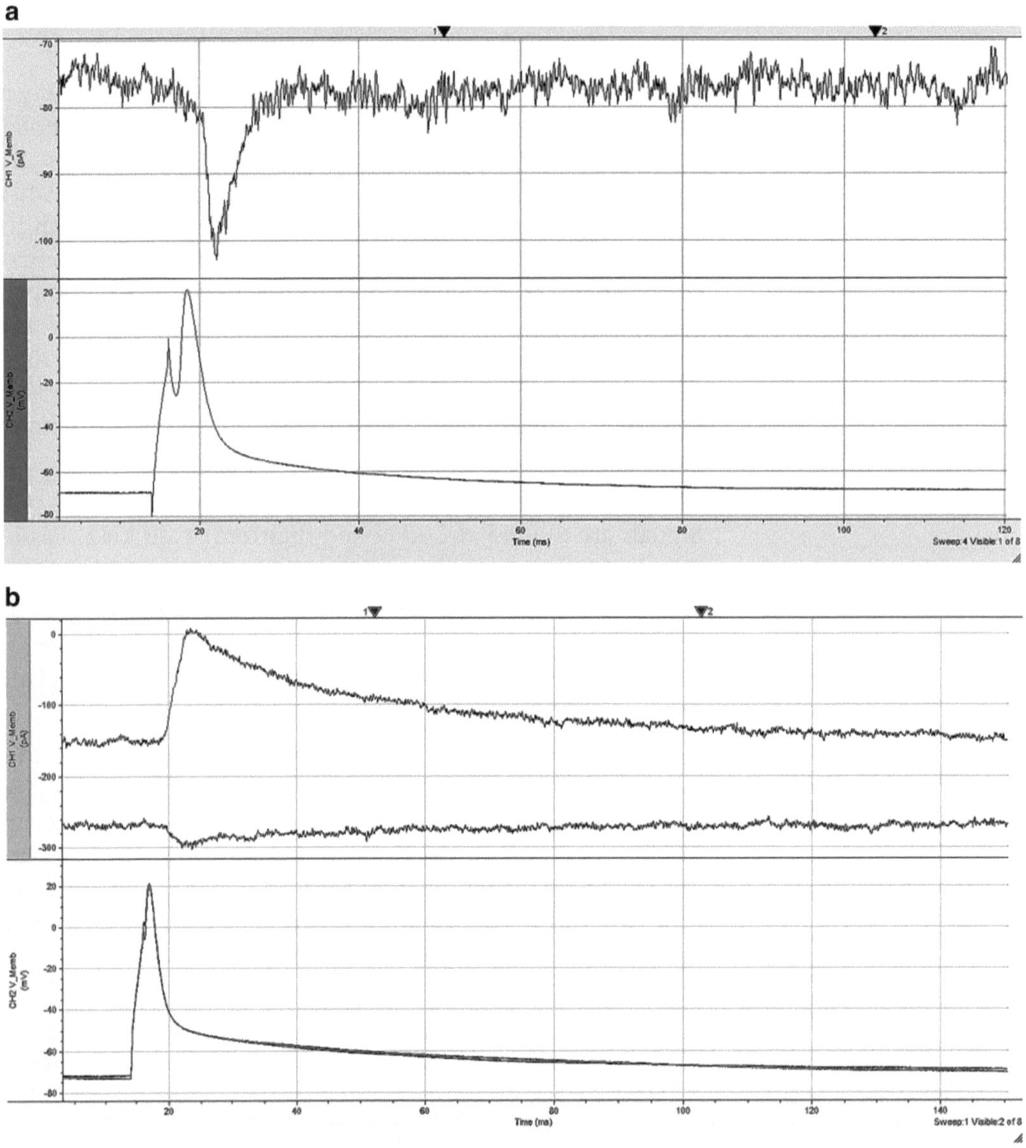

Fig. 3 Synaptic transmission in engineered networks. Action potentials are evoked by current injection in one of the cells, whereas postsynaptic currents are recorded from the other cell. (**a**) AP—lower trace, EPSC recorded at −70 mV holding—upper trace. (**b**) AP—lower trace, IPSC recorded at −70 and −30 mV holding—upper trace. Note the reversal of the current. Reprinted from [23] with permission

4. If the laser is not available, several other methods can be used for patterning, for example microcontact printing or inkjet printing [21].
5. For the photomask creation, several other programs can be used such as AutoCAD. Neuronal polarity can be determined by spatial clues [9].

6. The mask should be tightly pressed to the substrate. The UV wavelength and output power are important [22].
7. The timing of the 13F backfill is important, because 13F incorporates into the DETA layer. Backfilling should be optimized to result in a minimal incorporation and maximal coverage of the ablated surface. Incorporation can be tested by XPS.
8. In the last step of the electroless copper deposition, the timing is difficult; the coverslip must be checked regularly and removed from the solution when it is ready.
9. Gentle trituration of the tissue without bubble formation is important.

Acknowledgements

This work was supported by NIH Career Award K01 EB03465 and DOE grant DE-FG02-04ER46171.

References

1. Bhadriraju K, Chen CS (2002) Engineering cellular microenvironments to cell-based drug testing. Drug Discov Today 7:612–620
2. Singhvi R, Kumar A, Lopez GP et al (1994) Engineering cell shape and function. Science 264:696–698
3. Chen CS, Mrksich M, Huang S et al (1998) Micropatterned surfaces for control of cell shape, position, and function. Biotechnol Prog 14:356–363
4. Natarajan A, Stancescu M, Dhir V et al (2011) Patterned Cardiomyocytes on microelectrode arrays as a functional, high information content drug screening platform. Biomaterials 32:4267–4274
5. Sung JH, Esch MB, Prot JM et al (2013) Microfabricated mammalian organ systems and their integration into models of whole animals and humans. Lab Chip 13:1201–1212
6. Schober A, Fernekorn H, Singh S et al (2013) Mimicking the biological world: methods for the 3D structuring of artificial cellular environments. Eng Life Sci 13:352–367
7. Ravenscroft MS, Bateman KE, Shaffer KM et al (1998) Developmental neurobiology implications from fabrication and analysis of hippocampal neuronal networks on patterned silane- modified surfaces. J Am Chem Soc 120:12169–12177
8. Stenger DA, Pike CJ, Hickman JJ et al (1993) Surface determinants of neuronal survival and growth on self- assembled monolayers in culture. Brain Res 630:136–147
9. Stenger DA, Hickman JJ, Bateman KE et al (1998) Microlithographic determination of axonal/dendritic polarity in cultured hippocampal neurons. J Neurosci Methods 82: 167–173
10. Edwards D, Stancescu M, Molnar P et al. (2013) Two cell circuits of oriented adult hippocampal neurons on self-assembled monolayers for use in the study of neuronal communication in a defined system. ACS Chem Neurosci 4: 1174–1182
11. Natarajan A, DeMarse TB, Molnar P et al (2013) Engineered in vitro feed-forward networks. J Biotechnol Biomater 3:153–160
12. Craig AM, Boudin H (2001) Molecular heterogeneity of central synapses: afferent and target regulation. Nat Neurosci 4:569–578
13. Paoletti P, Bellone C, Zhou Q (2013) NMDA receptor subunit diversity: impact on receptor properties, synaptic plasticity and disease. Nat Rev Neurosci 14:383–400
14. MacDermott AB, Role LW, Siegelbaum SA (1999) Presynaptic ionotropic receptors and the control of transmitter release. Annu Rev Neurosci 22:443–485

15. Zimmermann M (2001) Pathobiology of neuropathic pain. Eur J Pharmacol 429:23–37
16. Yamada KA (1998) Modulating excitatory synaptic neurotransmission: potential treatment for neurological disease? Neurobiol Dis 5:67–80
17. Albuquerque EX, Santos MD, Alkondon M et al (2001) Modulation of nicotinic receptor activity in the central nervous system: a novel approach to the treatment of Alzheimer disease. Alzheimer Dis Assoc Disord 15:S19–S25
18. Nistico R, Pignatelli M, Piccinin S et al (2012) Targeting synaptic dysfunction in Alzheimer's disease therapy. Mol Neurobiol 46:572–587
19. Kang JF, Poeta M, Riedal L et al. (2004) Patterned neuronal networks for robotics, neurocomputing, toxin detection and rehabilitation. in 24th Army Conference. Orlando, FL
20. Schaffner AE, Barker JL, Stenger DA et al (1995) Investigation of the factors necessary for growth of hippocampal neurons in a defined system. J Neurosci Methods 62: 111–119
21. Kane RS, Takayama S, Ostuni E et al (1999) Patterning proteins and cells using soft lithography. Biomaterials 20:2363–2376
22. Hickman JJ, Bhatia SK, Quong JN et al (1994) Rational pattern design for in-vitro cellular networks using surface photochemistry. J Vac Sci Tech 12:607–616
23. Molnar P, Kang J-F, Bhargava N et al (2007) Synaptic connectivity in engineered neuronal networks. In: Patch clamp methods and protocols. Humana Press, Totowa, NJ, p 165–173

Chapter 16

Modeling of Action Potential Generation in NG108-15 Cells

Peter Molnar and James J. Hickman

Abstract

In order to explore the possibility of identifying toxins based on their effect on the shape of action potentials, we created a computer model of the action potential generation in NG108-15 cells (a neuroblastoma/glioma hybrid cell line). To generate the experimental data for model validation, voltage-dependent sodium, potassium and high-threshold calcium currents, as well as action potentials, were recorded from NG108-15 cells with conventional whole-cell patch-clamp methods. Based on the classic Hodgkin–Huxley formalism and the linear thermodynamic description of the rate constants, ion-channel parameters were estimated using an automatic fitting method. Utilizing the established parameters, action potentials were generated using the Hodgkin–Huxley formalism and were fitted to the recorded action potentials. To demonstrate the applicability of the method for toxin detection and discrimination, the effect of tetrodotoxin (a sodium channel blocker) and tefluthrin (a pyrethroid that is a sodium channel opener) were studied. The two toxins affected the shape of the action potentials differently, and their respective effects were identified based on the predicted changes in the fitted parameters.

Key words Action potential, Computer modeling, Hodgkin–Huxley, Linear thermodynamic description, Toxin detection, NG108-15, Parameter fitting

1 Introduction

Action potential generation and the shape of the action potential depends on the status of several ion channels located in a cell's membrane, which are regulated by receptors and intracellular messenger systems [1–4]. Changes in the extracellular or intracellular environment (receptor activation, gene expression) can be reflected in an alteration of spontaneous firing properties such as the frequency and firing pattern [1, 5–7] of excitable cells and in the changes of the action potential shape [8–13].

Experimental data-driven computer modeling has been an excellent tool to integrate our knowledge concerning elements of a complex biological system as well as to draw conclusions about the behavior of the complex system under different experimental conditions. These predictions can be correlated again with the experimental data. One typical example is the modeling of the

Marzia Martina and Stefano Taverna (eds.), *Patch-Clamp Methods and Protocols*, Methods in Molecular Biology, vol. 1183, DOI 10.1007/978-1-4939-1096-0_16, © Springer Science+Business Media New York 2014

electrophysiological behavior of an excitable cell. We have detailed knowledge about individual ion channels, as well as extensive knowledge about the behavior of the whole cell; but we know relatively little about the interaction and modulation of ionic currents that shape the action potentials. One of the most complex single cell models created simulates the electrophysiological behavior of human cardiac myocytes [13–15]. This model was validated based on electrophysiological experiments under physiological and pathophysiological conditions [16] and used for a deeper understanding of behavior of ion channels in disease models and for proteomic studies [17].

The most commonly used mathematical formalism which describes the action potential generation in excitable cells was developed by Hodgkin and Huxley in 1952 [18]. In our studies we have used this formalism, although for the description of the ion channels we utilized the linear thermodynamic approach, which, in our opinion, was more general and does not require "guessing" the form of the functions describing the voltage dependence of the state parameters [19, 20].

We used the NG108-15 neuroblastoma/glioma cell line in our studies [11], because these cells do not form synapses in culture, and thus they are ideal single cell sensors [21, 22].

2 Materials

2.1 NG108-15 Cultures

1. Culture medium: 90 % Dulbecco's modified Eagle's medium (DMEM) supplemented with 10 % Fetal Bovine Serum and 1 % HAT supplement.
2. Differentiating medium: DMEM + 2 % B27 supplement.
3. Poly-D-Lysine (PDL) solution: 5 mg PDL in 500 ml water, sterile filtered.

2.2 Patch-Clamp Recording of Ionic Currents and Action Potentials

1. Extracellular solution: 140 mM NaCl, 3.5 mM KCl, 2 mM $MgCl_2$, 2 mM $CaCl_2$, 10 mM Glucose, 10 mM HEPES, pH = 7.34.
2. Extracellular solution for the recording of potassium currents: Extracellular solution plus 1 μM Tetrodotoxin.
3. Extracellular solution for the recording of sodium currents: 50 mM NaCl, 100 mM TEA-Cl, 5 mM CsCl, 1 mM $CaCl_2$, 1 mM $CoCl_2$, 1 mM $MgCl_2$, 10 mM Glucose, 10 mM HEPES, pH = 7.34.
4. Extracellular solution for the recording of calcium currents: 100 mM NaCl, 30 mM TEA-Cl, 10 mM $CaCl_2$, 2 mM $MgCl_2$, 10 mM Glucose, 10 mM HEPES, 1 μM Tetrodotoxin, pH = 7.34.

5. Intracellular solution for action potential and potassium channel measurement: 130 mM potassium gluconate, 2 mM $MgCl_2$, 1 mM EGTA, 15 mM HEPES, 5 mM ATP, pH = 7.2, osmolarity = 276 mOsm.
6. Intracellular solution for sodium channel measurement: 130 mM CsF, 10 mM NaCl, 10 mM TEA-Cl, 2 mM $MgCl_2$, 1 mM EGTA, 10 mM HEPES, 5 mM ATP.
7. Intracellular solution for calcium channel measurement: 120 mM CsCl, 20 mM TEA-Cl, 2 mM $MgCl_2$, 1 mM EGTA, 10 mM HEPES, 5 mM ATP. For selecting L-type calcium channels, 1 μM ωCTxGVIA (Tocris) is added.
8. Pipette puller.
9. Glass pipettes.
10. Vibration isolation table with Faraday cage.
11. Upright microscope.
12. Patch clamp amplifier, A/D converter, data acquisition software.

2.3 Obtaining the Parameters Describing Ion Channel Currents and Action Potentials in NG108-15 Cells

1. MATLAB software (MathWorks).

2.4 Effect of Drugs on Action Potential Shape and Ion Channel Parameters

1. TTX.
2. Tefluthrin.

3 Methods

3.1 NG108-15 Cultures

1. The NG108-15 cell line frozen (passage number 16) in liquid N_2 in 1 ml vials (1 million cells/vial).
2. Cell stock is grown in a T-75 flask in culture medium at 37 °C with 10 % CO_2 (*see* **Note 1**).
3. Glass coverslips are cleaned utilizing concentrated nitric acid for 30 min, rinsed three times with water and sterilized by submerging in 100 % alcohol for 20 min.
4. Coverslips are placed in 6-well plates and incubated in poly-D-lysine solution for 1 h in a 37 °C incubator.
5. After confluence, the culture medium is replaced by 6 ml of differentiating medium. The cell layer is dislodged by knocking

the flask on the table. Cells are dissociated by trituration using a 5 ml pipette.

6. NG108-15 cells are plated at a density of 40,000 cells/35 mm culture dish in 2 ml differentiating medium on the PDL-coated coverslips and are cultured for 1 week at 37 °C with 5 % CO_2.

3.2 Patch-Clamp Recording of Ionic Currents and Action Potentials

1. Coverslips are transferred into a chamber on the microscope stage, which is continuously perfused with the appropriate extracellular solution. Experiments are performed at room temperature.
2. Glass pipettes, pulled with the electrode puller, are filled with the appropriate intracellular solution and should have a resistance of 4–6 MΩ.
3. Signals are filtered at 2 kHz and digitized at 20 kHz.
4. Sodium and potassium currents are measured in voltage-clamp mode using 10 mV voltage steps from a −85 mV holding potential. The high-threshold calcium currents are recorded at −40 mV holding potential.
5. Whole-cell capacitance and series resistance are compensated and a p/6 protocol is used.
6. Action potentials are evoked with short (2 ms) current injections in current-clamp mode either at resting membrane potential or at a −85 mV holding potential. Data are saved in text-format and imported into MATLAB for further analysis.
7. Tip potential is calculated and compensated by subtracting 15 mV from the membrane potential values.

3.3 Obtaining the Parameters Describing Ion Channel Currents in NG108-15 Cells

1. A computer program is created in MATLAB to fit the parameters to the recorded data according to the following equations. Basically, the experimental data is given as total ionic current vs. time (sodium, potassium, and calcium currents are measured independently in separate experiments) and membrane potential vs. time.
2. Total ionic current (sodium I_{Na}, potassium I_K calcium I_{Ca}, and leakage channels I_l included):

$$I_{ionic} = I_{Na} + I_K + I_{Ca} + I_l = \bar{g}_{Na} m^3 h (V - V_{Na}) + \bar{g}_K n^4 (V - V_K) + \bar{g}_{CaL} e^3 (V - V_{CaL}) + \bar{g}_l (V - V_l)$$

where $\bar{g}_{Na}, \bar{g}_K, \bar{g}_{CaL}, V_{Na}, V_K, V_{CaL}$ are parameters (maximum conductance of the channels and reversal potentials, respectively) and m, n, h, e are the state variables.

3. The dynamics of the state variables are:

$$\frac{dm}{dt} = \frac{m_\infty - m}{\tau_m}$$ where $m_\infty, n_\infty, h_\infty, e_\infty$ are the steady-state values of the state variables and the τ–s are their voltage-dependent time-constants.

4. The voltage dependence of the time constants and the steady-state state parameters are given according to the general thermodynamic formalism, an example was given for the *m* state parameter:

$$m_{\infty} = \frac{1}{1 + \exp^{-\frac{zF}{RT}\left(V_{\mathrm{m}} - V_{1/2}\right)}} \quad \text{and}$$

$$\tau_{\mathrm{m}} = \frac{A}{\exp^{\frac{zF}{RT}\xi\left(V_{\mathrm{m}} - V_{1/2}\right)} \cosh\left(\frac{zF}{2RT}\left(V_{\mathrm{m}} - V_{1/2}\right)\right)}$$

   where z, $V_{1/2}$, A and ξ are fitting parameters and V_{m} represents the membrane potential. As it can be seen from these equations $V_{1/2}$ corresponds to the half activation/inactivation potential of the channel and A is linearly related to the activation or inactivation time-constant. The meanings of z and ξ are not as obvious: z is related to the number of moving charges during the opening or closing of the channel, whereas ξ describes the asymmetric position of the moving charge in the cell membrane.
5. Sodium, potassium, and calcium channel mediated current data, which are recorded in voltage-clamp mode at different membrane potentials (5 mV increments, −40 mV, +30 mV range), are imported into MATLAB. Based on the above equations, simulated current traces are calculated (Fig. 1).
6. An error function is generated based on the difference between the recorded and the simulated traces through the whole voltage range (*see* **Notes 2** and **3**).
7. MATLAB's fminsearch routine is used to optimize the parameters to minimize the error function (*see* **Notes 4** and **5**).
8. Parameters obtained from different cells are averaged and considered as initial values for the action potential modeling.

3.4 Fitting Parameters to the Experimentally Recorded Action Potentials

1. Effect of the currents on the membrane potential:

$$\frac{\mathrm{d}V}{\mathrm{d}t} = \frac{I_{\text{external}} - I_{\text{ionic}}}{C_{\mathrm{M}}}$$

   where I_{external} is the externally injected current used to depolarize the membrane and evoke the action potential in current clamp mode.
2. The following parameters are obtained from the patch-clamp recordings and used in the modeling: membrane resistance, resting membrane potential, membrane capacitance, and injected current. The maximum conductance of the leakage current (g_{l}) is calculated from the ionic conductance and from the resting membrane potential.

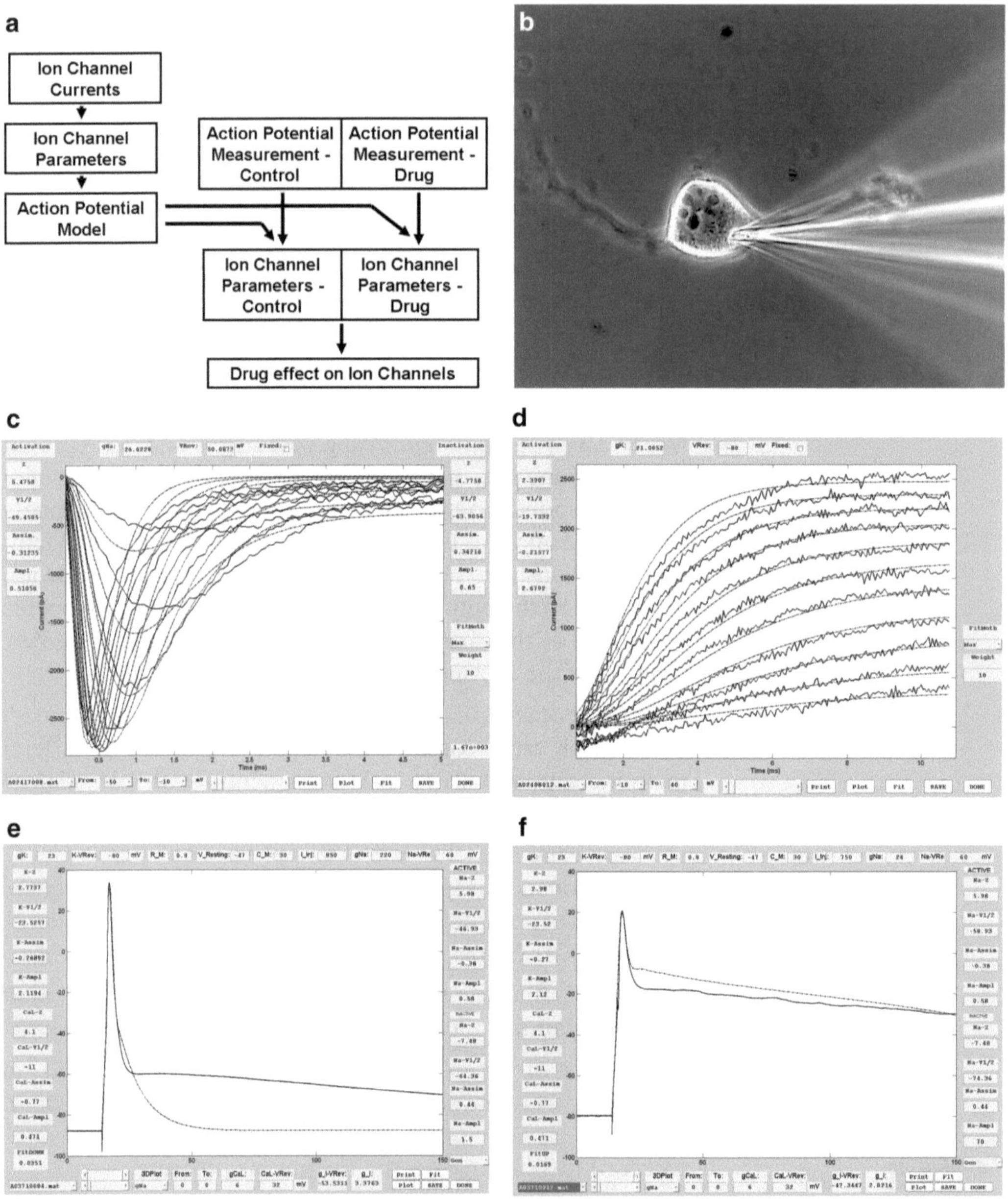

Fig. 1 Determination of the ion channel parameters obtained from action potential shapes. (**a**) Ion channel parameters are determined from action potential recordings based on a computer model of action potential generation and parameter fitting. Drug effects could be quantified based on changes of ion channel parameters. (**b**) Phase-contrast picture of an NG108-15 cell with a patch electrode. (**c**) Measured sodium channel currents (*solid line*) with simulated current traces (*dotted line*). Parameters for the simulation were determined by an automatic fitting routine. (**d**) Measured potassium channel currents (*solid line*) with simulated current traces (*dotted line*). (**e**) Control action potential (*solid line*) with simulated action potential trace (*dotted line*). (**f**) Action potential shape after 1 μM Tefluthrin (*solid line*) and simulated action potential trace (*dotted line*). Reprinted from [23] with permission

Table 1
Effect of TTX and tefluthrin on the action potential parameters

			Activation				Inactivation			
Channel	***g***	V_{Rev}	***z***	$V_{1/2}$	ξ	***A***	***z***	$V_{1/2}$	ξ	***A***
Control	**550**	60	6.2	-46	−0.38	**0.6**	-7.48	-64.36	0.53	**0.368**
TTX	**16.5**	60	6.0193	-45.992	−0.3917	**1.2963**	-7.4792	-64.364	0.51198	**6.0021**
Control	**220**	60	5.98	**-46.93**	−0.38	0.58	-7.48	**-64.36**	0.44	**1.5**
Tefluthrin	**24**	60	5.98	**-58.93**	−0.38	0.58	-7.48	**-74.36**	0.44	**70**

(Representative recordings out of 3–4 experiments with similar results. Only sodium channel parameters are shown, other ion-channel parameters did not change. Bold: more than 10 % change). Reprinted from [23] with permission

3. The differential equation above and the differential equations for the state parameters formed a first order coupled differential equation system, which is solved with MATLAB's ODE23 solver.
4. Parameters determining the simulated curve are fitted to the experimental data using the fminunc routine (*see* **Note 6**).

3.5 Effect of Drugs on Action Potential Shape and on Ion Channel Parameters

1. Drugs are applied by perfusing the experimental chamber with extracellular medium containing the drug.
2. Action potentials are recorded before and 10 min after the drug administration.
3. Ion channel parameters are calculated for the control and for the drug modified action potentials by fitting the simulated action potentials to the recorded experimental data (Table 1).
4. Drug effects are quantified as percentage changes in the ion channel parameters obtained before and after drug administration.

4 Notes

1. The cells also grow well at 5 % CO_2.
2. Curves were fitted after an initial 0.1 ms delay to eliminate the effect of experimental artifacts
3. To quantify the difference between the fitted curves and the recorded data the following error-functions were implemented: Maximum error: $E_{Max} = \text{Max}(\text{Abs}(R(t_n) - S(t_n)))$ where $R(t_n)$ is the recorded value and $S(t_n)$ is the simulated data at time t_n.
   Least Square: $E_{Lsquare} = \sum_n (R(t_n) - S(t_n))^2$.
4. Weighted Least Square: $E_{WLsquare} = E_{Lsquare}$ if $t_n < 30$ ms and $E_{WLsquare} = 5 \times E_{Lsquare}$ if $t_n \geq 30$ ms.

5. After several trials it was concluded that simulations that used the Weighted Least Square Error function gave the most satisfactory results because the other Error Functions occasionally obtained a non-inactivating sodium-current component in the simulated data.
6. Fminunc is used to find a minimum of a scalar function (the error function) for several variables, starting at an initial estimate. This method is generally referred to as unconstrained nonlinear optimization. Because it finds only local minimums, it is very important to start the optimization as close to the final result as possible. In our case, the averaged ion channel parameters obtained from voltage clamp experiments served the initial values for the parameter estimations.

Acknowledgements

This work was supported by NIH Career Award K01 EB03465 and DOE grant DE-FG02-04ER46171.

References

1. Gross GW, Harsch A, Rhoades BK et al (1997) Odor, drug and toxin analysis with neuronal networks in vitro: extracellular array recording of network responses. Biosens Bioelectron 12:373–393
2. Gross GW, Rhoades BK, Azzazy HM et al (1995) The use of neuronal networks on multielectrode arrays as biosensors. Biosens Bioelectron 10:553–567
3. Morefield SI, Keefer EW, Chapman KD et al (2000) Drug evaluations using neuronal networks cultured on microelectrode arrays. Biosens Bioelectron 15:383–396
4. Akanda N, Molnar P, Stancescu M et al (2009) Analysis of toxin-induced changes in action potential shape for drug development. J Biomol Screen 14:1228–1235
5. Amigo JM, Szczepański J, Wajnryb E et al (2003) On the number of states of the neuronal sources. Biosystems 68:57–66
6. Chiappalone M, Vato A, Tedesco MB et al (2003) Networks of neurons coupled to microelectrode arrays: a neuronal sensory system for pharmacological applications. Biosens Bioelectron 18:627–634
7. Xia Y, Gopal KV, Gross GW (2003) Differential acute effects of fluoxetine on frontal and auditory cortex networks in vitro. Brain Res 973:151–160
8. Akay M, Mazza E, Neubauer JA (1998) Nonlinear dynamic analysis of hypoxia-induced changes in action potential shape in neurons cultured from the rostral ventrolateral medulla (RVLM). FASEB J 12:2881
9. Clark RB, Bouchard RA, Salinas-Stefanon E et al (1993) Heterogeneity of action-potential wave-forms and potassium currents in rat ventricle. Cardiovasc Res 27:1795–1799
10. Djouhri L, Lawson SN (1999) Changes in somatic action potential shape in guinea-pig nociceptive primary afferent neurones during inflammation in vivo. J Physiol 520:565–576
11. Mohan DK, Molnar P, Hickman JJ (2006) Toxin detection based on action potential shape analysis using a realistic mathematical model of differentiated NG108-15 cells. Biosens Bioelectron 21:1804–1811
12. Muraki K, Imaizumi Y, Watanabe M (1994) Effects of noradrenaline on membrane currents and action-potential shape in smooth-muscle cells from guinea-pig ureter. J Physiol 481: 617–627
13. Nygren A, Fiset C, Firek L et al (1998) Mathematical model of an adult human atrial cell: the role of K^+ currents in repolarization. Circ Res 82:63–81
14. Bernus O, Wilders R, Zemlin CW et al (2002) A computationally efficient electrophysiological

model of human ventricular cells. Am J Physiol Heart Circ Physiol 282:H2296–H2308

15. Nygren A, Leon LJ, Giles WR (2001) Simulations of the human atrial action potential. Philos Trans R Soc Lond Ser A 359 (1783):1111–1125
16. Winslow RL et al (2001) Computational models of the failing myocyte: relating altered gene expression to cellular function. Philos Trans R Soc Lond Ser A 359:1187–1200
17. Winslow RL, Cortassa S, Greenstein JL (2005) Using models of the myocyte for functional interpretation of cardiac proteomic data. J Physiol 563:73–81
18. Hodgkin AL, Huxley AF (1952) A quantitative description of membrane current and its application to conduction and excitation in nerve. J Physiol 117:500–544
19. Destexhe A, Huguenard JR (2000) Nonlinear thermodynamic models of voltage-dependent currents. J Comput Neurosci 9:259–270
20. Weiss TF (1996) Cellular biophysics. MIT Press, Cambridge, MA
21. Kowtha VC, Quong JN, Bryant HJ et al (1993) Comparative electrophysiological properties of NG108-15 cells in serum-containing and serum-free media. Neurosci Lett 164: 129–133
22. Ma W, Pancrazio JJ, Coulombe M et al (1998) Neuronal and glial epitopes and transmitter-synthesizing enzymes appear in parallel with membrane excitability during neuroblastoma x glioma hybrid differentiation. Dev Brain Res 106:155–163
23. Molnar P, Kang J-F, Bhargava N, Das M, Hickman JJ (2007) Synaptic connectivity in engineered neuronal networks. Patch-clamp methods and protocols. Humana, Totowa, NJ, pp 165–173

Chapter 17

Whole-Cell Patch-Clamp Recordings in Freely Moving Animals

Albert K. Lee, Jérôme Epsztein, and Michael Brecht

Abstract

The patch-clamp technique and the whole-cell measurements derived from it have greatly advanced our understanding of the coding properties of individual neurons by allowing for a detailed analysis of their excitatory/inhibitory synaptic inputs, intrinsic electrical properties, and morphology. Because such measurements require a high level of mechanical stability they have for a long time been limited to in vitro and anesthetized preparations. Recently, however, a considerable amount of effort has been devoted to extending these techniques to awake restrained/head-fixed preparations allowing for the study of the input–output functions of neurons during behavior. In this chapter we describe a technique extending patch-clamp recordings to awake animals free to explore their environments.

Key words Intracellular recording, Awake-behaving, Hippocampus, CA1

1 Introduction

Understanding the role of individual identified cells in specific microcircuits underlying specific cognitive functions is a major goal of neuroscience. The whole-cell patch-clamp recording technique [1] provides a very efficient way to characterize the morphophysiological properties of individual neurons as it gives access to (1) their intrinsic electrophysiological properties (e.g., fast spiking/non-adapting versus regular spiking/adapting cells), (2) the dynamics of their membrane potential in response to inputs, (3) their output spiking activity, (4) their morphology following dye filling during the recording, and (5) the expression of specific proteins or neuropeptides when coupled with immunohistochemistry.

For a long time patch-clamp recording was limited to the in vitro culture or slice preparation because it requires a high level of mechanical stability to maintain physical contact between the

Albert K. Lee, and Jérôme Epsztein have contributed equally to this chapter.

Marzia Martina and Stefano Taverna (eds.), *Patch-Clamp Methods and Protocols*, Methods in Molecular Biology, vol. 1183, DOI 10.1007/978-1-4939-1096-0_17, © Springer Science+Business Media New York 2014

glass recording electrode and the neuron's membrane. While in vitro preparations have several advantages over in vivo preparations such as high spatial and temporal control of input stimulation and the ability to easily apply a wide variety of pharmacological agents to dissect the ligand-gated and voltage-gated channels involved in specific cell functions, in vitro preparations also suffer from inherent limitations such as the bathing of neurons in an artificial medium with artificial oxygenation levels, separation from major input structures and neuromodulators, and uncertainty about the physiological relevance of the applied stimuli.

To overcome these limitations, the whole-cell technique has been extended to in vivo preparations in both anesthetized [2–7] and awake, restrained or head-fixed [5, 8–16] animals (reviewed in ref. 17). These technical developments have allowed for the characterization of responses to more natural stimuli under more natural conditions. The combination of such recordings with two-photon microscopy allows for the targeting of specific classes of cells [18–20], and the development of optogenetic techniques allows for refinement in the types of stimulation which can be applied in vivo (reviewed in ref. 21). The recent use of virtual interfaces [22] driven by locomotion has enhanced the types of behavior which can be combined with whole-cell recordings, including spatial exploration of virtual visual environments in head-fixed animals running on treadmills [12, 14, 15]. However, the effect of the constraints imposed by head fixation or body restraint on neural coding when compared to an animal naturally exploring its environment is an area of active research [23, 24].

Here we present a technique allowing for whole-cell recordings in non-head-fixed animals free to explore their environment [25, 26]. The applicability of whole-cell recording to this problem is related to the biophysics of the patch-clamp technique. Upon seal formation a tight link between patch pipette and cell membrane is established, which results in a hose-like connection between recorded neuron and pipette for the duration of the recording. Mechanically favorable properties of whole-cell recordings such as the disappearance of heartbeat or locomotion-related mechanical artifacts derive from this configuration. These considerations may also explain why obtaining a high-resistance seal between neuron and patch pipette is a major predictor of the quality and stability of subsequent whole-cell recordings. The key technical principle for the adaptation of whole-cell recordings to freely behaving animals involves the rigid anchoring of the recording pipette relative to the skull of the animal. The procedure can be broken down into three major steps: the first is to get a high quality patch-clamp recording in an anesthetized head-fixed animal, the second is to fix the pipette with respect to the skull using an anchoring material (e.g., acrylic), and the third is to release and wake-up the animal to allow it to freely explore and interact with its environment. In the initial version of the technique [25], a head-mounted motor moved the pipette through the brain to obtain the gigaseal. Later, the use of a non-head-mounted motor and gripper improved the success rate [26], and it is this version that is presented here. This technique has allowed for the first intracellular recordings of hippocampal place cells [27], allowed for spatial analysis of small amplitude intracellular events called spikelets which are impossible to detect with extracellular recording or imaging [28], and revealed cellular differences between hippocampal place and silent cells [29]. A variant of this technique allowed for the discovery and manipulation of the nonlinear voltage-dependent basis of spatial inputs underlying place and silent cells [30]. A method using a head-mounted motor drive to fix a sharp electrode in place without a separate anchoring step has also led to intracellular recordings from freely moving animals, including singing birds [31].

While the steps below include specific procedures for targeting dorsal hippocampal CA1 pyramidal neurons [26], it can be easily adapted for other neurons. For instance, rigid fixation of the pipette also allows for whole-cell recording from the superficial layers of the motor cortex in freely moving rats [25].

2 Materials

2.1 Reagents

1. Intracellular recording solution: 135 mM K-gluconate, 10 mM HEPES, 10 mM Na_2-phosphocreatine, 4 mM KCl, 4 mM MgATP, 0.3 mM Na_3GTP (pH adjusted to 7.2 with KOH, target osmolality ~291 mmol/kg). Make in advance and store at ≤−20 °C. Thaw, then add ~0.05 % (wt/vol) biocytin to each aliquot and re-filter before each experiment (*see* **Note 1**).
2. Artificial cerebrospinal fluid (ACSF): 135 mM NaCl, 5.4 mM KCl, 5 mM HEPES, 1.8 mM $CaCl_2$, 1 mM $MgCl_2$ (pH adjusted to 7.2 with NaOH, target osmolality ~290 mmol/kg). Make in advance and store at 4 °C.
3. Prepare anesthetic mix. Initial dose is based on animal weight and consists of a mixture of dex medetomidine (112.5 μg/kg), midazolam (6 mg/kg), and fentanyl (7.5 μg/kg). Make enough mix in advance to allow for at least five doses of this size for each experiment (to be used for supplemental injections of ~1/3 dose each) and store at 4 °C.
4. Prepare local anesthetic for scalp incision: single dose of bupivacaine (2 mg/kg). Make 0.5 % (wt/vol) dilution of bupivacaine in advance and store at 4 °C.
5. Prepare antagonist ("wake") mix for anesthetics. Nominally, the initial dose of anesthetics is antagonized with a mixture of

atipamezole (1 mg/kg), flumazenil (600 μg/kg), and naloxone (180 μg/kg). Make enough mix in advance to allow for at least five doses of this size for each experiment and store at 4 °C.

6. Prepare agar. Add 2 % (wt/vol) agarose powder to ACSF, heat and stir until clear, then continue to heat and stir throughout the experiment to keep as liquid. Make in advance before each experiment.

2.2 Custom-Made Small Equipment for Head-Mounted Patching

1. Make a plastic ring that will surround craniotomy (Fig. 1c, e). Can be made by cutting off a piece from the cap of a microcentrifuge tube.
2. Make an acrylic base by pouring dental acrylic into a mold shaped like a "C" as in Fig. 1c (*see* **Note 2**).
3. Make a side support (Fig. 1d, e) from stainless steel tube, e.g., by cutting a 18 or 19 gauge hypodermic needle.
4. Make air tubing/pipette wire assembly (sideways "T"-shaped item in Fig. 1e) by gluing a signal wire and a small air tube (to apply air pressure to the inside of the pipette while attempting to get a gigaseal) with silicone glue to two ends of a 3-way small tubing connector. Connect a bigger tube (into which the pipette can be tightly inserted) to the remaining end of the connector (*see* **Note 3**).
5. Make a head-mountable head stage/LED assembly with a detachable post on the bottom (Fig. 1b, d) so that it can easily be attached through the post to the acrylic base (Fig. 1d) with dental acrylic and, after recording, can easily be removed from the base without damaging the equipment. It should also have a point of attachment on top so that it can be held and released by the manual manipulator (Fig. 1b). Prepare in advance (*see* **Note 4**).
6. Set up a counterweight system by attaching fishing lines to the top of the head stage/LED assembly, looping the ends over pulleys, then attaching weights onto the ends. Adjust the weights in advance so that the rat can easily move around the behavioral arena (located next to the stereotax) with the assembly on its head (*see* **Note 5**). Prepare in advance.
7. Prepare pipette electrodes. Pull whole-cell pipettes with resistance 5–7 MΩ from 1.5 mm o.d. thick-walled borosilicate capillaries. The distance between the tip and the point at which the 1.5 mm o.d. diameter capillary first starts narrowing should be ~5 mm. The tip should have an i.d. of ~1 μm. Prepare 5–10 electrodes in advance for each experiment because a new pipette must be used for each attempt at obtaining a GΩ seal, and each attempt is successful with a probability of ~1/3.

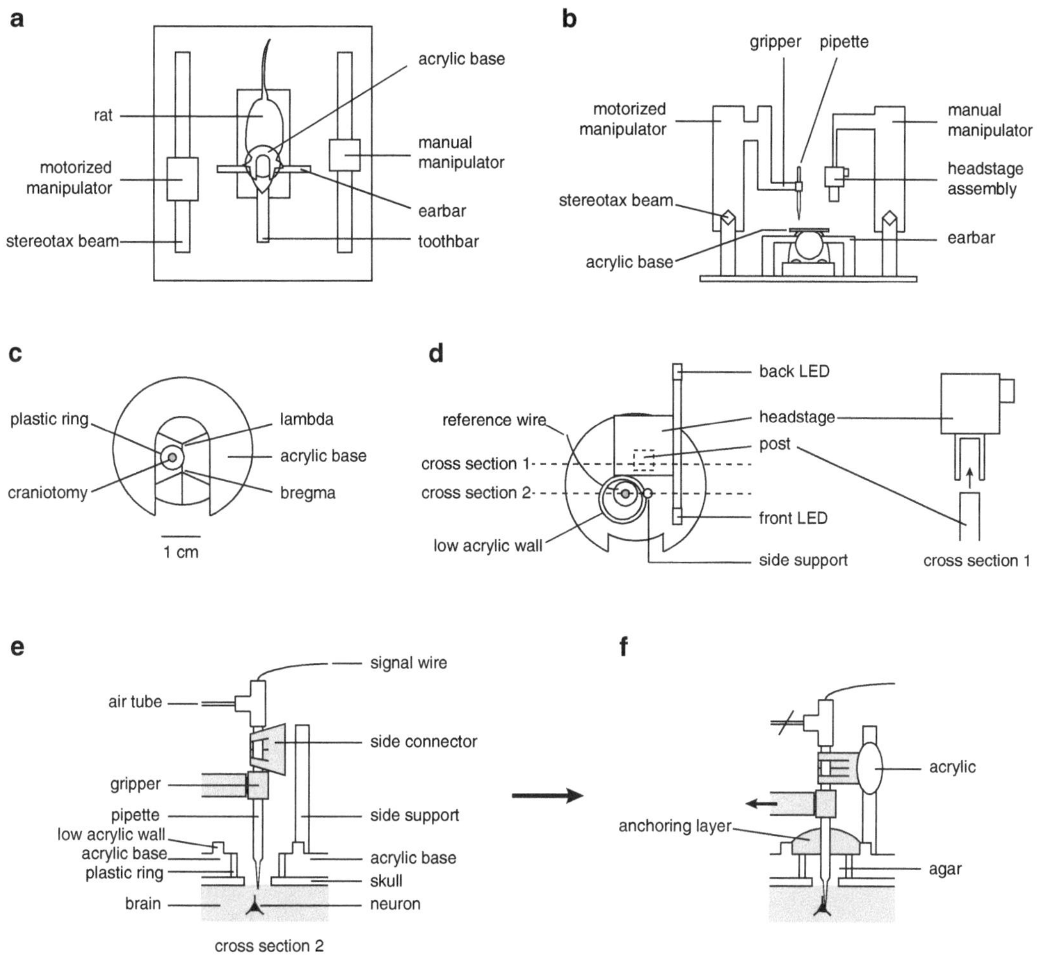

Fig. 1 Overview of the experimental setup, head-mounted and non-head-mounted equipment and anchoring process. (**a**) *Top view*: rat in the stereotaxic frame. (**b**) *Front view*: the gripper is attached to a motorized micromanipulator which is attached to the stereotaxic frame. The head stage/LED assembly is initially held by a manual manipulator. (**c**) Construction of the head-mounted assembly begins with the craniotomy and acrylic base (*top view*). (**d**) Completion of the head-mounted assembly. The head stage assembly is attached to the acrylic base via a post. A section showing the connection between the post and head stage/LED assembly is on the *right*. (**e**) A combined front view/section (corresponding to cross section 2 in (**d**)) showing the arrangement of items when the pipette starts to enter the brain to search for a neuron. (**f**) Same as in (**e**) after the whole-cell recording has been obtained and the pipette has been anchored to the animal's skull in 2 places: first by applying acrylic inside the low acrylic wall to form the "anchoring layer," and later by applying acrylic (marked as "acrylic") between the side support and side connector. Once the anchoring process is complete, the air tube is cut (*diagonal line*), the pipette is released from the gripper, the gripper is moved away from the pipette (*small arrow*), and the animal is released from the stereotax. Modified from [26] with permission from Nature Publishing Group

3 Methods

3.1 Exposure of the Brain Surface

1. Anesthetize a P25-28 male Wistar rat with isoflurane followed by intraperitoneal (i.p.) injection of the anesthetic mix (*see* **Note 6**). Supplemental doses of ~1/3 of the initial dose must be administered every 40–45 min after the initial dose to prevent the animal from waking up prematurely.
2. Place the animal on a heating pad, shave the fur on top of the head, then fix the head in a stereotaxic apparatus with earbars and toothbar (Fig. 1a).
3. Inject a local anesthetic (e.g., bupivacaine) subcutaneously (s.c.) into the scalp, apply eye ointment, then cover the eyes to protect them from the bright surgical light (which should be a cold light to prevent heating).
4. Make an incision along the midline of the scalp and clean the surface of the skull so that it is dry and free of any soft tissue for optimal attachment of the acrylic base.
5. Mark the position on the skull for the center of the craniotomy (e.g., 3.5 mm posterior and 2.5 mm lateral of bregma for the dorsal hippocampal CA1 target).
6. Perform craniotomy (1.5–2.0 mm diameter, *see* **Note 7**).
7. Remove the dura from a small area (~0.5 mm diameter) near the center of the craniotomy while avoiding blood vessels as much as possible (*see* **Note 8**). To begin, make a small incision in the dura away from any blood vessels using the tip of a new hypodermic needle (e.g., 30 gauge). Then, using very fine-tipped forceps (e.g., Dumont #5CO), grab the dura at the incision and gently pull it away until it is removed from the desired region. Once the dura is removed, keep the brain surface from drying out (i.e., with ACSF or agar) during the remainder of the procedure.

3.2 Construction of the Head-Mounted Assembly

1. Protect the craniotomy from bone particles with agar, drill 5–6 holes in the skull, screw a stainless steel screw into each of the holes, then remove the agar.
2. Position the plastic ring around the craniotomy, center it around the dura hole, then fill the ring with agar. Position the pre-made acrylic base around the skull (Fig. 1c) then add acrylic around the ring and screws to connect them to the base. Wait for ~5 min for the acrylic to harden.
3. Hook the end of an Ag/AgCl reference wire into the plastic ring then add a small amount of acrylic on top of the wire and base to connect it to the base, keeping the wire exposed inside the ring. Make a low acrylic wall (~1 mm high, Fig. 1d, e) around the plastic ring to prevent acrylic from flowing away

from the pipette during the anchoring step (Fig. 1f). Wait for ~5 min for the acrylic to harden.

4. Position a vertical side support for the pipette using the manual manipulator (Fig. 1b) then add acrylic to connect it to the base (Fig. 1d, e). Wait for ~5 min for the acrylic to harden (*see* **Note 9**).
5. Position the head stage/LED assembly with post above the base using the manual manipulator (Fig. 1b) then add acrylic to attach its detachable post to the base (Fig. 1d). Wait for ~5 min for the acrylic to harden.

3.3 Determination of the Target Recording Depth

1. This section is written specifically for targeting the dorsal hippocampal CA1 pyramidal layer, but can be adapted to target other cells in other brain areas (*see* **Note 10**).
2. Remove the agar with (non-fine-tipped) forceps, rinse the brain surface by adding ACSF into the ring using a transfer pipette and removing it with surgical swabs, then place an extracellular glass electrode (~2 MΩ of resistance, made by gently breaking the tip of a 5–7 MΩ whole-cell pipette) filled with ACSF into the gripper (which can be remotely opened and closed).
3. Position the electrode tip on the brain surface using the motorized manipulator that holds the gripper (Fig. 1b), zero the manipulator depth reading, then fill the ring with ACSF.
4. While in current or voltage clamp recording mode, advance the electrode into the brain relatively quickly to a depth of ~1,800 μm then advance slowly in ~5 μm steps while listening to the extracellular activity using an audio monitor. Hippocampal sharp-wave/ripple activity can be heard after entering the hippocampus and it becomes stronger as one approaches the CA1 pyramidal cell layer (~2,200 μm). Once the cell layer is reached, bursts of APs can be distinguished. Note this as the target depth.

3.4 Obtaining a Whole-Cell Recording

1. Insert a needle subcutaneously (under the animal's back skin) that is connected to a supply of the antagonist mix for the anesthetics in preparation for waking the animal up, but do not inject at this time.
2. Attach a side connector (Fig. 1e) to a new patch pipette filled with intracellular solution. Attach the pipette to the air tubing/pipette wire assembly then place the pipette into the gripper (Fig. 1e).
3. Apply high positive pressure (~500–800 mbar) to the pipette interior (*see* **Note 11**). Remove any ACSF in the ring, position the pipette tip on the brain surface, then zero the manipulator depth reading.

4. Advance the pipette into the brain relatively quickly to ~100 μm above the target depth then quickly reduce the pressure to the search value (25–35 mbar). Rotate the side connector so that it contacts the side support (Fig. 1f) then add liquid agar into the ring (*see* **Note 12**).
5. Use standard, blind in vivo patch-clamp techniques to obtain a whole-cell recording [5, 26]. A brief description of this procedure is as follows. While in voltage clamp mode, monitor the current response to applied voltage steps (~10 mV amplitude, ~10 ms duration, ~10 Hz) as one advances the electrode in small steps (~2 μm). When a consistent decrease in response amplitude over several (e.g., 3–4) consecutive steps is observed, quickly reduce the pressure to zero, then apply gentle suction to the inside of the pipette (as necessary) and hyperpolarize the command potential to ~−60 mV until the formation of a GΩ seal. Apply gentle suction to achieve the whole-cell configuration.
6. If a whole-cell recording is not obtained (*see* **Note 13**), retract the pipette from the brain, remove the pipette from the gripper, remove the agar from the ring, add ACSF into the ring, then repeat **steps 2–5** until a high-quality whole-cell recording is obtained.

3.5 Anchoring the Recording Electrode to the Head

1. After achieving the whole-cell configuration, switch the mode to current clamp and wait for a few minutes to assess the quality and stability of recording (e.g., based on the series resistance and baseline membrane potential), then apply acrylic (~0.03 ml or enough to yield a ~0.5-mm-thick layer) inside the low wall to completely surround the pipette and connect it to the base (Fig. 1f). Wait for ~5 min for the acrylic to harden (*see* **Note 14**).
2. For a second attachment point, apply acrylic (~0.03 ml) to the side connector to connect the pipette to the side support (Fig. 1f). Wait for ~5 min for the acrylic to harden (*see* **Note 14**).

3.6 Waking up the Animal

1. Cut the air tube, open the gripper, use the motorized manipulator to move the gripper away from the pipette (Fig. 1f), remove the eye cover, detach the head stage/LED assembly from the manual manipulator, release the animal from the stereotaxic apparatus (i.e., toothbar and earbars), then carefully carry the animal to the behavioral arena.
2. Inject the antagonist mix to counteract the anesthetics. The animal should wake up within ~3 min (*see* **Note 15**).

3.7 Recording the Data

1. Record simultaneous electrophysiological and behavioral (video and LED position tracking) data (Fig. 2).

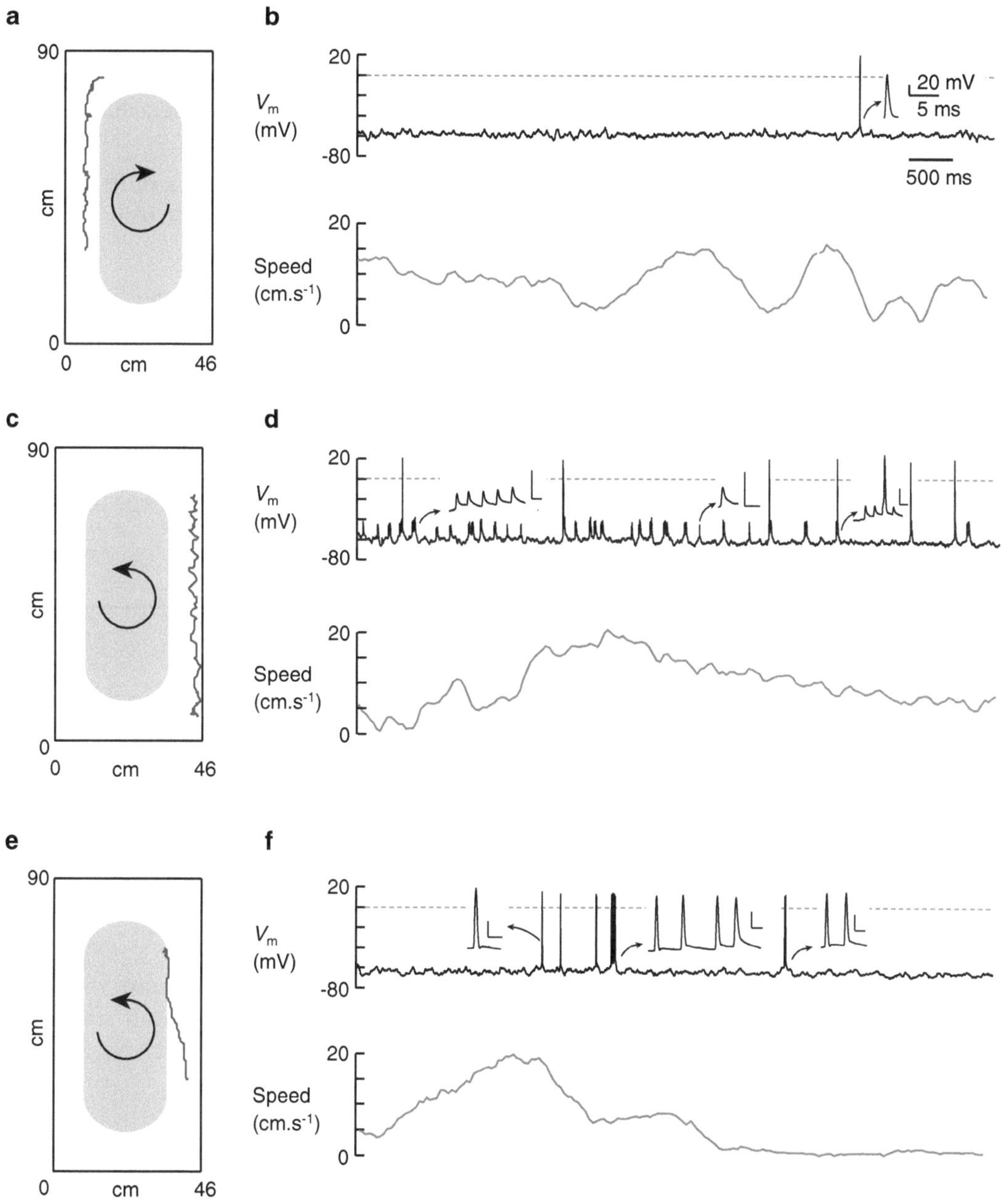

Fig. 2 Whole-cell recordings of hippocampal neurons in freely moving rats. (**a**, **c**, **e**) Trajectory (*dark gray*) of three different animals in an "O"-shaped maze (the *light gray area* at the center of the maze is not accessible) during the three recordings shown in (**b**, **d**, **f**). The *arrow* indicates the corresponding clockwise or counter-clockwise direction of animal movement. (**b**, **d**, **f**) Membrane potential (*black trace*, *top*) of three different CA1 pyramidal cells recorded in three different freely moving rats and corresponding speed of their head (*gray trace*, *bottom*). Single action potentials (**b**, **f**), isolated spikelets or bursts of spikelets with and without spikes (**d**) and bursts of spikes (**f**) are shown at expanded time scales (*arrows*). All scale bars in the insets are 20 mV and 5 ms. The 500 ms scale bar in (**b**) also applies to recordings in (**d**) and (**f**)

3.8 Recovering the Morphology of the Recorded Neuron

1. After the recording session (e.g., when the cell has been lost), inject an overdose of anesthetics (e.g., ketamine 200 mg/kg i.p.), wait until the animal is deeply anesthetized (administering more anesthetics as necessary), then detach the head stage/LED assembly from the head.
2. Perfuse the brain transcardially with phosphate-buffered saline followed by a 4 % (wt/vol) paraformaldehyde solution using standard methods.
3. Gently pull the pipette out from the brain and anchoring material along its axis (to minimize damage to the brain).
4. Extract the brain and store it overnight at 4 °C in a 4 % paraformaldehyde solution (*see* **Note 16**).
5. Slice the brain with a vibratome (75–150-μm-thick sections) then process the slices using the avidin–biotin–peroxidase method [32] to visualize the morphology of the recorded, biocytin-filled cell.

4 Notes

1. Alternatively, biocytin can be added when making the original internal solution but each aliquot should be refiltered before every experiment.
2. Prepare in advance to let the acrylic harden for at least 1 h before use.
3. Prepare in advance to allow the silicone glue time to cure overnight. This part is fragile so it is advisable to have a spare one prepared.
4. The whole-cell head stage is a miniature (~12 mm × 12 mm × 12 mm) version of a standard patch-clamp head stage. It is connected via a long cable to a standard patch-clamp amplifier and data acquisition system.
5. The weights can be properly set by placing the head stage/LED assembly (without being attached to an animal) in the behavioral arena and adjusting the weights so that the assembly hangs in the air at the height of an animal's head.
6. Appropriate regulations and guidelines for animal experiments must be followed for the entire procedure.
7. Perform craniotomy under a stereomicroscope. The goal is to remove the bone but leave the dura intact. The dura lies just below the bone so avoid drilling through the bone and dura all at once. Using a small drill bit (e.g., 0.45 mm diameter), thin the bone in a 1.5–2.0 mm diameter region centered on the target point, then drill gently around the edge until the bone begins to break. Using fine-tipped forceps (e.g., #5 forceps),

gently break the bone around the circumference then lift away the bone in the center in one piece.

8. Perform duratomy at high magnification (e.g., 60×) under the stereomicroscope. The dura must be completely removed to ensure that the patch pipettes can enter the brain cleanly or else the probability of forming a gigaseal will be much lower.
9. The side support must be (a) vertical so that it will be parallel to the pipette and (b) positioned at a specific distance from where the pipette will be so that the side connector attached to the pipette can reach it.
10. Without an accurate depth estimate, it will be much more difficult to obtain a whole-cell recording from a hippocampal pyramidal cell layer neuron. A similar procedure could be used for targeting other cells in other brain areas as long as there is a clear electrophysiological signature of these cells or the region where they are. However, if the goal is to obtain a superficial or deep layer cortical neuron, Subheading 3.3 could be skipped and the neuron identified post hoc through the biocytin staining procedure.
11. For a more shallow target than the hippocampus (e.g., superficial cortex), use a lower pressure such as ~200–300 mbar.
12. The agar should be applied as a liquid to the brain with as low a viscosity as possible provided that it is not too hot. Draw heated liquid agar with a syringe, wait for a short duration to allow it to cool down (the temperature can be tested by applying small amounts onto one's finger), then apply into the plastic ring. After obtaining the whole cell-recording the agar will harden and protect the brain surface from dental acrylic during anchoring of the pipette.
13. Difficulty in obtaining a high-quality whole-cell recording could be due to several reasons. First, pipettes may not be entering the brain cleanly (often indicated by a large decrease in the current response amplitude upon entry). In this case, clean the brain surface of any unremoved dura, blood, etc. Second, pipettes may not be clean when reaching the target depth (indicated by a smaller current response amplitude at the target depth when compared with the original value at the brain surface). In this case, place the pipette at a different position on the brain surface (since pipettes may have been passing through large blood vessels beneath the brain surface). Third, the criterion used for deciding when to reduce pressure to zero and attempt to form a gigaseal could be incorrect. Try changing the total distance covered by the pipette after observing the initial decrease in current response amplitude before reducing pressure to zero. Fourth, the pipette tip may not be optimal. Try increasing the resistance of pipettes or changing

the shape of the pipette tip. Fifth, the osmolality of the internal solution may not be in the target range. Try preparing a new internal solution. Finally, the animal may be unhealthy due to anesthetic administration. In this case, reduce the depth of anesthesia in future experiments.

14. Use a syringe and moderately wide hypodermic needle (e.g., 19 gauge) to apply acrylic precisely. Monitor the membrane potential (in current clamp mode) and series resistance (by injecting intermittent hyperpolarizing current steps) during the anchoring process to assess ongoing recording quality. During hardening of the acrylic, recordings can be lost (indicated by a large depolarization of the baseline membrane potential and/or a large increase in series resistance). This could be because the acrylic used for anchoring shrinks and moves the pipette. To reduce this problem one could decrease the amount of acrylic used for anchoring, increase the force with which the gripper holds the pipette, replace the gripper (if it appears that it is getting weaker due to mechanical rundown after use in many experiments), decrease the distance between the first anchoring layer and the bottom of the gripper, or use a light-cured adhesive system that yields less overall shrinkage [30].
15. Animals may exhibit a low level of behavioral activity after waking up. To address this, one could shorten the duration of anesthesia, though this would reduce the number of attempts to obtain a whole-cell recording. Also, the amount of antagonists to the anesthetics administered could be nonoptimal. Finding the right amount of antagonists to administer is somewhat complex. First, determine the total amount of anesthetic mix administered (i.e., the initial dose plus the supplemental doses), then calculate the amount of antagonist mix theoretically needed to fully antagonize it. Try varying the actual amount of antagonist mix administered between ~100 and 150 % of this theoretical amount.
16. The brain should not be stored in the paraformaldehyde solution for more than 1–2 days. Afterwards, it may be stored in phosphate-buffered solution for ~1 week.

Acknowledgments

This work was supported by an EMBO Long Term Fellowship and HHMI (to A. K. L.); a Human Frontier Science Program Long Term Fellowship, INSERM, Agence Nationale de la Recherche (grant ANR-09-BLAN-0259-01 and ANR-10-R11014AA), Région PACA (project EPIVIRT), A*MIDEX project (ANR-11-IDEX-0001-02) funded by the «Investissements d'Avenir» French

Government program (to J.E.); Humboldt Universität zu Berlin, the Bernstein Center for Computational Neuroscience Berlin, the German Federal Ministry of Education and Research (BMBF, Förderkennzeichen 01GQ1001A), NeuroCure, a European Research Council grant, and the Gottfried Wilhelm Leibniz Prize of the DFG (to M.B).

References

1. Hamill OP, Marty A, Neher E, Sakmann B, Sigworth FJ (1981) Improved patch-clamp techniques for high-resolution current recording from cells and cell-free membrane patches. Pflugers Arch 391:85–100
2. Pei X, Volgushev M, Vidyasagar TR, Creutzfeldt OD (1991) Whole cell recording and conductance measurements in cat visual cortex in-vivo. Neuroreport 2:485–488
3. Ferster D, Jagadeesh B (1992) EPSP-IPSP interactions in cat visual cortex studied with in vivo whole-cell patch recording. J Neurosci 12:1262–1274
4. Borg-Graham L, Monier C, Fregnac Y (1996) Voltage-clamp measurement of visually-evoked conductances with whole-cell patch recordings in primary visual cortex. J Physiol Paris 90: 185–188
5. Margrie TW, Brecht M, Sakmann B (2002) In vivo, low-resistance, whole-cell recordings from neurons in the anaesthetized and awake mammalian brain. Pflugers Arch 444:491–498
6. Wehr M, Zador AM (2003) Balanced inhibition underlies tuning and sharpens spike timing in auditory cortex. Nature 426:442–446
7. Bruno RM, Sakmann B (2006) Cortex is driven by weak but synchronously active thalamocortical synapses. Science 312:1622–1627
8. Covey E, Kauer JA, Casseday JH (1996) Whole-cell patch-clamp recording reveals subthreshold sound-evoked postsynaptic currents in the inferior colliculus of awake bats. J Neurosci 16:3009–3018
9. Aksay E, Gamkrelidze G, Seung HS, Baker R, Tank DW (2001) In vivo intracellular recording and perturbation of persistent activity in a neural integrator. Nat Neurosci 4:184–193
10. Wilson RI, Turner GC, Laurent G (2004) Transformation of olfactory representations in the Drosophila antennal lobe. Science 303: 366–370
11. Crochet S, Petersen CC (2006) Correlating whisker behavior with membrane potential in barrel cortex of awake mice. Nat Neurosci 9: 608–610
12. Harvey CD, Collman F, Dombeck DA, Tank DW (2009) Intracellular dynamics of hippocampal place cells during virtual navigation. Nature 461:941–946
13. Maimon G, Straw AD, Dickinson MH (2010) Active flight increases the gain of visual motion processing in Drosophila. Nat Neurosci 13: 393–399
14. Domnisoru C, Kinkhabwala AA, Tank DW (2013) Membrane potential dynamics of grid cells. Nature 495:199–204
15. Schmidt-Hieber C, Häusser M (2013) Cellular mechanisms of spatial navigation in the medial entorhinal cortex. Nat Neurosci 16: 325–331
16. Haider B, Häusser M, Carandini M (2013) Inhibition dominates sensory responses in the awake cortex. Nature 493:97–100
17. Long MA, Lee AK (2012) Intracellular recording in behaving animals. Curr Opin Neurobiol 22:34–44
18. Margrie TW, Meyer AH, Caputi A, Monyer H, Hasan MT, Schaefer AT, Denk W, Brecht M (2003) Targeted whole-cell recordings in the mammalian brain in vivo. Neuron 39: 911–918
19. Gentet LJ, Avermann M, Matyas F, Staiger JF, Petersen CC (2010) Membrane potential dynamics of GABAergic neurons in the barrel cortex of behaving mice. Neuron 65:422–435
20. Polack PO, Friedman J, Golshani P (2013) Cellular mechanisms of brain state-dependent gain modulation in visual cortex. Nat Neurosci 16:1331–1339
21. Fenno L, Yizhar O, Deisseroth K (2011) The development and application of optogenetics. Annu Rev Neurosci 34:389–412
22. Hölscher C, Schnee A, Dahmen H, Setia L, Mallot HA (2005) Rats are able to navigate in virtual environments. J Exp Biol 208: 561–569
23. Chen G, King JA, Burgess N, O'Keefe J (2013) How vision and movement combine in the hippocampal place code. Proc Natl Acad Sci U S A 110:378–383

24. Ravassard P, Kees A, Willers B, Ho D, Aharoni D, Cushman J, Aghajan ZM, Mehta MR (2013) Multisensory control of hippocampal spatiotemporal selectivity. Science 340: 1342–1346
25. Lee AK, Manns ID, Sakmann B, Brecht M (2006) Whole-cell recordings in freely moving rats. Neuron 51:399–407
26. Lee AK, Epsztein J, Brecht M (2009) Head-anchored whole-cell recordings in freely moving rats. Nat Protoc 4:385–392
27. Lee AK, Epsztein J, Brecht M (2008) Whole-cell recordings of hippocampal CA1 place cell activity in freely moving rats. Soc Neurosci Abstr 690:21
28. Epsztein J, Lee AK, Brecht M (2010) Impact of spikelets on hippocampal CA1 pyramidal cell activity during spatial exploration. Science 327:474–477
29. Epsztein J, Brecht M, Lee AK (2011) Intracellular determinants of hippocampal CA1 place and silent cell activity in a novel environment. Neuron 70:109–120
30. Lee D, Lin BJ, Lee AK (2012) Hippocampal place fields emerge upon single-cell manipulation of excitability during behavior. Science 337:849–853
31. Long MA, Jin DZ, Fee MS (2010) Support for a synaptic chain model of neuronal sequence generation. Nature 468:394–399
32. Horikawa K, Armstrong WE (1988) A versatile means of intracellular labeling: injection of biocytin and its detection with avidin conjugates. J Neurosci Methods 25:1–11

Part III

Biophysics

Chapter 18

Pressure-Polished Borosilicate Pipettes are "Universal Sealer" Yielding Low Access Resistance and Efficient Intracellular Perfusion

Marco Aquila, Mascia Benedusi, Anna Fasoli, and Giorgio Rispoli

Abstract

Whole-cell recording is the most widely used configuration of the patch recording technique, mainly because it allows to manipulate the intracellular environment while recording membrane current. However, the patch pipette tapered shank and the small tip opening give high access resistances and preclude efficient exchange between pipette solution and cell cytosol. Independently by the recording configuration, another problem of this technique is to gain consistently tight seals.

Here we describe a method to enlarge the pipette shank without affecting the tip opening diameter, through the calibrated combination of heat and air pressure, with a custom-made inexpensive setup. These pressure-polished pipettes give small access resistances and allow for the accommodation of pulled quartz or plastic perfusion tubes very close to the pipette tip (to deliver exogenous molecules into the cytosol with a controlled timing). Finally, we describe a method to consistently attain seals with pipettes made from just *one* glass type, for a wide variety of cell types, isolated from different amphibian, reptilian, fish, and mammalian tissues, and on artificial membranes composed of many different lipid mixtures.

Key words Electrophysiology, Patch clamp, Whole cell, Sensory cells, Muscle cells, Cell culture, Neurons, Artificial membranes

1 Introduction

Among the various recording configurations of the patch-clamp technique, whole-cell recording is the most widely used one, because it allows to record the ion currents through the cell membrane and the concurrent manipulation of the intracellular environment via the patch pipette. However, the pipette tapered shank and the small tip opening yield high access resistances, giving errors, in voltage clamp configuration, in membrane potential control and in the recording of waveform of current onset and offset (for instance in response to the application of a voltage jump or to an agonist), especially in the presence of large currents.

Marzia Martina and Stefano Taverna (eds.), *Patch-Clamp Methods and Protocols*, Methods in Molecular Biology, vol. 1183,
DOI 10.1007/978-1-4939-1096-0_18,

Unfavorable pipette geometry also slows down the solute exchange between pipette and cell cytosol [1]. This problem becomes dramatic with small cells, as the recording from outer segment of cone photoreceptors requires a small tip opening and consequently even longer and thinner shanks, both yielding an even higher access resistance.

Independently by the recording configuration, another problem of the patch-clamp recording is to attain consistently tight seals on the various membrane substrates. In fact, despite the worldwide use of this 40-year-old technique, the molecular interactions producing the seal between membrane and glass are still debated [2, 3]. Consequently, all sort of "laboratory tales" mushroomed, for instance which kind of glass could be the most appropriate for a certain cell type, how to fabricate pipettes, prepare cells, handle and formulate intracellular and extracellular solutions, etc., resulting in a day-to-day variability in attaining the seal (brilliantly described in ref. 4).

In this paper we describe a method to "swell" the pipette shank through the controlled application of heat to the shank to soften it, and air pressure in the pipette lumen. This method, which improves an existing one [5–7], is realized with a custom-made inexpensive setup allowing to fabricate pipettes with a shank as widened as desired, while the tip opening diameter could be independently increased, decreased, or left unchanged. "Pressure polished" pipettes as shown in Fig. 3e typically yield access resistances fourfold smaller in respect to the resistances obtained with conventional ones (Fig. 3b; [5]), even when recording from very small cells, such as the outer segment of cones mechanically isolated from the zebrafish retina [8]. Moreover, the shank geometry of these pipettes allows to insert in their lumen pulled quartz or plastic perfusion tubes very close to the tip, allowing the controlled and fast cytosolic incorporation of exogenous molecules. Finally, a method is described to fabricate, handle, and use pipettes of just *one* glass type, able to consistently attain seals virtually on any cell or artificial membrane target.

2 Materials

1. Patch pipettes: 50 or 100 μl borosilicate glass microcaps (Drummond, Broomall, PA, USA).
2. A commercially available puller.
3. A spare pipette holder with a side port, usually supplied with the patch clamp amplifier (Fig. 1, *right*).
4. A vertical bright light microscope (Fig. 1, *left*), equipped with a 10×, 20×, and a long working distance 32× objectives, and an xy microscope stage.

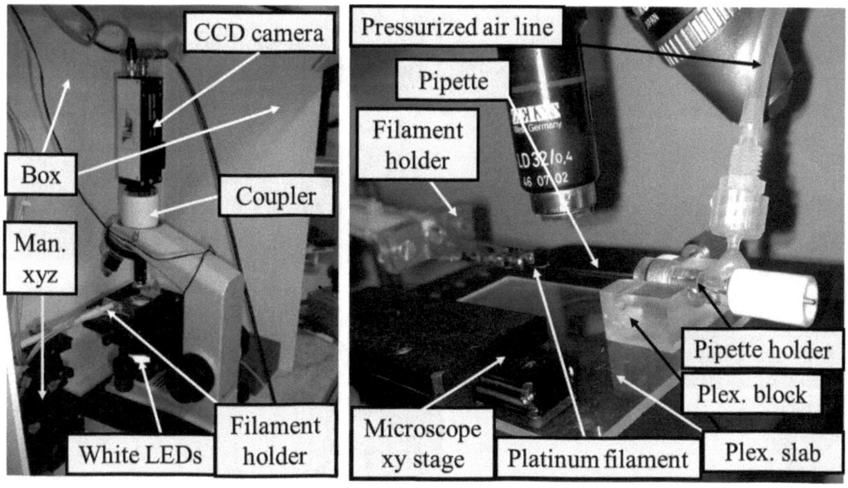

Fig. 1 Setup to fabricate the pressure-polished pipettes. *Left*, view of the microscope, the camera coupled to the objective nosepiece, the microscope stage, the filament holder, the LED illuminator enclosed in the box, and the micromanipulator (Man.); *right*, closer view of the microscope stage, the pressurized pipette holder, and the platinum filament; Plex.: plexiglass. Reproduced with permission from ref. 5

5. A plexiglass block with a groove (Fig. 1, *right*) to host the pipette holder that is firmly clamped to the block by a knob-ending screw; this block is glued to a transparent plexiglass slab, screwed to the microscope stage.
6. A platinum filament (50 μm), shaped in the form of an omega (Fig. 2) by using a needle of a 1 ml syringe and fine forceps: this shape ensures the homogeneous softening of the pipette shank, if it is placed into the central bend of this filament (Fig. 2, *bottom*), when the latter is heated to red color by passing a constant current through it.
7. Patch pipette glass powder, obtained by triturating few capillaries with a mortar and a pestle, in which to dip the filament heated to yellow color, to uniformly cover it by glass to avoid metal evaporation onto the pipette.
8. A two-prong copper holder (which also functions as a heat sink, Fig. 2, *top*), to which the filament is tin-soldered, mechanically coupled to a micromanipulator, and electrically connected to a variable current generator.
9. A conventional xyz micromanipulator to move the filament holder (Fig. 1, *left*);
10. A plastic box enclosing the entire pressure polishing setup to protect the filament from air currents (that could erratically reduce its temperature) and the entire setup from dust; this box be must have an opening on the front side to allow loading and unloading the pipette holder (with the pulled pipette screwed in).

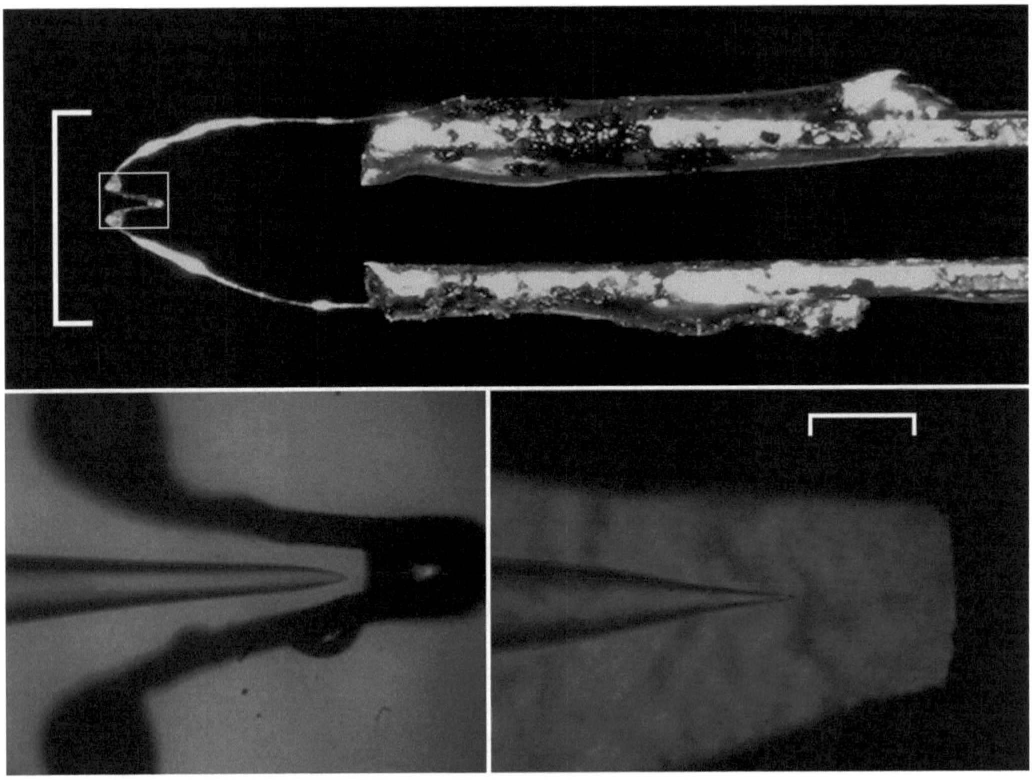

Fig. 2 Filament shape and pipette alignment. *Top*, shape of the glass-coated platinum filament tin-soldered to the holder; scale bar: 5 mm. The region within the white box is enlarged at *bottom left*. A pipette is also shown, properly aligned with the filament to ensure uniform pipette heating at the end of a mild pressure polishing. The "omega" shaped region of the filament is further enlarged at *bottom right*, where a pipette before pressure-polishing is also shown, correctly aligned with the filament; scale bar: 100 μm. Reproduced with permission from ref. 5

11. A variable high-current generator (0–2 A; custom-made on the basis of the circuit reported in a web site: [9]), remotely connected or not to the filament by a "push-to-make" switch, positioned outside the box.
12. To follow the pressure polish process, it necessary to view it on a LCD monitor positioned externally to the box: to this aim, the eyepiece is removed, and a contrast-intensified CCD camera is coupled to the objective nosepiece through an ocular, a 0.6 neutral density filter, and an infrared filter (the filters, positioned in the custom made optical "coupler" shown in Fig. 1, left, are necessary to cut the intense infrared light produced by the filament, which may saturate the camera). The video signal may be recorded by digitizing it online, stored it on a computer, and further process it offline.
13. A custom-made light source replacing the halogen lamp illuminator of the microscope to avoid the heat produced by the latter increases the temperature inside the box, preventing the fine control of the filament and the pipette temperature.

This light source is the top portion of a commercial 5-white-LEDs flashlight (Fig. 1, *left*), coupled to a frosted filter and the regulated low-current generator to even out the light intensity in space and time.

14. A variable low-current generators (0–10 mA), to regulate the intensity of the microscope light source (made with five white LEDs).
15. A pressurized air line from a compressor or a cylinder, equipped with an adjustable pressure adaptor and a 0.2 μm filter.
16. A three-way valve positioned outside the box: the center connected to the pipette, one end to the pressurized air line, and the other end left open (so to connect the pipette to the pressurized air or vent it to air).
17. A three-way valve positioned close to the patch-clamp setup: the center connected to the pipette, one end to a mouthpiece or to a 1 ml syringe (to deliver small controlled positive or negative pressures to clean the cell with the pipette solution and to gain the seal, respectively), the other end to a 50 ml syringe to keep a strong, positive pressure (used when the pipette enter in the bath and is moved close to the cell).

3 Methods

3.1 Pressure Polish Procedure

The pressure procedure is controlled by five parameters listed below:

1. The position of the pipette relatively to the omega-shaped, glass-coated platinum filament (Fig. 2).
2. The intensity of the current flowing in the filament (set to ~1.2 A for filaments as long as the one shown in Fig. 2, that would bring it to reddish color); the duration of this current flow.
3. The intensity of the pressure applied in the pipette lumen (~4 atm).
4. The duration of this pressure application as the current is flowing in the filament.
5. The most simple and efficient strategy to finely shape the shank profile and the tip opening section is to optimize and keep fixed the parameters *a*, *b*, and *d* (whose value is set in parenthesis), while the *c* and *e* ones are adjusted in order to obtain the desired pipette geometry. The pressure-polish procedure can be divided into the following steps:
   (a) Fabricate the patch pipette from borosilicate glass microcaps with the usual two-stage pulling procedure. To keep the shank length short, use 50 μl microcaps for recording from small cells (having the longer side <20 μm, as the sensory cells) or 100 μl microcaps for larger ones (up to ~300 μm, as the muscle fibers).

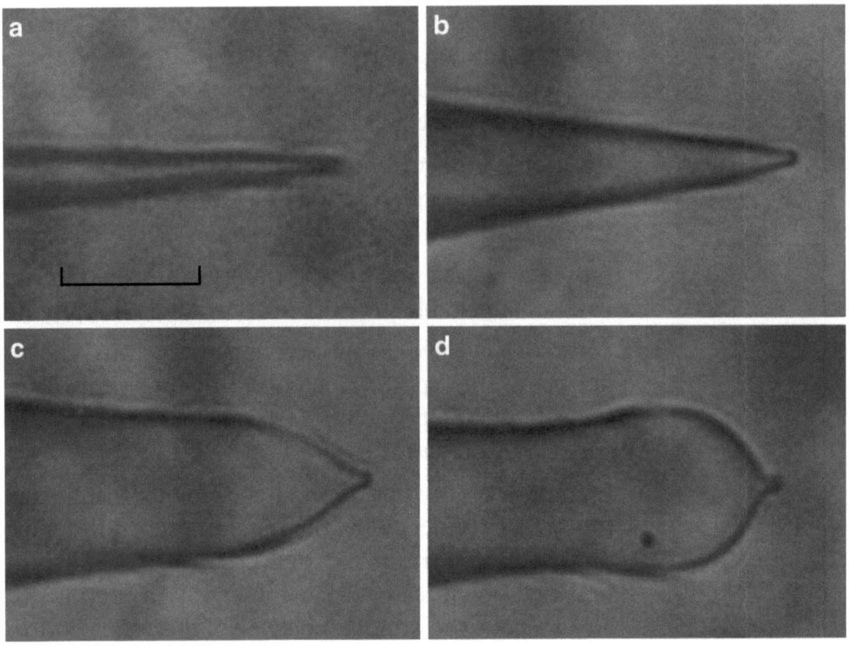

Fig. 3 Different pipette shanks obtained with the pressure fire polishing. (**a**) pulled pipette; the same pipette fire polished without pressure is undistinguishable from the pulled one at this magnification; (**b**–**d**) pulled pipettes similar to (**a**), fire polished with progressively longer pressure durations; scale bar = 20 μm for all panels. Reproduced with permission from ref. 5

(b) Carefully tighten the pulled pipette into the holder (Fig. 1, *right*) to avoid it to be displaced when the pressure is applied to the pipette lumen. Clamp the holder to the microscope stage (Fig. 1, *right*).

(c) Move the pipette holder by means of the xy manipulator of the microscope stage, and the filament with the xyz micromanipulator, to align each other as shown in Fig. 2.

(d) Set the three-way valve so as to connect the pipette to the pressurized air line.

(e) Turn on the current generator with the "push-to-make switch" and keep it on, while following the shaping process of the pipette shank on the LCD monitor, until the desired shape is attained (Fig. 3). A calibration grating superimposed on the monitor would help to precisely and reproducibly shape the shank. As expected, long heating applications (while applying the pressure) give very enlarged shanks. If the tip opening section results too large (as judged by the pipette electrical resistance), then the tip can be heated again in the absence of pressure until the desired section is achieved. If it is instead too small, then apply that pressure duration to pipettes pulled with larger tips, by reducing the current intensity of the puller filament during the second stage pull.

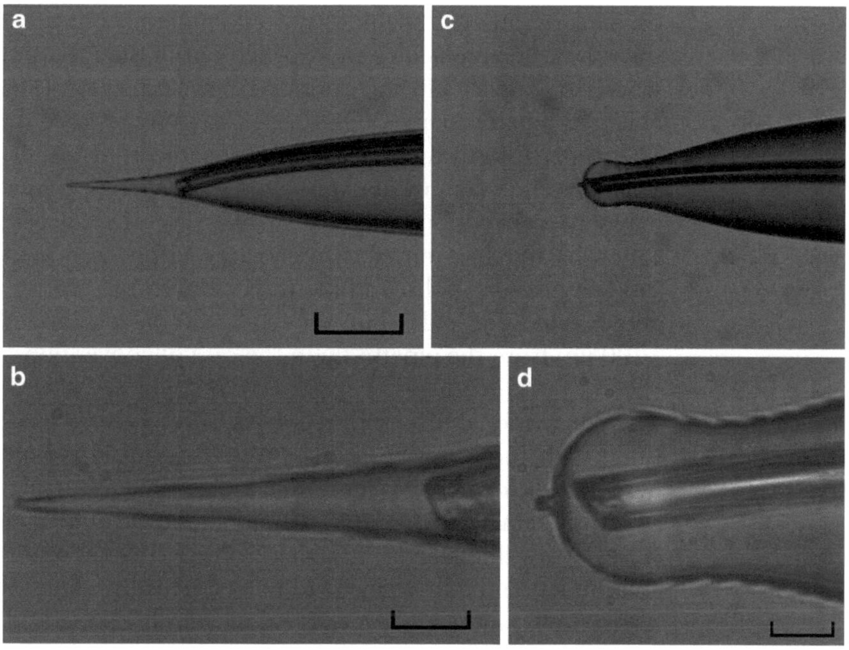

Fig. 4 Controlled intracellular perfusion. (**a**) A pulled quartz tube is positioned inside the lumen of a conventional pipette as close as possible to its tip, magnified in (**b**); (**c**) the same tube is inserted in a pressure polished pipette, magnified in (**d**). Scale bar = 100 μm in (**a**) and (**c**), and 20 μm in (**b**) and (**d**). Reproduced with permission from ref. 5

(f) Set the three-way valve so to vent the pipette to air and remove the pipette, that is now ready to be used, otherwise it must be stored in a clean and air-tight jar.

3.2 Intracellular Perfusion

With respect to conventional pipettes, the pressure-polished ones yield a dramatic improvement of the electrical recordings given by the huge access resistance decrease, and they doubled the rate of molecular diffusion between the pipette and the cell interior [5]. Many experiments require the intracellular delivery of dyes, proteogenic molecules, drugs, etc. during the whole-cell recording, but often they cannot be added to the pipette solution, because they may stick to the glass tip and reduce the probability or even prevent the seal formation [10]. In these cases, it is required to add these molecules to the pipette solution at a desired time, at least any time after attaining the seal. In any case, even the time-controlled intracellular delivery of molecules not affecting the seal formation is extremely important: this delivery may be necessary for instance after having performed some control experiments. Molecules dissolved in the intracellular solution could be pressure-ejected, with a controlled timing, out of pulled quartz or plastic tubes (which can be easily softened with a Bunsen flame, as polyethylene or polypropylene) placed into the pipette lumen. These tubes could be accommodated very close to the tip (i.e., very close to the cell cytosol; Fig. 4) of these pipettes fabricated with a particularly enlarged shank: this would minimize the delay between the time when molecules are expelled out of the tube and the time when they are actually incorporated in the cell cytosol. Indeed, the enlarged pipette shank (Fig. 4b, d) allows to advance the smallest quartz perfusion tube that could be handled (~10 μm of internal diameter and ~18 μm of external one) by further 95 μm as compared to conventional pipette shanks (Fig. 4a, c). More robust tubes, having an external diameter of ~45 μm, can be positioned inside a pressure-polished pipette as close to the pipette tip as the above one (Fig. 4d), but they have to be positioned at least ~200 μm from the pipette tip in the case of conventional pipettes.

3.3 Sealing on Any Membrane Target with Just One Glass Type

The conventional fire-polished or pressure-polished pipettes fabricated with the borosilicate glass microcaps, handled and used as described below (*see* **Notes 1** and **2**), are able to attain tight seals on an extremely wide range of membrane substrates, as sensory cells (labyrinthic hair cells, retinal rods and cones from reptilian, amphibian, and fish retinae) [11–14], neurons (horizontal cells, bipolar cells, and ganglion cells enzymatically dissociated from the turtle retina) [15], culture cells (HEK 293, NB4, MDA-MB-231, COS, and HeLa) [16], cardiac myocytes and skeletal muscle fibers, and erythrocytes from amphibian, reptilian, fish, and human blood. These pipettes sealed perfectly on giant unilamellar vesicles (GUVs) as well, prepared by electroformation [17] using any combination of these lipids: egg phosphatidylcholine, egg phosphatidylglycerol, cholesterol, dipalmitoylphosphatidylcholine, nitrobenzoxadiazole labeled phosphatidylethanolamine, and dansyl labeled phosphatidylethanolamine. Given the "sealing power" of the borosilicate glass microcaps, the pressure polishing technique described above was optimized just for this glass type only.

In order to have a very high rate of sealing on virtually *any* membrane surface with just *one* type of glass, the pipette must be kept thoroughly clean, as follows:

1. Carefully clean the microscope stage, the puller, the filament, and the pipette storage jar with a "cleaning solution" containing 50 % distilled water and 50 % ethanol, and blow the dust away from them with the filtered compressed air used for pressure polish.
2. Periodically, completely disassemble the pipette holder (used for the pressure polish and for the patch-clamp experiments) and sonicate the pieces submersed in a beaker containing the cleaning solution.
3. Clean the microcaps thoroughly with the cleaning solution, avoiding to touch them with bare hands especially in their center, i.e., where there will be thinned to form the pipette tips, and blow the microcaps interior with filtered compressed air.

4. Try to minimize the amount of organic debris in the bath, and keep the pipette exposed to air with no protection as little as possible.
5. Follow this protocol to attain the seal:
   (a) Focus the high magnification objective on the membrane targets (cells, GUVs, etc.) laying on the recording chamber bottom, while the pipette is still in air.
   (b) Move the chamber away with the xy microscope stage.
   (c) Move the pipette (still in air) to the same focus of the cell layer (do not move the microscope focusing knob).
   (d) Lift the pipette by using the coarse vertical manipulator only (z), just enough to allow to bring the chamber back on the objective without hitting the pipette tip (*see* **Note 3**).
   (e) Search a suitable membrane target.
   (f) Apply a steady strong pressure to the pipette lumen using the 50 ml syringe.
   (g) Enter the pipette into the liquid, by using the coarse z manipulator only, and bring it just above the chosen membrane target as fast as possible.
   (h) Release the strong pressure, but keep a positive pressure by mouth once the pipette is brought close to the membrane target (by using the fine xyz manipulator knobs), to clean it from debris but avoiding at the same time to blow away the target, especially if it is a small isolated cell or a GUV. This "pressure cleaning" does not suffice to remove connective tissue or other organic material sticking to the membrane target (which may be for instance a cell inserted in a tissue slice): in this case, it is necessary to use specific proteolytic enzymes, following known protocols described in literature [20].
   (i) Press the pipette tip on the membrane target and only at this point apply a gentle suction: the seal should form almost instantly if the membrane target is clean.

Summing up, a failure in attaining the seal is due to a soiled patch pipette tip, and/or because the membrane target is contaminated with cell debris, connective tissue, or other organic material (*see* **Note 4**).

4 Notes

1. At difference with what reported by many authors, but in agreement with Priel et al. [2], it is found that at physiological pH, the presence or the lack of Ca^{2+} in the patch pipette is not

necessary to promote the seal, which can be routinely obtained with solutions containing up to 30 mM of BAPTA or 30 mM Ca^{2+} [18, 19].

2. The conventional pipette fire polishing smooths the surface of the pipette tip and thickens its wall, resulting in a significant increase of the probability to gain the seal on all the membrane target tested: this is expected, because it allows a larger membrane surface to seal on the fire-polished tip in respect to the surface sealing on the rough and thin tip of a pipette just pulled.
3. The procedure of **step 5**, Subheading 3.3 can be greatly accelerated by memorizing with a step-motor manipulator the two positions, "home" (i.e. outside the bath) and "just above the membrane target."
4. It is sometimes possible to gain a seal with the same pipette more than once, especially when the cell extracellular matrix was broken down with enzymatic treatment [20], providing to move as quickly as possible while going from one cell to the other, to avoid soiling the pipette. Although the seal resistance is constantly decreasing from one trial to the other, it is possible to have seal resistances above 1 GΩ up to the fourth consecutive seal with the *same* pipette.

References

1. Pusch M, Neher E (1988) Rates of diffusional exchange between small cells and a measuring patch pipette. Pflugers Arch Eur J Physiol 411:204–211
2. Priel A, Gil Z, Moy VT et al (2007) Ionic requirements for membrane-glass adhesion and giga seal formation in patch-clamp recording. Biophys J 92:3893–3900
3. Suchyna TM, Markin VS, Sachs F (2009) Biophysics and structure of the patch and the gigaseal. Biophys J 97:738–747
4. Miller C (2009) Everything you always wanted to know about Sachs' seals. Biophys J 97:687
5. Benedusi M, Aquila M, Milani A et al (2011) A pressure-polishing set-up to fabricate patch pipettes that seal on virtually any membrane, yielding low access resistance and efficient intracellular perfusion. Eur Biophys J 40: 1215–1223
6. Goodman MB, Lockery SR (2000) Pressure polishing: a method for re-shaping patch pipettes during fire polishing. J Neurosci Methods 100:13–15
7. Johnson BE, Brown AL, Goodman MB (2008) Pressure-polishing pipettes for improved patch-clamp recording. J Vis Exp. doi:10.3791/964
8. Fasoli A, Benedusi M, Aquila M et al (2013) Modulation of cone photoresponse by whole-cell delivery of zGCAP3 and its monoclonal antibody. Eur Biophys J 42(Suppl 1):S47
9. Regan T (2005) Current sense circuit collection. Making sense of current. Application note 105, current control. http://www.linear.com/ad/19-current_control.pdf
10. Mathias RT, Cohen IS, Oliva C (1990) Limitations of the whole cell patch clamp technique in the control of intracellular concentrations. Biophys J 58:759–770
11. Martini M, Rossi ML, Rubbini G et al (2000) Calcium currents in hair cells isolated from semicircular canals of the frog. Biophys J 78:1240–1254
12. Aquila M, Benedusi M, Koch KW et al (2013) Divalent cations modulate membrane binding and pore formation of a potent antibiotic peptide analog of alamethicin. Cell Calcium 53:180–186
13. Aquila M, Benedusi M, Milani A et al (2012) Enhanced patch-clamp technique to study antimicrobial peptides and viroporins, inserted in a cell plasma membrane with fully inactivated endogenous conductances. In: Kaneez FS (ed) Patch clamp technique. InTech, Rijeka, pp 329–356

14. Moriondo A, Rispoli G (2010) The contribution of cationic conductances to the potential of rod photoreceptors. Eur Biophys J 39: 889–902
15. Tatsukawa T, Hirasawa H, Kaneko A et al (2005) GABA-mediated component in the feedback response of turtle retinal cones. Vis Neurosci 22:317–324
16. Navangione A, Rispoli G, Gabellini N et al (1997) Electrophysiological characterization of ionic transport by the retinal exchanger expressed in human embryonic kidney cells. Biophys J 73:45–51
17. Stella L, Pallottini V, Moreno S et al (2007) Electrostatic association of glutathione transferase to the nuclear membrane. Evidence of an enzyme defense barrier at the nuclear envelope. J Biol Chem 282:6372–6379
18. Rispoli G (1998) Calcium regulation of phototransduction in vertebrate rod outer segments. J Photochem Photobiol B 44:1–20
19. Rispoli G, Navangione A, Vellani V (1995) Transport of K+ by Na(+)-Ca2+, K+ exchanger in isolated rods of lizard retina. Biophys J 69: 74–83
20. Corey D, Stevens C (1982) Science, and technology of patch recording electrodes. In: Sakmann B, Neher E (eds) Single-channel recording, 1st edn. Plenum, New York, pp 53–68

Chapter 19

Current Recordings at the Single Channel Level in Adult Mammalian Isolated Cardiomyocytes

Romain Guinamard, Thomas Hof, and Laurent Sallé

Abstract

This chapter describes appropriate methods to investigate mammalian cardiac channels properties at the single channel level. Cell isolation is performed from new born or adult heart by enzymatic digestion on minced tissue or using the Langendorff apparatus. Isolation proceeding is suitable for rabbit, rat, and mouse hearts. In addition, isolation of human atrial cardiomyocytes is described. Such freshly isolated cells or cells maintained in primary culture are suitable for patch-clamp studies. Here we describe the single channel variants of the patch-clamp technique (cell-attached, inside-out, outside-out) used to investigate channel properties. Proceedings for the evaluation of biophysical properties such as conductance, ionic selectivity, regulations by extracellular and intracellular mechanisms are described. To illustrate the study, we provide an example by the characterization of a calcium-activated non-selective cation channel (TRPM4).

Key words Single channel, Patch clamp, Cell isolation, Mammalian cardiomyocytes, Langendorff, Calcium-activated non-selective cation channel, TRPM4

1 Introduction

Determination of ion currents in cardiomyocytes primarily requires isolated cells. Isolated myocytes have the advantages to provide a uniform population of a single cell type free of neuronal and humoral influences in a controllable environment. These cells are suitable for electrophysiological studies, but also are used in metabolic, receptor and ligand binding, pharmacological or immunological studies, and evaluation of intracellular dynamics using fluorescence. The highest difficulties to obtain isolated cells come from the facts that adult cardiomyocytes are highly differentiated then do not multiply anymore and are strongly associated together in the cardiac tissue, thus complicating cell dissociation and culture.

Concerning laboratory mammals, adult cardiomyocytes dissociation is achieved using either heart mincing and enzymatic digestion for atrial cells or the Langendorff method, described by the end of the nineteenth century for ventricular cells [1].

Marzia Martina and Stefano Taverna (eds.), *Patch-Clamp Methods and Protocols*, Methods in Molecular Biology, vol. 1183,
DOI 10.1007/978-1-4939-1096-0_19,

This technique allows for the perfusion of the isolated heart by retrograde flow into the aorta with different buffers and enzymatic solutions that reach the coronary vascular system, providing enzymes at the depth of the tissue. Enzymes induce the degradation and loosing of the extracellular and collagen matrixes that hold the myocytes in the myocardium. As the aortic valve prevents the retrograde entrance of solutions, the ventricles do not fill with the perfusate, and then do not perform pressure-volume work, preventing cell injury due to ischemia. After isolation, cells can be used as freshly isolated cells or maintained in culture to follow their modifications in controlled conditions. The technique, which gives routinely high yields of calcium-tolerant myocytes from different parts of the heart, appears to be convenient and reliable for providing electrophysiologically useful cardiac single cells. It can be commonly used by any investigator and easily applied to the heart dissociation of small animals [2].

Dissociated cells are suitable for both whole-cell and single-channel electrophysiological recordings, including the physiological cell-attached configuration and both inside-out and outside-out excised configurations. While whole-cell recordings provide currents from thousands of channels, single channel proceedings give currents from a unique channel protein and thus provide information at the molecular level. If experiments are well conducted, single channel results can be considered as free of contamination by other currents. Thus, it is possible to identify precisely a channel type through its biophysical properties such as single channel conductance, rectification, open and close time constants, ionic selectivity, and voltage dependency. According to these results, a rectification observed at the whole-cell level will be attributed to a variation of single channel conductance and/or open probability upon voltage. In addition, the intracellular and extracellular channel modulators can be easily investigated since solutions bathing both sides of the membrane can be rapidly changed in the excised configurations. Also, the cell-attached configuration, considered as the more physiological condition, allows for the description of intracellular pathways involved in channel regulation. These approaches are suitable for most of the channel types with sufficiently large (few pS) single channel conductance to allow for individual events detection.

This chapter focuses on the proceedings to obtain freshly isolated adult cardiomyocytes using the Langendorff apparatus, their surviving in culture and the use of the patch-clamp technique to investigate channels properties at the unitary level. In the patch-clamp section, neither the patch-clamp setup nor software used for analysis is presented since both are basic apparatus. However, the proceedings to characterize channel biophysical and regulatory properties are described. To illustrate the study, the authors provide an example by the characterization of a calcium-activated non-selective cation channel (TRPM4) [3].

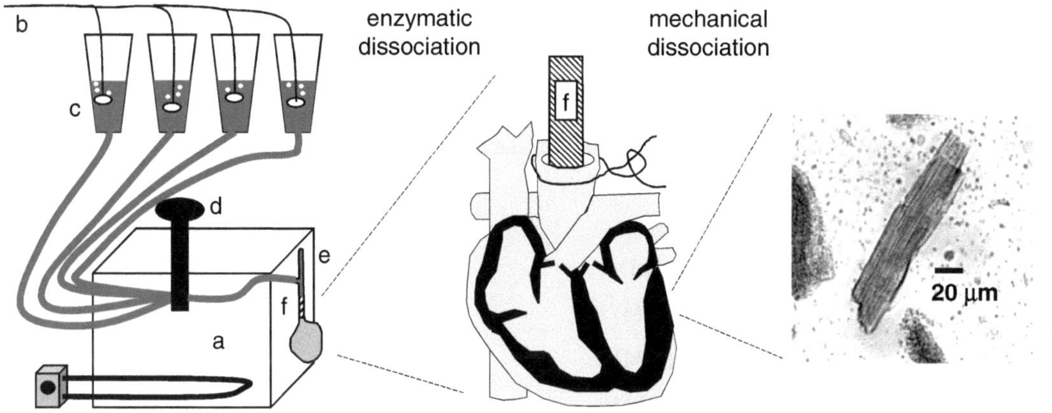

Fig. 1 Cardiomyocytes isolation using a Langendorff apparatus. The apparatus is composed of (*a*) water bath, (*b*) oxygenation system, (*c*) tank for solutions and tubulures, (*d*) faucet, (*e*) bubble trap, and (*f*) cannula. Introduction of the cannula into the aorta is schematized in the central magnification. *Right panel* is a phase-contrast micrograph of a freshly isolated adult rat ventricular cardiomyocyte (republished with permission from ref. 23)

2 Materials

2.1 Cell Dissociation and Culture

1. Langendorff apparatus.
   To perform cell-dissociation, we used the isolated perfused heart preparation, according to the Langendorff technique [1, 4] that is suitable for most warm-blooded animals including, rabbit, mouse, rat, ferret, and guinea pig. The aim of the technique is to perfuse the coronary vessels with solutions that allow for the digestion of the ventricular extracellular matrix. Simplistic apparatus could easily be home made. However, companies provide complete setups. The Langendorff apparatus is drawn in Fig. 1 and composed of:
   (a) A water bath to warm solutions at 37 °C.
   (b) A system to oxygenate the solutions.
   (c) Tubulures going from solutions tanks through the water bath to a faucet which allow for selecting the perfused solution.
   (d) A bubble trap.
   (e) An aortic cannula. It can be made in stainless steel (for example a sectioned needle), glass (hematocrit tube or patch-clamp pipette glass) or polyethylene tube with section and length that depend on the animal. The external diameter is typically similar or slightly larger than the aorta. For example: 2–4 cm in length and 1–2 mm in diameter for mouse, 2–3 mm for rat and 4–5 mm for rabbit hearts. Several small circumferential grooves are machined into the distal end of the cannula to prevent the aorta from slipping off.
   (f) A chamber to protect the heart.

Table 1
Ionic solutions used for atrial cardiomyocytes isolation

Solution	Low Ca^{2+}	Enzymatic	KB
NaCl (mM)	140	140	
KCl (mM)	5.4	5.4	70
$CaCl_2$ (mM)		0.07	0.08
$MgCl_2$ (mM)	1.8	1.8	
EGTA (mM)	0.1	0.1	0.5
Taurine (mM)			20
Glucose (mM)	10	10	10
HEPES (mM)	10	10	10
KH_2PO_4 (mM)			20
$MgSO_4$ (mM)			5
Na_2-ATP (mM)			5
K-glutamate (mM)			5
Creatine (mM)			5
Collagenase type I (i.u./ml)		100	
Protease type XIV (i.u./ml)		0.3	
pH (with NaOH or KOH for KB)	7.4	7.4	7.4

2. Cell isolation.
   The complete compositions of solutions used for cell isolation are given in Tables 1 and 2.
3. Cell culture.
   (a) The medium used for cardiomyocytes culture is M199 supplemented with 0.2 % BSA, 10-7 M insulin, 1 % antibiotics (penicillin G 100 i.u./ml and streptomycin 50 i.u./ml), and 10 μM cytosine 1-β-d-arabinofuranoside, an inhibitor of cell proliferation, including fibroblasts.
   (b) Cells are seeded in 35 mm petri dishes preincubated for 30 min with laminin (20 μg/ml).

2.2 Patch-Clamp

1. Patch-clamp setup.
   The patch-clamp setup used for the electrophysiological recordings includes an inverted microscope and a perfusion system connected to the petri dishes containing cardiomyocytes. Currents are recorded using a patch-clamp amplifier connected to a head stage and controlled by a computer connected via an A/D converter. A computer is also used for data

Table 2
Ionic solutions used for ventricular cardiomyocytes isolation using a Langendorff apparatus

Solution	Tyrode	Tyrode 0 Ca^{2+}	Enzymatic	Low Ca^{2+}	KB
NaCl (mM)	130	130	130	130	
KCl (mM)	5.4	5.4	5.4	5.4	70
$CaCl_2$ (mM)	1		0.1	0.1	0.08
$MgCl_2$ (mM)	1.4	1.4	1.4	1.4	
NaH_2PO_4 (mM)	0.4	0.4	0.4	0.4	
EGTA (mM)		0.1			0.5
Taurine (mM)	20	20	20	20	20
Creatine (mM)	10	10	10	10	
Glucose (mM)	10	10	10	10	10
HEPES (mM)	10	10	10	10	10
KH_2PO_4 (mM)					20
$MgSO_4$ (mM)					5
Na_2-ATP (mM)					5
K-glutamate (mM)					5
Creatine (mM)					5
Collagenase type I (i.u./ml)			100–150		
Protease type XIV (i.u./ml)			0.3		
pH (with NaOH or KOH for KB)	7.4	7.4	7.4	7.4	7.4

acquisition, storage, and analysis. Data are acquired using pClamp software (Molecular Devices, Sunnyvale, CA, USA).

2. Perfusion system.
   The perfusion system can differ for cell-attached and cell-free configurations experiments. For cell-attached recordings it is preferred a unique perfusion tube into the recording chamber connected to all the experimental solution tanks. At the opposite, for the cell-free configurations it is useful to have a superfusion tube for each solution. Such a system minimizes blending of the different solutions by putting the pipette at the entrance of the concerned tube.

3. Patch pipettes.
   Patch pipettes are made from borosilicate microhematocrits glass tubes pulled in two stages and coated with the hydrophobic silicon elastomer SYLGARD 184 (Dow Corning, Seneffe, Belgium) to reduce pipette capacitance and noise. After application on the

pipette, SYLGARD is polymerized using a heated wire. Before patching, the tip of the pipettes is heat-polished using a fire-polisher before filling with pipette solution (*see* **Note 1**). Good seals and recordings are achieved with a pipette tip resistance of 7–10 MΩ (in 140 mM NaCl solution).

4. Solutions for patch-clamp experiments.
   Standard 140 mM NaCl solution used for bath and pipette: 140 mM NaCl, 4.8 mM KCl, 1.2 mM $MgCl_2$, 10 mM glucose, and 10 mM HEPES. Pipette and bath solutions usually contain 1 and 1.8 mM $CaCl_2$, respectively.

3 Methods

3.1 Cells Preparation from Atria

Atrial wall is sufficiently thin to allow enzymes to reach the depth of the tissue. Thus, a simplistic enzymatic digestion can be achieved after mincing of the tissue by scissors. Note that such a proceeding is also suitable for ventricles from new born small mammals. Moreover, this proceeding can also be applied to human atrial specimens after merge modifications (*see* **Note 2**) [5] or even mouse sinus node cells [6].

1. Animal preparation.
   The proceeding is suitable for most laboratory mammals but is described here for rats. Adult animals of either sex are used; however, better results are obtained from young animals (approximately 1–2 months old weighing 0.25–0.3 kg for rats) (*see* **Note 3**).

   Animals are injected with heparin (5,000 i.u./kg, intraperitoneal). Anesthesia can be performed 10 min after injection. Inhalation of methoxyflurane or intraperitoneal injection of pentobarbital is suitable for anesthesia. If the last proceeding is applied, it is possible to inject heparin and pentobarbital at the same time. In some experiments, cervical dislocation is preferred, in accordance with guidelines for animal experimentations, to avoid potential side effects of anesthetics on electrophysiological results.

2. Heart excision.
   Once the animal is anesthetized, a transabdominal incision is performed. The thorax is opened and the thoracic cage is reflected over the animal head, exposing the heart. The aorta is pinched with forceps as high as possible and aorta, vena cava, and pulmonary vessels are incised. The heart is immersed in an ice-cold Tyrode solution previously bubbled with oxygen. Low temperature stops heart beating and protects cells.

3. Enzymatic dissociation for atrium.
   (a) Atria are isolated from the heart and finely minced with iridectomy scissors in an oxygenated low-calcium physiological solution (Table 1).

(b) After 4 min, the solution is changed to the enzymatic solution (Table 1) for digestion at 37 °C for 30–40 min.

(c) Pieces are then rinsed with Kraft-Brühe medium (KB) (Table 1).

(d) Pieces are gently triturated in KB using a fire-polished Pasteur pipette to free the cardiomyocytes.

(e) Cells are stored in KB in the petri dishes until use for patch-clamp experiments at the day of dissociation. Cells may also be prepared to be maintained in culture, following the proceeding that will be described for ventricular cells (*see* **step 4** in Subheading 3.2).

3.2 Cell Preparation from Adult Ventricle

After animal anesthesia and excision described in Subheading 3.1 (**steps 1** and **2**), the whole heart (atria and ventricles) is perfused in a retrograde way. Therefore five steps have to be performed for cells preparation: (a) washing of the preparation to clear the blood; (b) perfusion of the tissue with a calcium-free solution in order to weaken intercellular adhesion; (c) enzymatic digestion of extracellular matrix; (d) mechanical agitation to free the cells from the tissue; (e) cell seeding (*see* **Note 4**).

1. Aorta cannulation.

(a) Immediately after excision, the heart is immersed in an ice-cold Tyrode solution (Table 2) to stop beating and limit any ischemic injury during the period between the excision and the restoration of vascular perfusion.

(b) The cannula is mounted on a 10 ml syringe containing Tyrode solution (at 4 °C). One must be sure that no air bubbles are trapped in the cannula, to avoid the chance of air emboli.

(c) The cannula is introduced into the aorta. At this point it should be taken care that the position of the cannula is not too low to impede the aortic valves or the coronary ostia.

(d) The aorta is fixed on the cannula using a thread or a clamp.

(e) The syringe content is perfused gently into the coronary vessels to empty any traces of blood.

(f) Any surplus tissue can be trimmed away.

(g) The cannula is removed from the syringe and mounted on the Langendorff apparatus. Once again, a great care should be taken to avoid any air bubble go into the heart.

Isolation success highly depends on the time from heart excision to recirculation that should be reduced to only few minutes.

2. Heart perfusion.

(a) The heart is retrogradely perfused with the 37 °C Tyrode solution (about 3–4 min) at a perfusion rate of 10 ml/min using a peristaltic pump. Because of warming the heart recovers its spontaneous activity (*see* **Note 5**).

(b) Tyrode solution is switched to the calcium-free Tyrode solution (Table 2) for 4 min which causes total Ca^{2+} depletion, cessation of heart beat within a minute, and flimsiness of intercellular cements.

(c) The heart is then perfused for 8–12 min with the low calcium enzymatic solution containing collagenase and protease (Table 2). By the time of enzymes action, the heart becomes flaccid and its color lightens (*see* **Note 6**).

3. Mechanical isolation.

(a) Once flaccid, the heart is removed from the cannula. Ventricles are excised from the heart, minced, and bathed in 10 ml enzymatic solution (Table 2) supplemented with 1 % bovine serum albumin (BSA) in an Erlenmeyer, under oxygenation. The preparation is gently shaken at 37 °C for 5 min (*see* **Note 7**).

(b) Ventricular cells are filtered from this solution using a 200 μm mesh nylon gauze to remove undissociated pieces. Undissociated pieces undergo a new digestion cycle, as described in step 1, while filtered cardiomyocytes are submitted to a gentle centrifugation (400 rpm). The supernatant containing solution and fibroblasts is removed and cardiomyocytes are resuspended in the low Ca^{2+} solution (Table 2) for BSA and enzymes washing. Cells are leaved 15 min for sedimentation, then the supernatant is removed and cells are resupsended in the Tyrode solution containing 1 mM $CaCl_2$ (*see* **Notes 8** and **9**).

(c) The second batch that was resubmitted to 5 min enzymatic digestion is thus treated as described in step 2. Once again, dissociated cells are saved and undissociated pieces submitted to an additional cycle. This proceeding is reproduced 4–5 times. It allows batch of cells with different times of digestion to be obtained, increasing the probability to get cells suitable for patch-clamp use. At this point, cells might be directly used for patch-clamp experiments or prepared to be maintained in culture.

In some protocols, it is recommended to mechanically dissociate cells in the K^+ rich Kraft-Brühe solution medium to protect the cells [7] (*see* **Note 10** and Table 2).

4. Cell culture.

Culture method is derived from the protocols used by Jacobson and Piper [8] and Volz et al. [9].

(a) After cell isolation (**steps 2** and **3** in Subheading 3.2), isolated cells are left for 15 min to allow for sedimentation. The supernatant is then removed and cells are resuspended in culture medium containing 1.8 mM Ca^{2+} (*see* **step 3** in Subheading 2.1).

(b) Cells are seeded in petri dishes (35 mm) preincubated for 30 min with laminin (20 μg/ml, dissolved in sterile water and store in freeze aliquots) to facilitate cell adhesion and incubated for 2 h at 37 °C for platting.

(c) The culture medium is changed once. At this point cells are ready for electrophysiological experiments or cell culture (*see* **Note 11**).

(d) If cells have to be maintained in culture, the medium is renewed every 2 days while dishes are kept at 37 °C with a constant moist 95 % air and 5 % CO_2 (*see* **Note 12**).

3.3 Patch-Clamp Experiments

In this chapter, we focus the description of patch-clamp experiments on single channels recordings. However, cardiomyocytes dissociations described above are also suitable for whole-cell currents recordings [10, 11].

Proceeding for patch-clamp recordings, including ionic solutions highly depends on the type of channel that is expected (*see* **Note 13**). In the following **steps** (**1** through **5** in Subheading 3.3), we will take the example of a non-selective cation channel expressed in mammalian cardiomyocytes. Non-selective cation channels are a family of channels hard to investigate in native cells as several members coexist in the same cell and only few specific pharmacological modulators are known. That is particularly true for the TRPM4 channel, a calcium-activated non-selective cation channel (NSC_{Ca}) widely expressed in tissues including heart [3]. Recordings of this channel at the whole-cell level are compromised by the presence of a large variety of other cationic channels, in particular in the heart. Thus, an approach at the single channel level was necessary to investigate biophysical and regulatory properties of this channel in native cardiomyocytes. The study was done on several preparations such as human and mouse atrial myocytes [5, 12], rat ventricular myocytes [13], and mouse sinus node cells [6]. In these preparations, an NSC_{Ca} was sharply described at the single channel level. Its properties are similar to those described for the TRPM4 protein expressed in cell lines. Thus, the work unmasked the role of the TRPM4 channel in heart physiology and occurrence of arrhythmias [3].

1. Recording dishes preparation.

   In most of the studies the bath solution is the physiological saline solution (so-called 140 mM NaCl solution) described in Subheading 2, with a pH adjusted to 7.4. Before patching, the culture medium is removed and the bath solution is added gently to avoid cell wasting (*see* **Note 14**). Several solution changes are necessary to totally remove culture medium components such as BSA that would prevent seal formation.

Experiments can be carried at room temperature that do not modify cell surviving by the time of experiment or at 37 °C using a microchamber coupled to a cooler/heater with a temperature controller.

2. Seal formation.
   (a) Before introducing the patch pipette into the bath, a small positive pressure is applied on the pipette. This pushes back all impurities present at the surface of the dish. The patch amplifier is switched to the Vtrack mode.
   (b) The pipette resistance is estimated using a voltage pulse and measure of the resulted current.
   (c) Liquid junction potential is adjusted to zero.
   (d) The pipette is approached to the cell under microscopic observation, and then the final approach is achieved following pipette resistance variations. When the cell membrane is touched, the pipette resistance increases of few MΩ, then the suppression of the positive pressure gives a second weak increase in resistance. At that time, a small suction is applied in the pipette, inducing a drop of resistance to 1 GΩ. While the cell-attached seal is reached, the suction is suppressed and the patch amplifier switched to the Vclamp mode. Seal formation and duration highly depend on cell quality. Thus, depending of the dissociation, seal formation success is in the range of 50–10 % of trials.
3. Cell-attached recordings.
   (a) In the cell-attached configuration [14], the cell remains intact and so its resting membrane potential (Em) is not known. Accordingly, the applied potential (ΔVm = –pipette potential) corresponding to the difference between bath and pipette potentials is superimposed on the spontaneous cell membrane potential for the patch-membrane. In these conditions, the real potential applied to the patch-membrane (Vm) corresponds to the sum of ΔVm and Em. This has to be taken in account for protocols and interpretation of the results (*see* **Note 15**).
   (b) The first approach to investigate channel properties is the realization of a current/voltage relationship (*I*/*V*). This can be performed by long voltage steps (30 s to few minutes) with increase of 10–20 mV after each sweep from –100 mV to +100 mV. Channel openings and closings are observed as rectangular changes in current traces. In some cases, voltage ramp protocols can be used.
   (c) Current changes are analyzed with classical patch-clamp analysis software providing histograms amplitude and thus unitary current values and open probabilities (Fig. 2).

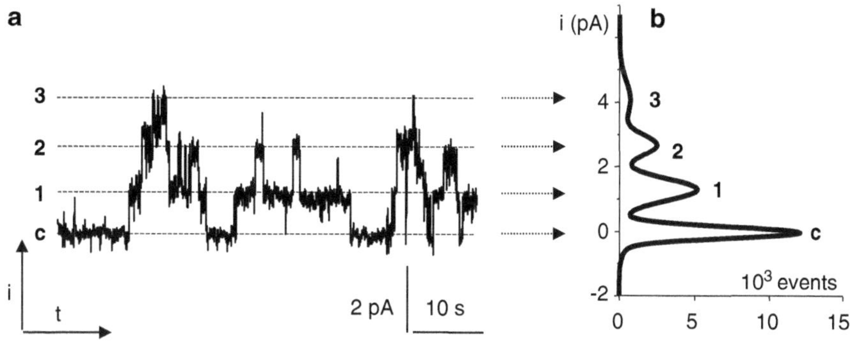

Fig. 2 Single channel trace and amplitude histogram construction. Single channel current trace recorded from a rat ventricular cardiomyocyte. (**a**) Changes in current (*i*) recorded during a constant voltage pulse. Three identical channels are present in the patch membrane, labelled "c" for the closed state and *1*, *2*, and *3* indicating the current opening level for the three channels. (**b**) Corresponding amplitude histogram. Each peak corresponds to one level of the current. Multi Gaussian equation was fitted to the distribution, allowing for the estimation of the single channel current amplitude for the corresponding voltage and also estimation of the channel open probability using ratios from areas under curve for each level (republished with permission from ref. 23)

These data are used for I/V construction and estimation of reversal potential, unitary conductance and variation of open probability in function of voltage (Fig. 3). Also, kinetics studies can be achieved, providing open and close lifetime distribution.

(d) Investigation of channel selectivity in the cell-attached configuration will be more qualitative than quantitative as neither Em nor the precise intracellular ionic composition is known. However, an approach can be the realization of I/V with different ionic solutions into the pipette. For example, replacing a part (120 mM) of Cl^- in the pipette by gluconate shifts the I/V, changing reversal potential, in case of a chloride channel. The same can be done for cations, replaced by *N*-methyl-d-glucamine (NMDG). The disadvantage is that these experiments are performed on different patch membranes. Lastly, unknown Em can be shifted to zero by bathing the cells in a 145 mM KCl rich solution that totally depolarizes them. Figure 3 illustrates such recordings from a non-selective cation channel.

(e) Channel regulation can be investigated in cell-attached configuration by perfusing a molecule in the bath. In that case, the channel is not directly accessible, so active molecules have to permeate the cell membrane before reaching the channel or to interact with an extracellular receptor that modifies channel activity. Then channel modulation and particularly washing can be slow. An example is provided in Fig. 3.

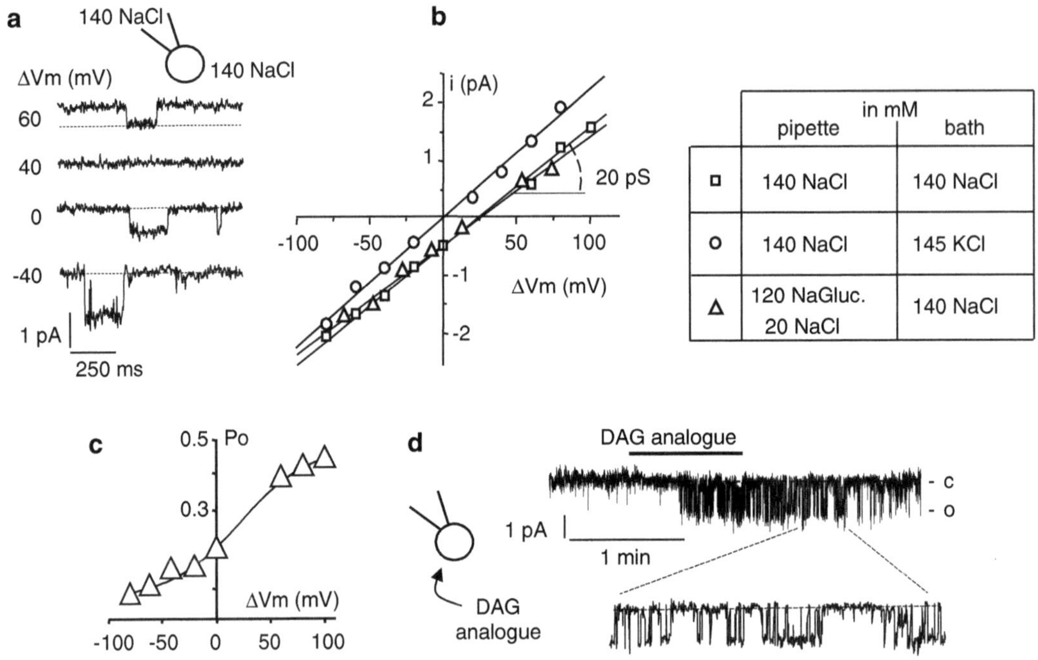

Fig. 3 Properties of a non-selective cation channel in the cell-attached configuration. (**a**) Single-channel tracings recorded at various voltages from a cell-attached patch of an adult rat cardiomyocyte after 8 days in culture. Pipette and bath contained the standard 140 mM NaCl solution. ΔVm corresponds to the clamp potential between the pipette and the bath. (**b**) *Channel conductance and selectivity*. Current–voltage relationships in ionic conditions as indicated in the *right table*. In the control condition (140 mM NaCl in pipette and bath, *squares*), the *I*/*V* is linear with a slope conductance of 20 pS and a reversal potential at 24 mV. This corresponds to the opposite of resting membrane potential (Em) of such cell [15]. Thus, it indicates that ions going through the channel are at the equilibrium when the membrane potential (Vm = ΔVm + Em) is zero. That is the case only for non-selective cation channels if one considers that the cell contains approximately 140 mM K^+ and 5 mM Na^+. Depolarizing the cell using a bathing solution with KCl instead of NaCl (*circles*) annuls Em, and thus Vm = ΔVm. In these conditions, the *I*/*V* reversal potential is close to zero, that confirms the previous statement. In addition, replacing most of the Cl^- in the pipette by gluconate (*triangles*) does not modify reversal potential, indicating that Cl^- does not permeate through the channel and thus that it is a cationic non-selective channel. (**c**) *Voltage dependence*. Open probability (Po), determined at various voltages using amplitude histograms, is plotted against imposed voltage. Fitting the points indicates that the channel is activated by membrane depolarization. (**d**) *Channel regulation*. During a recording, the bath is perfused with a diacylglycerol analogue (ΔVm = −40 mV) [15]. It produces a long activation of the channel. c and o indicate the close and open states of the channel, respectively (republished with permission from ref. 23)

   (f) While delicate to use, some apparatus were designed to allow for changes in pipette solution during a recording. This can be useful to investigate channel properties in the cell-attached configuration [16].

4. Inside-out recordings.

   (a) The inside-out configuration [14] is obtained by lifting up the pipette after the cell-attached seal was obtained.

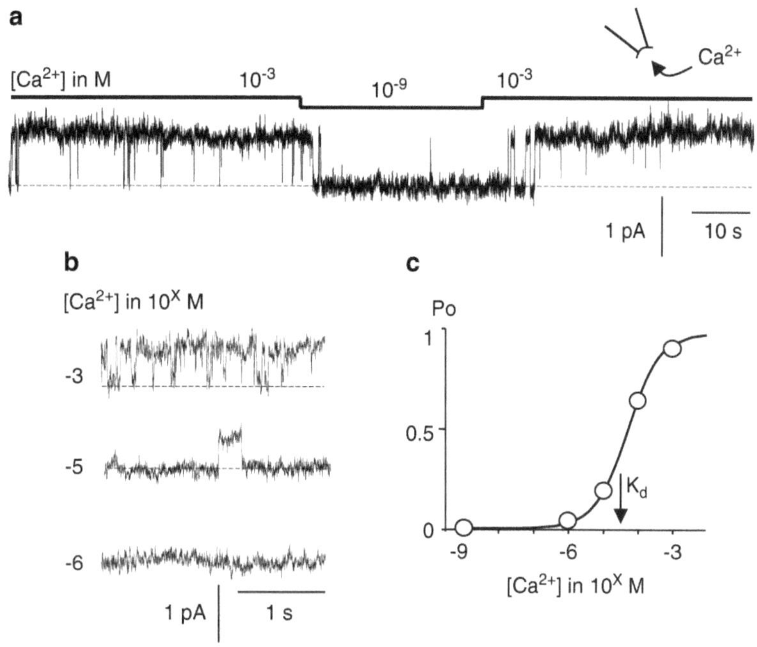

Fig. 4 Inside-out recordings. The effect of $[Ca^{2+}]_i$ on cardiac NSC_{Ca} activity is investigated on rat ventricular cardiomyocytes in the inside-out configuration [13]. Pipette and bath contained the standard 140 mM NaCl solution. Vm = +40 mV. (**a**) Single-channel tracing showing the rapid and reversible change in channel activity when $[Ca^{2+}]_i$ is strongly reduced. (**b**) Current tracings from the same patch membrane with variable $[Ca^{2+}]_i$. (**c**) Corresponding values of Po in function of $[Ca^{2+}]_i$. Fitting of the Hill equation to the experimental data points yields apparent dissociation constant (K_d) and Hill coefficient (republished with permission from ref. 23)

(b) Free of the cell, the pipette can be approached very close to the perfusion system. As the intracellular side is now on the bath, the pH of perfused solutions is adjusted to 7.2.

(c) Recordings of single channel traces resemble those of the cell-attached recordings (Fig. 4). Yet realization of *I*/*V* allows for the determination of single channel conductance, rectification properties, and channel dependence upon voltage. This has to be compared with the properties observed in the cell-attached configuration. However, it is possible that channels appear only in the inside-out configuration or inversely, since intracellular regulators are affected by excision.

(d) Since solutions bathing both sides of the membrane are now controlled, the ionic channel selectivity can be sharply investigated by replacing ions in the perfusing solution. *I*/*V* reversal potentials are used to estimate permeability ratios, according to the Goldman-Hodgkin-Katz equation [17] (*see* **Note 16**).

(e) During analysis, the membrane potential has to be corrected for the liquid junction potentials arising from changes in bathing solutions at the inner surface of the membrane patch. Variation in liquid junction potentials can be measured beforehand with a pipette containing 2.7 M KCl by measuring the zero-current voltage deflections induced by NaCl-substituted solutions [18]. Junction potentials can also be calculated according to Henderson equation using the JPCalc program [19, 20] in pClamp software.

(f) In the inside-out configuration, direct channel activation can be investigated by adding molecules or modifying ionic concentrations in the perfusing solutions. Contrary to the cell-attached configuration, the intracellular machinery is now totally washed out and only direct effects can be observed. However, it is conceivable that proteins associated to the channel are present. Figure 4 illustrates the direct regulation of the cardiac NSC_{Ca} by intracellular calcium.

5. Outside-out recordings.

(a) The outside-out configuration is obtained after the formation of the Giga seal in the cell-attached configuration. Cell membrane is ruptured by a strong suction in the pipette to reach the whole-cell configuration. It is detectable by the appearance of capacitive components at the onset and end of the pulse, on the oscilloscope. The pipette is then lifted up to reach the outside-out configuration.

(b) Investigations of channels properties are identical to those described in **step 4**, Subheading 3.3. However, active molecules will now act at the outside of the channel as the extracellular side of the membrane faces the perfusion system.

4 Notes

1. Although it is feasible to pull several pipettes in advance, it is recommended to polish the pipette at the time of patching.
2. *See* Benardeau et al. [21] and Guinamard et al. [5] for complete proceedings. Right atrial appendages can be obtained from patients undergoing cardiac surgery. Specimens are finely minced with iridectomy scissors in Krebs buffer containing 30 mM 2,3-butanedione monoxime (BDM) and 0.5 mM EGTA. Digestion is performed in Krebs supplemented with 0.5 % BSA and 6 i.u./ml protease type XXIV, then 165 i.u./ml collagenase type V for 15 min. After a further two to four 10 min steps in the presence of collagenase, samples are triturated with a Pasteur pipette. Cells are filtered and gradually

resuspended in DMEM (Dulbecco's modified Eagle's medium) culture medium supplemented with 1 % antibiotics (100 i.u./ml penicillin G; 50 i.u./ml streptomycin), 1 nM insulin, and 10 % fetal calf serum. Cells suitable for patch-clamp experiments are seeded and maintained in culture as described in **step 4**, Subheading 3.2, for rat cardiomyocytes.

3. It is possible to use older animals until 1 year old, but the success of cell dissociation is reduced due to the high level of fibroblasts and extracellular matrix.
4. For further reading and examples of the use of isolated adult ventricular cells, *see* refs. 2, 10, 11.
5. When reintroduced in 37 °C Tyrode solution, arrhythmic activities reveal that the perfusion is not efficient and the dissociation process compromised. Also, when calcium is removed the heart beat should stop within a minute and do not start again, indicating that the perfusion is correct.
6. The end-point of enzymes perfusion is one of the major points of the process. Unfortunately, the time of perfusion highly depend on the age of animal, quality of perfusion, and efficiency of the enzymes. If the heart is not sufficiently digested, very few isolated cells are obtained. On the contrary a strong digestion deteriorates cells membrane. One trick is to remove a small sample of the ventricle during the perfusion of the enzymatic solution, quickly triturate with a Pasteur pipette and examine, using a microscope, to see if some cells are already isolated, indicating time to stop the digestion. Sometimes the heart recovers spontaneous beating in the enzymatic solution, indicating that the solution is contaminated with calcium. That is injurious for the rest of the experiment.
7. To reduce proceedings cost, it is feasible to reuse enzymatic solution recovered after perfusion using the Langendorff apparatus.
8. Before filtering the pieces of ventricle, it might be benefit to release additional cells by using a fire-polished Pasteur pipette to gently triturated pieces by up and down movements.
9. In some cases, it could be preferable to reintroduce calcium in the solution by increasing concentration steps. This proceeding allows for a gentle restoration of calcium concentration. In a calcium free medium, the cells are highly calcium sensitive and any strong variation of calcium is noxious.
10. The high K^+ concentration of the KB solution which depolarize the cells helps their protection during mechanical dissociation [7]. KB composition in Table 2.
11. Freshly isolated cells can be used rapidly after dissociation proceeding. If so, it is not necessary to reintroduce calcium in the

solution, using the culture medium. A Tyrode solution can be used. However, due to the traumatic injury induced by cell dissociation proceeding, it is recommended to let cells recover few hours before using for experiments.

12. One should take care of the differentiating process that occurs on cardiomyocytes in culture. After 3–4 days, the morphology of the ventricular cells changes and so is their protein expression. The freshly isolated rod-shaped myocytes progress to a round shape; after 1 week, most of the cells are largely flattened and considerably bigger with apparent multinucleation, and then show spontaneous activity. Interestingly, these dedifferentiated cells have similarities with cardiomyocytes obtained after cardiac hypertrophy [8, 10, 13, 22].
13. For single channel recordings it is important to minimize the number of currents types present in the patch membrane. A first approach to eliminate undesirable currents is the use of known specific inhibitors. However, one has to be sure that these molecules do not interfere with the expected channel. A second approach is the use of solutions with specific ionic compositions, for example lacking the ion permeating through undesirable channels.
14. A simple Pasteur pipette can be used on well attached cells. However, if cells are isolated since only few hours, it can be useful to use a perfusion and suction system during few minutes that has a lower flow than a manually operate Pasteur pipette.
15. One should consider that in the cell-attached configuration, the junction potential that was corrected before sealing disappears after sealing. Thus, this has to be taken in account during the analysis if bath and pipette solutions have different compositions.
16. A useful method is the use of solutions containing low NaCl concentrations in the perfusing system (e.g., 14 mM, with sucrose added to maintain osmolarity) and a 140 mM NaCl standard solution in the pipette. Thus, Na^+ and Cl^- have opposite equilibrium potentials. In these conditions, if several currents types are present, the unexpected one can be eliminated by realizing measurements at the reversal potential for its permeating ion.

Acknowledgments

The author is in indebted to Pr. Patrick Bois and Pr. Jacques Teulon for helpful comments on this manuscript and teaching in both tissue preparation and patch-clamp proceedings.

References

1. Langendorff O (1895) Untersuchungen am uberlebenden Saugethierherzen. Pflügers Arch 61:291–332
2. Bouron A, Potreau D, Besse C, Raymond G (1990) An efficient isolation procedure of Ca-tolerant ventricular myocytes from ferret heart for applications in electrophysiological studies. Biol Cell 70(3):121–127
3. Guinamard R, Sallé L, Simard C (2011) The non-selective monovalent cation channels TRPM4 and TRPM5. Adv Exp Med Biol 704: 147–171
4. Bell RM, Mocanu MM, Tellon DM (2011) Retrograde heart perfusion: the Langendorff technique of isolated heart perfusion. J Mol Cell Cardiol 50:940–950
5. Guinamard R, Chatelier A, Demion M et al (2004) Functional characterization of a Ca^{2+}-activated non-selective cation channel in human atrial cardiomyocytes. J Physiol 558: 75–83
6. Demion M, Bois P, Launay P, Guinamard R (2007) TRPM4, a Ca2+-activated nonselective cation channel in mouse sino-atrial node cells. Cardiovasc Res 73:531–538
7. Isenberg G, Klockner U (1982) Calcium tolerant ventricular myocytes prepared by preincubation in a "KB medium". Pflugers Arch 395:6–18
8. Jacobson SL, Piper HM (1986) Cell cultures of adult cardiomyocytes as models of the myocardium. J Mol Cell Cardiol 18:661–678
9. Volz A, Piper HM, Siegmund B, Schwartz P (1991) Longevity of adult ventricular rat heart muscle cells in serum-free primary culture. J Mol Cell Cardiol 23:161–173
10. Fares N, Gomez JP, Potreau D (1996) T-type calcium current is expressed in dedifferentiated adult rat ventricular cells in primary culture. C R Acad Sci III 319:569–576
11. Brette F, Blandin E, Simard C et al (2013) Epac activator critically regulates action potential duration by decreasing potassium current in rat adult ventricle. J Mol Cell Cardiol 57: 96–105
12. Simard C, Hof T, Keddache Z et al (2013) The TRPM4 non-selective cation channel contributes to the mammalian atrial action potential. J Mol Cell Cardiol 59:11–19
13. Guinamard R, Rahmati M, Lenfant J, Bois P (2002) Characterization of a Ca^{2+}-activated nonselective cation channel during dedifferentiation of cultured rat ventricular cardiomyocytes. J Membr Biol 188:127–135
14. Hamill OP, Marty A, Neher E et al (1981) Improved patch-clamp techniques for high-resolution current recording from cell-free membrane patches. Pflugers Arch 391:85–100
15. Guinamard R, Chatelier A, Lenfant J, Bois P (2004) Activation of the Ca^{2+}-activated nonselective cation channel by diacylglycerol analogues in rat cardiomyocytes. J Cardiovasc Electrophysiol 15:342–348
16. Tang JM, Wang J, Quandt FN, Eisenberg RS (1990) Perfusing pipettes. Pflugers Arch 416: 347–350
17. Hille B (1984) Selective permeabilities. Ionic channels of excitable membranes. Sinauer Associated Inc Publishers, Sunderland, MA
18. Nastuk WI, Hodkin AL (1950) The electrical activity of single muscles fibers. J Cell Comp Physiol 35:39–73
19. Barry PH (1994) JPCalc, a software package for calculating liquid junction potential corrections in patch-clamp, intracellular, epithelial and bilayer measurements and for correcting junction potential measurements. J Neurosci Methods 51:107–116
20. Ng B, Barry PH (1995) The measurement of ionic conductivities and mobilities of certain less common organic ions needed for junction potential corrections in electrophysiology. J Neurosci Methods 56:37–41
21. Benardeau A, Hatem SN, Rucker-Martin C et al (1996) Contribution of Na^{+}/Ca^{2+} exchange to action potential of human atrial myocytes. Am J Physiol 271:H1151–H1161
22. Hefti MA, Harder BA, Eppenberger HM, Schaub MC (1997) Signaling pathways in cardiac myocyte hypertrophy. J Mol Cell Cardiol 29:2873–2892
23. Guinamard R (2007) Determination of channel properties at the unitary level in adult mammalian isolated cardiomyocytes. In: Molnar P, Hickman JJ (eds) Patch-clamp methods and protocols. Series: methods in molecular biology, 1st edn. Humana, Totowa, NJ, pp 195–209

Chapter 20

Dynamic Clamp as a Tool to Study the Functional Effects of Individual Membrane Currents

Géza Berecki, Arie O. Verkerk, Antoni C.G. van Ginneken, and Ronald Wilders

Abstract

Today, the patch-clamp technique is the main technique in electrophysiology to record action potentials or membrane current from isolated cells, using a patch pipette to gain electrical access to the cell. The common recording modes of the patch-clamp technique are current clamp and voltage clamp. In the current clamp mode, the current injected through the patch pipette is under control while the free-running membrane potential of the cell is recorded. Current clamp allows for measurements of action potentials that may either occur spontaneously or in response to an injected stimulus current. In voltage clamp mode, the membrane potential is held at a set level through a feedback circuit, which allows for the recording of the net membrane current at a given membrane potential.

A less common configuration of the patch-clamp technique is the dynamic clamp. In this configuration, a specific non-predetermined membrane current can be added to or removed from the cell while it is in free-running current clamp mode. This current may be computed in real time, based on the recorded action potential of the cell, and injected into the cell. Instead of being computed, this current may also be recorded from a heterologous expression system such as a HEK-293 cell that is voltage-clamped by the free-running action potential of the cell ("dynamic action potential clamp"). Thus, one may directly test the effects of an additional or mutated membrane current, a synaptic current or a gap junctional current on the action potential of a patch-clamped cell. In the present chapter, we describe the dynamic clamp on the basis of its application in cardiac cellular electrophysiology.

Key words Action potential, Membrane current, Patch clamp, Current clamp, Voltage clamp, Dynamic clamp, Dynamic action potential clamp, Coupling clamp, Cardiac myocytes, Computer simulation

1 Introduction

In neurophysiology, "dynamic clamp" was introduced by Sharp et al. [1–3], whereas similar techniques were introduced by Robinson and Kawai [4], named "conductance injection," and by Hutcheon et al. [5], named "reactive current clamp," all with the purpose to introduce artificial conductances into biological neurons, thus simulating a synaptic input, an additional membrane

Marzia Martina and Stefano Taverna (eds.), *Patch-Clamp Methods and Protocols*, Methods in Molecular Biology, vol. 1183, DOI 10.1007/978-1-4939-1096-0_20, © Springer Science+Business Media New York 2014

ionic current or an electrical synapse. At that time, as described in detail elsewhere [6], related techniques had already been applied in cardiac electrophysiology by Scott [7] and by Joyner and coworkers [8, 9], who designed an "Ersatz Nexus" and a "coupling clamp" circuit, respectively, both with the purpose to simulate intercellular electrical coupling through gap junctions. Although dynamic clamp thus exists in cardiac electrophysiology for well over 30 years now, it is less common than in neurophysiology. However, there is an emerging interest in the potential of dynamic clamp methodology in heart research.

Recently, dynamic clamp has been used in cardiac cellular electrophysiology to study the role of the L-type calcium current ($I_{Ca,L}$) and the transient outward potassium current (I_{to}) [10, 11]. Madhvani et al. [10] blocked the native $I_{Ca,L}$ of isolated rabbit ventricular myocytes by nifedipine and introduced a virtual $I_{Ca,L}$, which was computed in real time based on a mathematical model. Thus, it was possible to examine the role of the steady-state half-activation and half-inactivation potential of $I_{Ca,L}$ on action potential characteristics, including early afterdepolarizations (EADs). Similarly, Workman et al. [11] computed a virtual I_{to} in real time that was either used to enhance (addition) or inhibit (subtraction) the native I_{to} of rabbit and human atrial myocytes. Thus, it was possible to determine the effects of I_{to} on action potential characteristics, including afterdepolarizations. In both studies, dynamic clamp was used to introduce a virtual membrane current into a single isolated cardiomyocyte. This "conventional" application of the dynamic clamp technique, which has been designated as "model clamp" [12] or "model reference current injection" [13, 14] because a model based current is injected into the myocyte, is described in Subheading 3.7.

Instead of being computed, the aforementioned virtual current may also be recorded from a HEK-293 cell, or a similar type of cell, that expresses this current and is voltage-clamped by the free-running action potential of the myocyte. This "dynamic action potential clamp" is experimentally challenging, but it has the advantage that it does not rely on a mathematical model to represent the ionic current of interest. This application of the dynamic clamp technique has been used to compare the functional effects of wild-type and mutant currents on the action potential of isolated cells [15, 16] and is described in Subheading 3.8.

A further application of the dynamic clamp technique, which is described in Subheading 3.9, is to create intercellular electrical coupling through a virtual gap junction between two isolated cardiac myocytes that are not physically coupled ("coupling clamp"). We have used this configuration to study the synchronization of two spontaneously active sinoatrial node pacemaker cells [17]. In a more recent study by Nguyen et al. [18], this configuration was used to assess the interactions between an isolated patch-clamped ventricular myocyte and one or more surrounding (simulated) myofibroblasts.

2 Materials

2.1 Plasmids

1. pSP64T vector.
2. pCGI mammalian expression vector (provided by Drs. David Johns and Eduardo Marbán, Johns Hopkins University, Baltimore, MD, USA) [19].

2.2 Human Embryonic Kidney-293 Cell Culture Medium

Minimal essential medium (MEM, stored at +4 °C) supplemented with:

1. 10 % (v/v) fetal bovine serum (FBS).
2. 1 % (v/v) of 10,000 IU/mL penicillin and 10,000 μg/mL streptomycin stock solution, stored in aliquots at −20 °C, and then added to the culture dishes as required (*see* **Note 1**).
3. 1 % (v/v) nonessential amino acids.

2.3 Human Embryonic Kidney-293 Cell Transfection

1. Phosphate-buffered saline (PBS) without Ca^{2+} and Mg^{2+}.
2. Trypsin–versene: 500 mg/L trypsin, 200 mg/L ethylenediaminetetraacetic acid (EDTA) (*see* **Note 2**).
3. 2.0 mg/mL Lipofectamine.

2.4 Isolation of Rabbit Ventricular Myocytes

1. Heparin (5,000 IU/mL, 0.2 mL) and pentobarbital (60 mg/mL, 5 mL).
2. "Normal" Tyrode's solution: 140 mM NaCl, 5.4 mM KCl, 1.8 mM $CaCl_2$, 1.0 mM $MgCl_2$, 5.0 mM HEPES, and 5.5 mM glucose (pH adjusted to 7.4 with NaOH) (*see* **Note 3**).
3. "Ca^{2+}-free" Tyrode's solution: 140 mM NaCl, 5.4 mM KCl, 0.5 mM $MgCl_2$, 1.2 mM KH_2PO_4, 5.0 mM HEPES, and 5.5 mM glucose (pH adjusted to 7.2 with NaOH).
4. "Low-Ca^{2+}" Tyrode's solution: nine parts "Ca^{2+}-free" Tyrode's solution and one part "normal" Tyrode's solution, yielding a final Ca^{2+} concentration of approximately 180 μM.
5. Myocyte dissociation solution: 150 mL Ca^{2+}-free Tyrode's solution with 10 μM $CaCl_2$ + 22 mg collagenase type B + 7 mg collagenase type P + 30 mg hyalonuridase + 15 mg trypsin inhibitor (*see* **Note 4**).

2.5 Isolation of Rabbit SA Nodal Myocytes

1. "Normal" Tyrode's solution as for the isolation of rabbit ventricular myocytes: 140 mM NaCl, 5.4 mM KCl, 1.8 mM $CaCl_2$, 1.0 mM $MgCl_2$, 5.0 mM HEPES, and 5.5 mM glucose (pH adjusted to 7.4 with NaOH) (*see* **Note 3**).

2. "Nominally Ca^{2+}-free" Tyrode's solution: "normal" Tyrode's solution without $CaCl_2$.
3. Myocyte dissociation solution: nominally Ca^{2+}-free Tyrode's solution with 0.22–0.25 U/mL liberase III (Roche Diagnostics, Mannheim, Germany), 0.21–0.24 U/mL elastase (Serva, Heidelberg, Germany), and 0.92 U/ml pronase E (final 6 min only; Serva, Heidelberg, Germany) (*see* **Note 4**).
4. Modified "Kraft-Brühe" (KB) solution: 85 mM KCl, 30 mM K_2HPO_4, 5.0 mM $MgSO_4$, 20 mM glucose, 5.0 mM pyruvic acid, 5.0 mM creatine, 30 mM taurine, 5.0 mM β-hydroxybutyric acid, 5.0 mM succinic acid, bovine serum albumin (BSA) 1 %, 2.0 mM Na_2ATP (pH adjusted to 6.9 with KOH) (*see* **Note 3**).

2.6 Electrophysiology

1. Extracellular solution for cardiac myocytes as well as human embryonic kidney-293 (HEK-293) cells expressing the *KCNH2*-encoded HERG current: "normal" Tyrode's solution (*see* Subheading 2.4).
2. Pipette solution for recording action potentials from rabbit ventricular myocytes in the whole-cell configuration of the patch-clamp technique: 125 mM K-gluconate, 20 mM KCl, 1.0 mM $MgCl_2$, 5.0 mM MgATP, and 10 mM HEPES (pH adjusted to 7.2 with KOH) (*see* **Note 3**).
3. Pipette solution for recording action potentials from rabbit SA nodal myocytes in the perforated-patch configuration of the patch-clamp technique: 125 mM K-gluconate, 20 mM KCl, 10 mM NaCl, 0.22 mM amphotericin-B, and 10 mM HEPES (pH adjusted to 7.2 with KOH) (*see* **Note 3**).
4. Pipette solution for whole-cell patch-clamp recording of HERG current expressed in HEK-293 cells: 125 mM K-gluconate, 20 mM KCl, 1.0 mM $MgCl_2$, 5.0 mM ethyleneglycol-bis-*N*,*N*,*N*′,*N*′-tetraacetic acid (EGTA), 5.0 mM MgATP, and 10 mM HEPES (pH adjusted to 7.2 with KOH) (*see* **Note 3**).
5. E-4031 (specific inhibitor of I_{Kr}) to be added to the extracellular solution from a 5.0-mM stock solution to a final concentration of 5.0 μM.

2.7 Dynamic Clamp

1. Moderately fast PC, for example an Intel Core i5 or i7 CPU based desktop computer, with a high-speed 16-bit data acquisition (DAQ) board with at least two analog in and two analog out channels, for example the NI PCI-6052E, PCIe-6251, or PCIe-6351 data acquisition boards (National Instruments, Austin, TX, USA) (*see* **Note 5**).
2. Real-time operating system, for example Real-Time Linux (RT-Linux) [20] (*see* **Note 6**).

3. Dedicated software that implements the dynamic clamp technique, for example the Microsoft Windows-based open-source "Spike timing dependent plasticity Clamp" (StdpC) [21] or the RT-Linux based open-source "Real-Time eXperiment Interface" (RTXI) with an appropriate dynamic clamp module [22] (*see* **Note 7**).
4. Mathematical model of the virtual ionic current to be injected into the myocyte, for example the L-type calcium current $I_{Ca,L}$ [10] or the transient outward potassium current I_{to} [11] (*see* **Note 8**).

3 Methods

3.1 Plasmid Construction

1. Clone the coding sequence (cDNA) of the ion channel subunit(s) of interest into an appropriate vector (e.g., pSP64T). Subclone the ion channel cDNA into the pCGI vector for bicistronic expression of the channel protein and green fluorescent protein (GFP) reporter in HEK-293 cells (*see* **Note 9**).
2. For co-expression studies, use an additional bicistronic vector: pCCI (cyano-fluorescent protein reporter, CFP instead of GFP).

3.2 HEK-293 Cell Culture

1. HEK-293 cells are passaged when approaching 80 % confluence (check confluence level on an inverted microscope). Remove all medium from the culture flask (Falcon, canted neck 50 mL, polystyrene). Wash the cells with 5 mL PBS (2×). Remove PBS. Add 0.5 mL trypsin–versene to the 50-mL culture flask. After 1–2 min incubation, round-shaped cells detach from the bottom of the flask.
2. Add 4.5 mL fresh culture medium and resuspend the medium 1× with the 5-mL pipette. Add 0.2 mL of this suspension to a new 50-mL cell culture flask and then add 4.8 mL cell culture medium. Place the cell culture flask in the incubator (5 % CO_2, 37 °C, 8 % relative humidity).

3.3 HEK-293 Cell Transfection and Preparation for Recording

1. Grow cells to 50 % confluence in culture medium described in Subheading 2.2.
2. In a 15-mL centrifuge tube add 0.5–2 μg DNA (plasmid with the cDNA of the ion channel subunit), 20 μL Lipofectamine, and 0.5 mL FBS-free culture medium. Mix gently and incubate for 40 min.
3. Remove the culture medium from the culture flask and wash the cells with 5 mL PBS (2×). Add the content of the centrifuge tube and 2 mL FBS-free culture medium to the cells. Incubate the flask in the incubator for 6 h.

4. Remove transfection medium and add 5 mL culture medium (with FBS) to the flask. Incubate the cells overnight.
5. Wash the transfected cells with 5 mL PBS (2×). Remove PBS. Add 0.5 mL trypsin–versene to the flask. After 1–2 min of incubation (round-shaped cells detach from the bottom of the flask) add 4.5 mL fresh culture medium. Gently resuspend the medium with a 5-mL pipette to detach all cells from the bottom of the flask.
6. Place 10–20 μL of the above suspension in the recording chamber mounted on the stage of an inverted microscope (*see* **Note 10**); fill the chamber with the extracellular solution (*see* **Note 11**).

3.4 Isolation of Rabbit Ventricular Myocytes

Single rabbit ventricular myocytes are isolated using an enzymatic dissociation procedure, modified from Tytgat [23]. The experimental protocol should comply with the "Guide for the Care and Use of Laboratory Animals" published by the US National Institutes of Health Office of Laboratory Animal Welfare (Eighth Edition, 2011).

1. Apply heparin and then kill New Zealand White rabbits by injection of pentobarbital.
2. Remove the heart rapidly and mount it on a Langendorff perfusion apparatus. Perfuse it retrogradely through the aorta with normal Tyrode's solution for 10 min, followed by an additional 10 min of perfusion with Ca^{2+}-free Tyrode's solution. Switch perfusion to myocyte dissociation solution for 15 min. During the isolation procedure, the enzymatic solution is recirculated. All solutions are saturated with 100 % O_2 and temperature is maintained at 37 °C.
3. Cut the ventricles into pieces and collect these in a beaker with myocyte dissociation solution. Shake the beaker in a bath at 60 and 100 rpm for 10 and 5 min, respectively. Remove supernatant, add fresh myocyte dissociation solution with 1 % fatty acid-free BSA, and shake the beaker at 120 rpm for 10 min. Collect supernatant with isolated myocytes. Further digest the remaining pieces of tissue (120 rpm for 10 min) and collect myocyte suspension. Combine both supernatants and allow myocytes to settle for 5–10 min.
4. Remove supernatant and add low-Ca^{2+} Tyrode's solution to the cells.
5. After 10 min, replace approximately 75 % of the low-Ca^{2+} Tyrode's solution by normal Tyrode's solution to obtain stepwise increases of the calcium concentration. Repeat this procedure four times at intervals of 10 min. Store the cells at room temperature (20–22 °C) and use them within 8 h.

6. Place 10–20 μL of myocyte suspension in the recording chamber mounted on the stage of an inverted microscope; fill the chamber with the extracellular (Tyrode's) solution (*see* **Note 12**). Select single, rod-shaped myocytes with smooth surfaces and clear cross-striations for electrophysiological measurements.

3.5 Isolation of Rabbit SA Nodal Myocytes

Single cells are isolated from the SA nodal region by enzymatic dissociation.

1. Follow **steps 1** and **2** of the isolation procedure for ventricular myocytes (*see* Subheading 3.4).
2. Excise the SA nodal region, cut it into small strips of 0.7–1 × 2 mm perpendicular to the crista terminalis, and let these equilibrate in normal Tyrode's solution for 15 min (20 °C).
3. Place the strips in nominally Ca^{2+}-free Tyrode's solution (20 °C) and refresh twice. Incubate the strips in myocyte dissociation solution for 30 min and triturate the strips through a pipette with a tip diameter of 2.0 mm during this period.
4. Stop the dissociation by transferring the strips into modified Kraft-Brühe (KB) solution (37 °C) and refresh the KB solution three times (at 20 °C). Store the strips for 30 min and then triturate (pipette tip diameter 1.2 mm) in modified KB solution (20 °C) for 2 min to obtain single cells. Store single cells in modified KB solution at room temperature for at least 45 min.
5. Place 10–20 μL of myocyte suspension in the recording chamber mounted on the stage of an inverted microscope; fill the chamber with the extracellular (Tyrode's) solution (*see* **Note 12**). Select spindle or elongated spindle-like cells displaying regular contractions for electrophysiological measurements.

3.6 Electrophysiological Recording

1. Superfuse myocytes with extracellular solution at 36 °C.
2. Pull electrodes from borosilicate glass with thin filament inside the lumen, using conventional one-stage or two-stage pullers. When backfilled with the appropriate pipette solution, electrodes should have resistances of 1–3 MΩ.
3. For recording large amplitude currents from HEK-293 cells (whole-cell configuration; voltage clamp mode), backfilled electrodes should exhibit resistances of 1–1.3 MΩ to obtain series resistance, R_s, values of typically 2–2.5 MΩ, which may be compensated for 90–95 %. To further minimize possible voltage errors, small HEK-293 cells of 8–12 pF membrane capacitance should be selected, because these are expected to express less current than larger cells (*see* **Note 13**).

4. Adjust the potential between the pipette and bath solution (liquid junction potential) to zero before establishing a high-resistance seal between the pipette and the cell. All potentials should be corrected for the change in liquid junction potential, which occurs when contact with the cells is made [24].
5. Record action potentials in current clamp mode, using the whole-cell (ventricular myocytes) or perforated-patch (SA nodal myocytes) configuration of the patch-clamp technique. In case of the intrinsically quiescent ventricular myocytes, elicit action potentials by 2-ms current pulses of 1.5× threshold, i.e., 1.5 times the amount of current required to reach firing threshold.
6. When recording action potentials or membrane current, use low-pass filtering with a cutoff frequency of 5 kHz and digitizing at 5–20 kHz (*see* **Note 14**).
7. For data acquisition and analysis use patch-clamp amplifiers and software of personal preference (*see* **Note 15**).

3.7 "Conventional" Dynamic Clamp

In its conventional configuration, dynamic clamp is used to simulate the presence of an additional conductance in the membrane of a single isolated myocyte. The design and use of this configuration are illustrated in Fig. 1. In this configuration, the membrane potential V_m of the myocyte is continuously sampled into a personal computer (PC), equipped with an appropriate data acquisition board, which calculates the V_m dependent current I_x and sends a command potential V_{cmd} to the patch-clamp amplifier to inject this current into the myocyte (Fig. 1a).

A relatively simple application of the "conventional" dynamic clamp is shown in Fig. 1b, c. The computed current I_x is an ohmic current with a reversal potential of −40 mV and an adjustable conductance G_x), which is injected into an isolated ventricular myocyte to represent the "injury current" that may flow from normal myocardium to ischemic tissue, which is known to contain inexcitable cells that are depolarized to ≈−40 mV [25] (*see* **Note 16**). In the absence of injury current, i.e., with $G_x = 0$, I_x is zero and the action potential of the ventricular myocyte is characterized by a resting potential of −82 mV and a duration of ≈420 ms (Fig. 1b, c, blue dashed lines). Upon the introduction of injury current with a conductance of 10 nS, the resting potential depolarizes by 3–4 mV and the action potential duration shortens to ≈180 ms (Fig. 1b, c, red solid lines). Thus, the dynamic clamp experiment directly shows the effect of injury current on the ventricular action potential.

1. Patch the isolated myocyte and, after setting up correction for liquid junction potential and series resistance compensation, determine its basic characteristics, like stimulus threshold, input resistance, and membrane capacitance (*see* **Note 17**).
2. Switch to the dynamic clamp software to further record the data and control the experiment (*see* **Notes 18** and **19**).

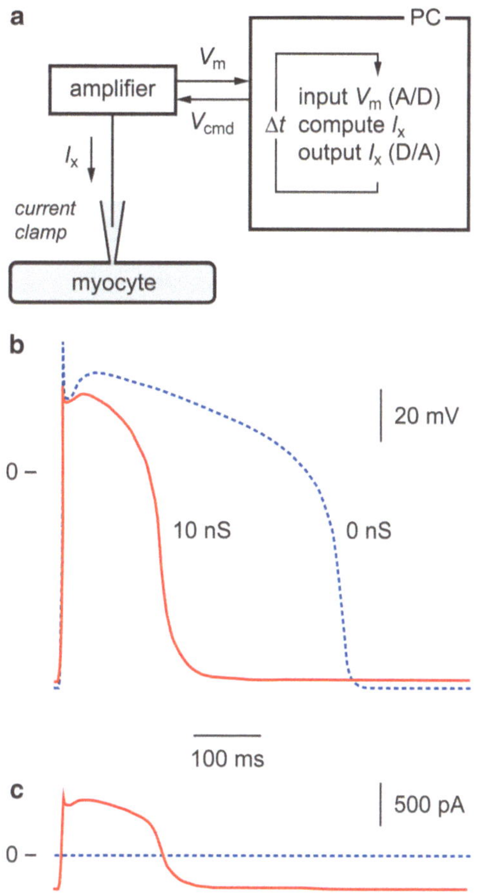

Fig. 1 Design and use of the "conventional" dynamic clamp technique. (**a**) The free-running membrane potential V_m of a patch-clamped cardiac myocyte in current clamp mode is sampled into a personal computer (PC) at high rate (time step Δt typically <50 μs). A virtual ionic current I_x is computed in real time, based on the sampled value of V_m, and injected into the cell as an additional membrane current through the recording pipette. V_{cmd} is the command potential sent out to the patch-clamp amplifier. (**b**) Action potential of an isolated ventricular myocyte (in this particular example of human left ventricular origin) in the absence and presence of a simulated "injury current" (*blue dashed line* labelled "0 nS" and *red solid line* labelled "10 nS," respectively). The injury current I_x that would flow from normal myocardium to a depolarized ischemic region, thus shortening the action potential of the normal region and depolarizing its resting potential, was computed as an ohmic current with a conductance of 10 nS and a reversal potential of −40 mV. (**c**) Simulated "injury current" I_x associated with the action potentials of panel **b**

3.8 Dynamic Action Potential Clamp

In its "dynamic action potential clamp" (dAPC) configuration, the current associated with the additional conductance in the membrane of a single isolated myocyte (*see* Subheading 3.7 and Fig. 1) is not computed, but recorded in real time from a transfected HEK-293 cell that expresses the ionic current of interest. The design and use of this configuration are illustrated in Fig. 2.

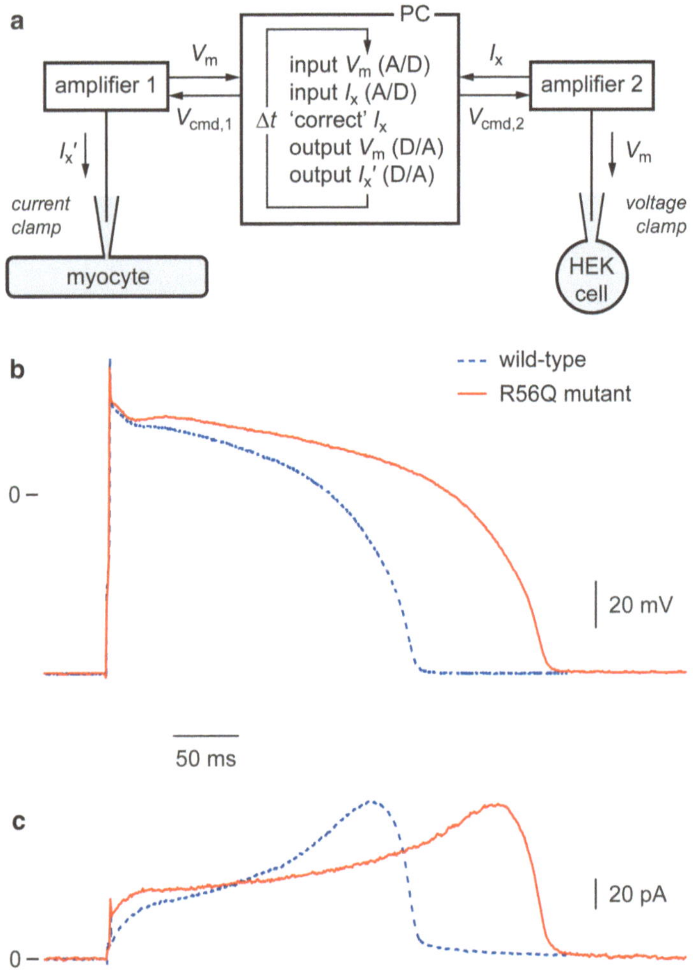

Fig. 2 Design and use of the "dynamic action potential clamp" (dAPC) technique. (**a**) The free-running membrane potential V_m of a single isolated myocyte is recorded in current clamp mode and used to voltage-clamp a HEK-293 cell (HEK cell, command potential $V_{cmd,2}$), in which a specific ion current is expressed. This ion current (I_x) is recorded, fed back into the personal computer (PC), and—after online "correction" (i.e., subtraction of endogenous background current of the HEK-293 cell and appropriate scaling of the remaining current)—injected into the myocyte as additional membrane current (I_x', command potential $V_{cmd,1}$). The time step for updating input and output values is Δt. (**b**) Action potential of an isolated rabbit left ventricular myocyte in the presence of either wild-type or R56Q mutant *KCNH2* (HERG) current expressed in a HEK-293 cell (*blue dashed line* and *red solid line*, respectively). The native *KCNH2*-encoded delayed rectifier potassium current (I_{Kr}) of the myocyte was blocked by E-4031 (5 μM). The HEK-293 cell was voltage-clamped by the membrane potential of the myocyte, so that the native I_{Kr} of the myocyte was effectively replaced with its *KCNH2*-encoded counterpart expressed in the HEK-293 cell. The myocyte was stimulated at 1 Hz to elicit action potentials. (**c**) Wild-type or R56Q mutant *KCNH2* current I_x' associated with the action potentials of panel **b**

A patch-clamp amplifier ("amplifier 1") is used to make recordings from a single isolated myocyte in current clamp mode. The recorded membrane potential V_m is continuously sampled into a personal computer (PC), where it is used to send out a command potential $V_{cmd,2}$ to a second patch-clamp amplifier ("amplifier 2") (*see* **Note 20**). This second amplifier operates in voltage clamp mode and is used to clamp the membrane potential of the transfected HEK-293 cell at the membrane potential of the myocyte. The expressed current I_x is sampled into the PC, where it may be "corrected" (subtraction of background current, scaling) before a command potential $V_{cmd,1}$ is sent out to the first amplifier to inject the corrected current I_x' into the myocyte as if this current were actually present as a native ionic current in the membrane of the myocyte (Fig. 2a).

An application of the dynamic action potential clamp is shown in Fig. 2b, c. The ventricular myocyte has its native *KCNH2*-encoded I_{Kr} blocked by E-4031 (5 μM). This current is effectively replaced with either wild-type or R56Q mutant *KCNH2* current expressed in the HEK-293 cell, which resides in a second patch-clamp setup and is voltage-clamped by the membrane potential of the myocyte (*see* **Note 21**). To this end, the recorded current is scaled down in the PC to produce a *KCNH2* current density similar to the native I_{Kr} density of the myocyte, as estimated from its E-4031-sensitive current (*see* **Notes 22** and **23**). This dAPC experiment [15] directly demonstrates that the R56Q mutation in *KCNH2* results in significant action potential prolongation (Fig. 2b). Also, it shows the alterations in the time course of the *KCNH2* current associated with this prolongation (Fig. 2c) (*see* **Notes 24** and **25**).

1. On one setup, patch the isolated myocyte and, after setting up correction for liquid junction potential and series resistance compensation, determine its basic characteristics, like stimulus threshold, input resistance, and membrane capacitance (*see* **Note 17**).
2. On another setup, let a second electrophysiologist simultaneously patch the HEK-293 cell expressing the current of interest and, after setting up correction for liquid junction potential and series resistance compensation, determine the amplitude of the expressed current (*see* **Note 22**).
3. Switch to the dAPC module of the dynamic clamp software to further record the data and control the experiment (*see* **Notes 18** and **19**).

3.9 Coupling Clamp

Dynamic clamp may also be used to simulate intercellular electrical coupling through gap junctions, known as "coupling clamp" [8, 9]. As in the dAPC configuration of Fig. 2, recordings are made from two isolated cells, using two separate patch-clamp amplifiers. However, both amplifiers now operate in current clamp mode.

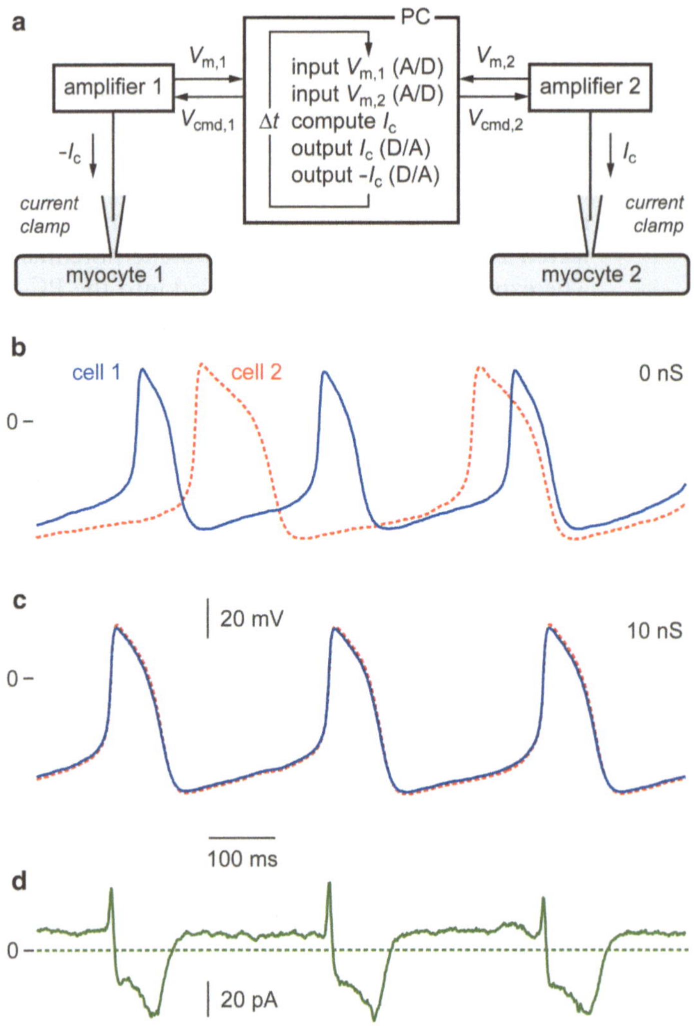

Fig. 3 Design and use of the "coupling clamp" technique. (**a**) The membrane potentials of two isolated myocytes ($V_{m,1}$ and $V_{m,2}$), both in current clamp mode, are sampled into a personal computer (PC) and an intercellular coupling current (I_c) is computed, based on $V_{m,1}$ and $V_{m,2}$. Command potentials ($V_{cmd,1}$ and $V_{cmd,2}$) are then sent back to the patch-clamp amplifiers to inject this current into the myocytes as additional membrane current through the recording patch pipettes. The time step for updating input and output values is Δt. (**b**). Intrinsic spontaneous action potentials simultaneously recorded from two single isolated rabbit SA nodal pacemaker cells, residing in two different patch-clamp setups. There is zero coupling conductance between the cells, as indicated by the "0 nS" label. (**c**) Synchronization of the action potentials of the two cells of panel **b** upon simulating an ohmic gap junctional current with a conductance of 10 nS. (**d**) Intercellular coupling current I_c associated with the action potentials of panels **b** and **c** (*dashed zero-current line* and *solid line*, respectively)

The membrane potentials $V_{m,1}$ and $V_{m,2}$ of two myocytes not in physical contact with each other are continuously sampled into a PC (*see* **Note 26**). At each time step, a virtual coupling current I_c, by definition positive if flowing from myocyte 1 to myocyte 2, is computed from $V_{m,1}$ and $V_{m,2}$. Command potentials $V_{cmd,1}$ and $V_{cmd,2}$ are then sent out to the patch-clamp amplifiers to make this current effectively flow between the two myocytes (Fig. 3a).

An example of a coupling clamp experiment is shown in Fig. 3b–d. The spontaneous action potentials of two isolated sino-atrial node pacemaker cells are recorded. The intrinsically different action potentials of the two cells are shown in Fig. 3b as blue solid and red dashed lines. If an ohmic coupling conductance of 10 nS is applied, the two cells synchronize and their action potentials become almost identical (Fig. 3c) as a result of the coupling current flowing between the two cells (Fig. 3d, green solid line). This coupling clamp allowed for studying the synchronization process of the two cells, from "frequency entrainment" at a coupling conductance as small as 0.2 nS to full "waveform entrainment" at 10 nS [17].

1. On one setup, patch the first isolated myocyte and, after setting up correction for liquid junction potential and series resistance compensation, determine its basic characteristics, like stimulus threshold, input resistance, and membrane capacitance (*see* **Note 17**).
2. On a second setup, let another electrophysiologist simultaneously patch the second isolated myocyte and determine its characteristics.
3. Switch to the coupling clamp module of the dynamic clamp software to further record the data and control the experiment (*see* **Notes 18** and **19**).

4 Notes

1. Use sterile technique: all handling in laminar flow hood with sterile disposable plasticware.
2. Small aliquots of this solution can be stored at −20 °C.
3. The pH of solutions is temperature dependent and should therefore be corrected for the physiological (working) temperature if this differs from the temperature at which the solutions are prepared.
4. The yield of myocytes may depend on the batch of enzymes used and on Ca^{2+} concentration. It may prove necessary to adjust the amount of enzymes and/or Ca^{2+} concentration, which should not exceed 50 μM. Of note, the suggested isolation protocol has been optimized with enzymes from the specified suppliers. Enzyme activity can differ slightly between suppliers, resulting in isolation protocols that require optimization.
5. A fast 16-bit data acquisition board is required to cover the wide dynamic range of the cardiac sodium current. For less demanding currents, a moderately fast 12-bit board may be sufficient. Still, a fast 16-bit board minimizes discretization errors and latencies in the feedback loop between voltage and current (*see* **Note 16**). Here, "fast" refers to the handling of single A/D and single D/A samples in a time-efficient manner

and with reproducible time steps (microsecond resolution), i.e., with minimal jitter. Note that whereas buffering of A/D input and D/A output samples helps to increase the advertized speed of data acquisition boards, it is not of interest for the real-time single sample in, single sample out process of dynamic clamp. Effects of time step, jitter, and latency have been documented in detail by Bettencourt et al. [26]. Make sure that the selected data acquisition board is supported by the dynamic clamp software of choice.

6. RT-Linux is available as open-source software and has been used as the real-time operating system in several dynamic clamp studies [10, 15, 16, 18]. As an exception, the StdpC dynamic clamp software [21], which does not rely on fixed length time step intervals in the dynamic clamp cycle, does not require a real-time operating system.
7. In RTXI and related dynamic clamp systems, the experiment is controlled by a single PC that runs under RT-Linux, which in simple wordings means that this PC runs under standard Linux, but switches to real-time mode during the run of an experiment. As an alternative configuration, one may use a Microsoft Windows-based host computer that controls a second target computer through a local Ethernet connection. This target computer, which is equipped with the data acquisition board, then runs under a dedicated real-time operating system and thus acts as—or may actually be—an embedded processor under control of the host. Such approach has successfully been applied in combination with LabVIEW-RT (LabVIEW with Real-Time Module extension; National Instruments) [27] and in combination with Simulink, MATLAB and xPC Target (The MathWorks, Natick, MA, USA) [28]. Less flexible than these custom configurations but more user friendly for the common electrophysiologist is the dynamic clamp configuration provided by the Windows-based CED Signal 5 electrophysiological recording software in conjunction with the CED Power1401-3 data acquisition interface (Cambridge Electronic Design, Cambridge, UK), which features an embedded processor.
8. It should be realized that the virtual current is injected through the pipette so that "chemical effects" of this current (local increases in ion concentration, activation of second messenger systems through conformational changes of the ion channel) are absent. This may pose a drawback as well as a unique opportunity to separate electrical effects from chemical effects.
9. The mutated insert and ligation regions should be verified by sequencing.

10. Ionic currents are recorded within 1–2 days from HEK-293 cells exhibiting cyan and/or green fluorescence in the whole-cell configuration of the patch-clamp technique.
11. The composition of this solution should be adequate for the expressed ion channel protein (*see* Subheading 2.6).
12. Allow myocytes to settle for 5 min before superfusion with "normal" Tyrode's solution is started.
13. If the expressed current is of large amplitude, there may be a noticeable voltage error in the membrane potential relative to the voltage clamp command potential. However, in case of a 10-nA current and a residual uncompensated $R_s < 200$ kΩ, this voltage error is <2 mV.
14. Selection of cutoff and digitizing frequency may vary with the current of interest. In dAPC experiments with *KCNH2* current (*see* Subheading 3.8), membrane currents may be low-pass filtered with a cutoff frequency of 2 kHz and digitized at 5 kHz.
15. In dynamic clamp experiments, one may prefer to use the dynamic clamp software to record the data and control the experiment. Storage of the acquired A/D samples and the chosen parameter settings is sufficient to "replay" the experiment with an "offline" version of the dynamic clamp software. The role of the standard recording software is then limited to determination of cell characteristics (membrane capacitance, input resistance), compensation of pipette resistance, and correction for liquid junction potential at the start of the experiment (and monitoring these during the experiment).
16. If the injected current is a simple ohmic current, the dynamic clamp may readily be accomplished with analog circuitry [8, 9], but in general a PC-controlled digital implementation is highly preferred because of its flexibility and ease of use. One should, however, be aware that a digital implementation introduces discretization errors and latencies in the feedback loop between voltage and current (*see* **Note 5**). Also, the actual time step may show considerable jitter.
17. It is important to determine membrane capacitance as a measure of cell size. In the dynamic clamp experiment, the virtual conductance can then be scaled to membrane capacitance to allow for direct comparison between experiments on cells of different size.
18. In intrinsically quiescent cells, action potentials may be elicited through the dynamic clamp software by adding a stimulus current to the injected current, thus using this software to control the experiment (*see* **Note 15**).

19. Always make sure that a zero current signal is sent out through the D/A channel of the data acquisition board before starting up a dynamic clamp experiment. Improper termination of a previous experiment or an awkward default setting of a driver may result in the unplanned and possibly destructive injection of a large current into the myocyte.
20. It is also possible to feed the recorded membrane potential V_m directly into the second patch-clamp amplifier, but then the PC cannot be used for subtraction of a V_m dependent background current or any other correction based on V_m. Furthermore, the PC then loses its role as the "control center" of the experiment (*see* **Note 15**).
21. Instead of a real isolated myocyte, a mathematical model may be used to simulate the myocyte in real time. Thus, the membrane potential is not obtained by patch-clamp recording, but by solving the differential equations of the model at each time step Δt, adding I_x' to the net membrane current of the model cell. Such approach has been used when comparing the effects of several mutations in the *SNC5A* gene [16].
22. Wild-type and mutant *KCNH2* currents were both scaled down to produce a *KCNH2* current density similar to the native I_{Kr} density of the myocyte. This scaling procedure seemed appropriate because the mutation of interest was associated with altered gating properties rather than impaired trafficking from the endoplasmic reticulum to the cell membrane. In general, the scaling procedure requires data on the expression level of the current in the myocyte.
23. If the recorded current needs to be scaled down, as with the *KCNH2* current, any potentially contaminating background current is scaled down too and readily becomes negligible. However, if the dAPC experiment requires the recorded current to be scaled up, as with the *SCN5A* encoded fast sodium current [16], correction for the background current may become necessary. If the characteristics of this background current are known, for example from recordings on non-transfected HEK-293 cells, an estimate of this current can be computed and subtracted from the recorded current before injection into the myocyte [16].
24. The dAPC experiment shows that action potential prolongation occurs despite the initially larger *KCNH2* current in the mutant case. This demonstrates that the observed prolongation is not only due to alterations in the time course of *KCNH2* current but also of other membrane currents, through their dependence on membrane potential.
25. A dAPC experiment in which the effects of wild-type and mutant currents are compared is experimentally challenging

because it requires simultaneous stable recordings on multiple patch-clamp setups. To allow for direct comparison, the same myocyte must consecutively be connected to two different HEK-293 cells, one expressing wild-type current and one expressing mutant current. Higher success rates are achieved if the real myocyte is replaced with a real-time simulation of such myocyte (*see* **Note 21**).

26. As with the dAPC configuration of Fig. 2, one of the myocytes may be replaced with a computer model that simulates the myocyte in real time (*see* **Note 21**). Such approach has been used to probe the action potential transfer between an isolated guinea pig ventricular myocyte exposed to conditions of acute ischemia and a non-affected myocyte [29]. The real cell may even be embedded in a virtual sheet of surrounding atrial or ventricular myocytes [30] or (myo)fibroblasts [18].

Acknowledgements

We thank Mr. Berend de Jonge and Mr. Jan G. Zegers for expert technical assistance.

References

1. Sharp AA, Abbott LF, Marder E (1992) Artificial electrical synapses in oscillatory networks. J Neurophysiol 67:1691–1694
2. Sharp AA, O'Neil MB, Abbott LF et al (1993) Dynamic clamp: computer-generated conductances in real neurons. J Neurophysiol 69: 992–995
3. Sharp AA, O'Neil MB, Abbott LF et al (1993) The dynamic clamp: artificial conductances in biological neurons. Trends Neurosci 16: 389–394
4. Robinson HPC, Kawai N (1993) Injection of digitally synthesized synaptic conductance transients to measure the integrative properties of neurons. J Neurosci Meth 49:157–165
5. Hutcheon B, Miura RM, Puil E (1996) Models of subthreshold membrane resonance in neocortical neurons. J Neurophysiol 76: 698–714
6. Wilders R (2006) Dynamic clamp: a powerful tool in cardiac electrophysiology. J Physiol 576:349–359
7. Scott S (1979) Stimulation simulations of young yet cultured beating hearts. PhD thesis, State University of New York at Buffalo, New York
8. Tan RC, Joyner RW (1990) Electrotonic influences on action potentials from isolated ventricular cells. Circ Res 67:1071–1081
9. Joyner RW, Sugiura H, Tan RC (1991) Unidirectional block between isolated rabbit ventricular cells coupled by a variable resistance. Biophys J 60:1038–1045
10. Madhvani RV, Xie Y, Pantazis A et al (2011) Shaping a new Ca^{2+} conductance to suppress early afterdepolarizations in cardiac myocytes. J Physiol 589:6081–6092
11. Workman AJ, Marshall GE, Rankin AC et al (2012) Transient outward K^+ current reduction prolongs action potentials and promotes afterdepolarisations: a dynamic-clamp study in human and rabbit cardiac atrial myocytes. J Physiol 590:4289–4305
12. Wilders R, Verheijck EE, Kumar R et al (1996) Model clamp and its application to synchronization of rabbit sinoatrial node cells. Am J Physiol 271:H2168–H2182
13. Butera RJ Jr, Wilson CG, Delnegro CA et al (2001) A methodology for achieving high-speed rates for artificial conductance injection in electrically excitable biological cells. IEEE Trans Biomed Eng 48:1460–1470

14. Raikov I, Preyer A, Butera RJ (2004) MRCI: a flexible real-time dynamic clamp system for electrophysiology experiments. J Neurosci Methods 30:109–123
15. Berecki G, Zegers JG, Verkerk AO et al (2005) HERG channel (dys)function revealed by dynamic action potential clamp technique. Biophys J 88:566–578
16. Berecki G, Zegers JG, Bhuiyan ZA et al (2006) Long-QT syndrome-related sodium channel mutations probed by the dynamic action potential clamp technique. J Physiol 570: 237–250
17. Verheijck EE, Wilders R, Joyner RW et al (1998) Pacemaker synchronization of electrically coupled rabbit sinoatrial node cells. J Gen Physiol 111:95–112
18. Nguyen TP, Xie Y, Garfinkel A et al (2012) Arrhythmogenic consequences of myofibroblast-myocyte coupling. Cardiovasc Res 93:242–251
19. Johns DC, Nuss HB, Marbán E (1997) Suppression of neuronal and cardiac transient outward currents by viral gene transfer of dominant-negative Kv4.2 constructs. J Biol Chem 272:31598–31603
20. Barabanov M, Yodaiken V (1997) Introducing real-time Linux. Linux J 34:19–23
21. Kemenes I, Marra V, Crossley M et al (2011) Dynamic clamp with StdpC software. Nat Protoc 6:405–417
22. Lin RJ, Bettencourt J, White JA et al (2010) Real-Time eXperiment Interface for biological control applications. Conf Proc IEEE Eng Med Biol Soc 2010:4160–4163
23. Tytgat J (1994) How to isolate cardiac myocytes. Cardiovasc Res 28:280–283
24. Barry PH, Lynch JW (1991) Liquid junction potentials and small cell effects in patch clamp analysis. J Membr Biol 121:101–117
25. Verkerk AO, Veldkamp MW, de Jonge N et al (2000) Injury current modulates afterdepolarizations in single human ventricular cells. Cardiovasc Res 47:124–132
26. Bettencourt JC, Lillis KP, Stupin LR et al (2008) Effects of imperfect dynamic clamp: computational and experimental results. J Neurosci Methods 169:282–289
27. Kullmann PHM, Wheeler DW, Beacom J et al (2004) Implementation of a fast 16-bit dynamic clamp using LabVIEW-RT. J Neurophysiol 91:542–554
28. Clausen C, Valiunas V, Brink PR et al (2013) MATLAB implementation of a dynamic clamp with bandwidth of >125 kHz capable of generating I_{Na} at 37 °C. Pflügers Arch 465: 497–507
29. Wilders R, Verheijck EE, Joyner RW et al (1999) Effects of ischemia on discontinuous action potential conduction in hybrid pairs of ventricular cells. Circulation 99:1623–1629
30. Joyner RW, Wang Y-G, Wilders R et al (2000) A spontaneously active focus drives a model atrial sheet more easily than a model ventricular sheet. Am J Physiol Heart Circ Physiol 279:H752–H763

Chapter 21

Dynamic Clamp in Cardiac and Neuronal Systems Using RTXI

Francis A. Ortega, Robert J. Butera, David J. Christini, John A. White, and Alan D. Dorval II

Abstract

The injection of computer-simulated conductances through the dynamic clamp technique has allowed researchers to probe the intercellular and intracellular dynamics of cardiac and neuronal systems with great precision. By coupling computational models to biological systems, dynamic clamp has become a proven tool in electrophysiology with many applications, such as generating hybrid networks in neurons or simulating channelopathies in cardiomyocytes. While its applications are broad, the approach is straightforward: synthesizing traditional patch clamp, computational modeling, and closed-loop feedback control to simulate a cellular conductance. Here, we present two example applications: artificial blocking of the inward rectifier potassium current in a cardiomyocyte and coupling of a biological neuron to a virtual neuron through a virtual synapse. The design and implementation of the necessary software to administer these dynamic clamp experiments can be difficult. In this chapter, we provide an overview of designing and implementing a dynamic clamp experiment using the Real-Time eXperiment Interface (RTXI), an open-source software system tailored for real-time biological experiments. We present two ways to achieve this using RTXI's modular format, through the creation of a custom user-made module and through existing modules found in RTXI's online library.

Key words Dynamic clamp, RTXI, Cardiac electrophysiology, Neural electrophysiology, Neuronal networks, Artificial conductance block, Reciprocal neuronal coupling

1 Introduction

1.1 Dynamic Clamp

The dynamic clamp technique is an extension of traditional intracellular recording techniques that allows for the introduction of an artificial conductance, controlled by the experimentalist, into a biological cell [1–5]. Dynamic clamp uses closed-loop feedback control between a cell's recorded membrane potential and an injected current calculated by a mathematical model (Fig. 1). This feedback loop, when iterated quickly and accurately enough, is sufficient to mimic one or more biological conductances in the targeted cell. Consequently, a versatile dynamic clamp system allows

Marzia Martina and Stefano Taverna (eds.), *Patch-Clamp Methods and Protocols*, Methods in Molecular Biology, vol. 1183,
DOI 10.1007/978-1-4939-1096-0_21, © Springer Science+Business Media New York 2014

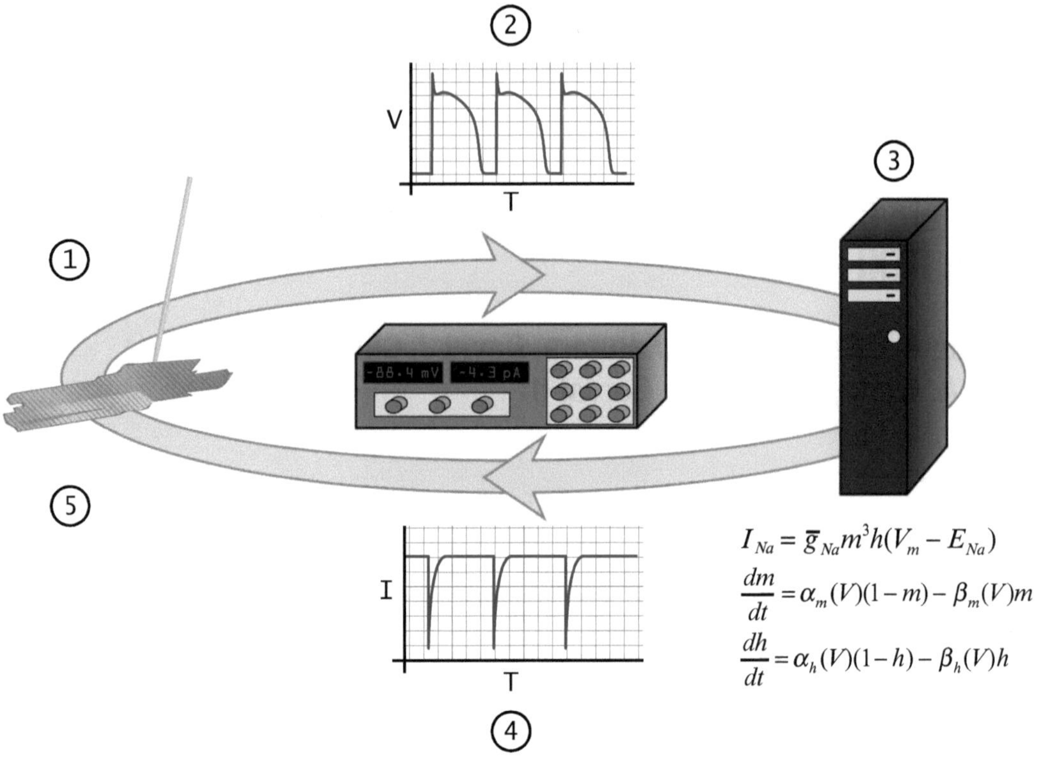

Fig. 1 The fundamental steps of the dynamic clamp feedback loop when injecting an artificial conductance. When run at high enough rates (10–50 kHz), this feedback loop can sufficiently mimic a biological conductance. (*Step 1*) Membrane voltage from a cell is measured using the patch clamp technique and (*Step 2*) sent from the amplifier to the computer's data acquisition system. (*Step 3*) Voltage data is used in a computational current model to calculate current, represented here by a sodium current. (*Step 4*) Calculated current data is sent to the patch clamp amplifier and (*Step 5*) injected into the cell

investigators to simultaneously control many electrophysiological parameters within a single experiment.

1.2 Real-Time eXperiment Interface

At the core of the dynamic clamp experimental setup is the software, which acts as the interface between the mathematical models and the patch-clamp hardware used to measure the parameters of the biological cell. Rapid sampling of membrane potential and fast computation of the conductance is required to adequately recapitulate a conductance [6]. Due to these requirements, real-time control is required to accomplish the dynamic clamp technique. For numerous technical reasons, real-time control is not possible on standard operating systems and software. Thus, a specialized operating system and software are required. RTXI is a fast and versatile software system adopted by many labs across many fields of electrophysiology to administer dynamic clamp experiments. The software provides real-time control using a real-time Linux operating system (Fig. 2). Free and open-source, the platform has been

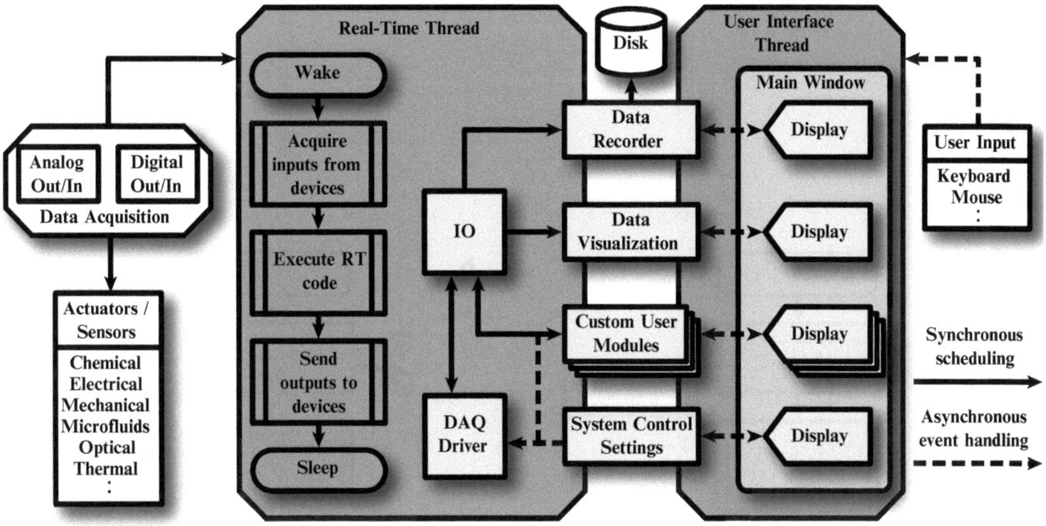

Fig. 2 Diagram of the RTXI software architecture. RTXI uses a two thread architecture where modules communicate and share data in real-time through a signal and slot system. On every clock cycle, a prioritized real-time thread wakes and sequentially runs its tasks—data acquisition, execution of real-time instructions of loaded modules, and signal output. Once all tasks are complete, the thread sleeps until the following cycle. Data acquisition and output is performed using the open-source COMEDI (www.comedi.org) drivers, which allows RTXI to interface with many multifunction data acquisition cards. A separate user interface thread is responsible for all graphical components and user input. This approach allows for all real-time instructions to take priority over graphical components during computation

tailored to fit the needs of the biological research community. RTXI uses a modular architecture, which allows investigators to construct complex experimental protocols using a combination of built-in, community-shared, and self-made modules. More information on the software, installation, and use can be found on its Web site (www.rtxi.org).

1.3 Simulating Artificial Conductance

In this chapter, we detail two different ways to build a dynamic clamp protocol using RTXI. In the first case, dynamic clamp will be used to artificially block a potassium conductance in a cardiac myocyte (Fig. 3a). By injecting a conductance with the opposite magnitude of a biologically present current, an artificial block can be achieved. This type of experiment is useful for testing the validity of a computational model in a biological setting, and when no available drugs block a current specifically. In this chapter, the described experimental protocol will use a custom module developed based on the equations describing the cardiac inward rectifier current (IK1). We will detail the steps required to convert the mathematical equations of a current model into a RTXI module capable of implementing this dynamic clamp experiment. In the second case, dynamic clamp will be used to construct a hybrid

$$t_k = t_{k-1} + \Delta t \qquad t_{k+1} = t_k + \Delta t$$

$$I_{K1} = \overline{G}_{K1} * K1_{\infty}(V - E_{K1})$$

$$I_{stim} = 0 \qquad I_{stim} = 1.5 * I_{th}$$

$$I_{app} = (c * I_{K1}) + I_{stim}$$

$$G = G_{max} * \frac{t}{\tau} * e^{[\frac{-(t-\tau)}{\tau}]}$$

$$I_{syn} = G * (V_m - E_{syn}) \qquad V_m > V_{th}$$

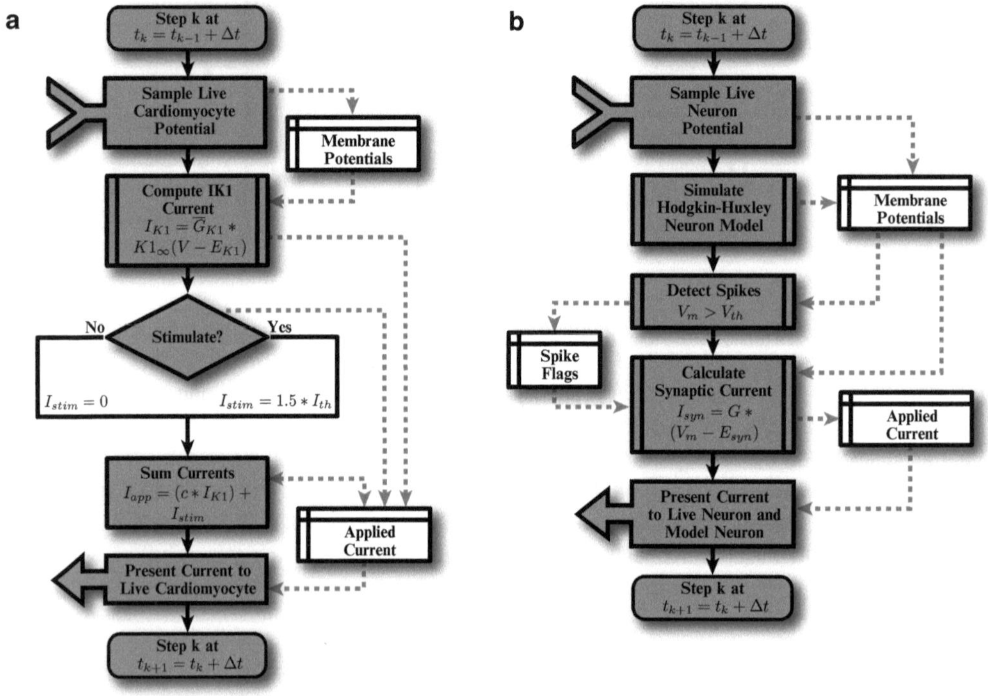

Fig. 3 Flowcharts depicting dynamic clamp and RTXI module activity for a time step. The time step Δt is a parameter that would ideally be at most 1/20 the smallest time constant in the current model. (**a**) Overview of the dynamic clamp procedure to block the IK1 current in cardiomyocytes. Voltage is sampled from the myocyte and passed to a module that computes the IK1 current. Stimulation is provided at a set cycle length, typically for 1 ms at 1.5× the determined threshold to invoke action potentials. The stimulation and the IK1 current, scaled by the parameter c (for full block, $c = -1$), is then injected into the cardiomyocyte. (**b**) Overview of the dynamic clamp procedure to reciprocally couple a live neuron and a virtual neuron by virtual alpha synapses to construct a neural network. Voltage is sampled from a live neuron and a simulated Hodgkin–Huxley neuron model. When the membrane potential of either the model or virtual neuron passes a certain threshold (V_{th}) indicating a spike has occurred, the spike flag is updated. Using the spike information and membrane potential, the synaptic current is computed and injected into the live neuron and model neuron

network by reciprocally coupling two neurons through alpha synapses (Fig. 3b). This hybrid network will consist of one real patch-clamped neuron, and a virtual Hodgkin–Huxley neuron model. Rather than manually constructing a module to accomplish this experiment, the protocol will be developed using already available modules: bundled with RTXI or available through its online library.

This chapter focuses on the setup, customization, and use of the RTXI platform to implement a dynamic clamp experiment. As such, it is assumed the reader is familiar with the example applications, cardiomyocyte conductances and hybrid neuronal networks, and with the more general techniques of patch clamping.

2 Materials

In this list of materials, Subheadings 2.1–2.3 are collections of the absolute essential materials required to prepare and patch-clamp isolated cardiomyocytes or neurons in slice. Subheading 2.4 describes the required patch clamp solutions needed in both experimental settings, cardiac and neural. Subsequent sections deal with the materials specific to dynamic clamp. Subheading 2.5 lists the required components, hardware and software, for a digital dynamic clamp system run on a personal computer.

2.1 Cardiomyocyte Isolation Equipment

1. Anesthetic appropriate for animal model (e.g., sodium pentobarbital for guinea pigs).
2. Heparin 1,000 units/ml.
3. Langendorff perfusion system containing water bath, heat exchanger, peristaltic pump, bubble trap, and tubing.
4. Gas dispersion tube with fritted cylinder and medical grade oxygen.
5. Aortic cannula, appropriately sized for animal model.
6. Surgical scissors, forceps, two fine bent forceps, scalpel, bulldog clip, and surgical suture.
7. 50 ml conical tubes and corresponding centrifuge.
8. 250 μm cell strainer.
9. Shaking water bath.
10. Collagenase, Type II and Protease Type XIV.
11. Common Tyrode's solution, calcium-free Tyrode's solution, 1-Enzyme solution, 2-Enzyme solution, and wash solution (Table 1) made with endotoxin-free water.

2.2 Brain Tissue Slice Preparation Equipment

1. Air tight anesthesia chamber with isoflurane.
2. Appropriately sized animal guillotine.
3. Surgical scissors, Scoopula, scalpel.
4. Horizontal serial slicer (vibratome) with walled slice chamber.
5. Filter paper, fine artist's paint brush.
6. Bubble chamber with tubing and oxygen.
7. Artificial cerebrospinal fluid (ACSF) (Table 2; *see* **Note 1**).

Table 1
Cardiomyocyte isolation solutions

Solution	Common Tyrode's	Calcium-free Tyrode's	1-Enzyme	2-Enzyme	Wash
$CaCl_2$ (mM)	2		0.025	0.025	0.375
EGTA (μM)		100			
Collagenase (mg/ml)			1	1	
Protease XIV (mg/ml)			0.05		
BSA (mg/ml)					5

In addition, all solutions are buffered to a pH of 7.2 with NaOH and contain the following: 130 mM NaCl, 5.4 mM KCl, 3.5 mM $MgCl_2$, 0.4 mM NaH_2PO_4, 5 mM HEPES, and 10 mM Dextrose
EGTA ethyleneglycol-bis-N,N,N′, N′-tetraacetic acid, *BSA* bovine serum albumin

Table 2
Patch clamp solutions

Solution	Superfusion Tyrode's	ACSF	IC cardiomyocyte	Artificial neuroplasm
NaCl (mM)	137	126		2
KCl (mM)	4.5	3	142	
$MgCl_2$ (mM)			2	2
$MgSO_4$ (mM)	1	2		
NaH_2PO_4 (mM)		1.25		
$NaHCO_3$ (mM)		26		
K-Gluconate (mM)				135
Dextrose (mM)	10	10		
$CaCl_2$ (mM)	2	2	3.8	
HEPES (mM)	10		10	10
EGTA (mM)			10	1.25
Tris–ATP (mM)				2
Tris–GTP (mM)				0.4
Na_2ATP (mM)			5	
pH	7.35	7.4	7.1	7.25
Osmolarity (mOsm)	300	300	290	290

EGTA ethyleneglycol-bis-N,N,N′, N′-tetraacetic acid

2.3 Patch Clamp Hardware

1. Inverted microscope or infrared differential interference contrast microscope depending on application (*see* **Note 2**).
2. Stage chamber with temperature controller and influx/efflux tubing for extracellular solutions.
3. One micromanipulator.
4. One current-clamp amplifier (Axon Instruments MultiClamp 700B; Molecular Devices, USA) with head stage (*see* **Note 3**).
5. Borosilicate glass electrodes drawn on a horizontal puller. Pipette specifications are dependent on application (*see* **Note 4**).

2.4 Patch Clamp Solutions

1. Extracellular solution: superfusion Tyrode's solution for cardiomyocyte application or ACSF for brain tissue application (Table 2).
2. Intracellular solution: intracellular cardiomyocyte solution or artificial neuroplasm (Table 2).

2.5 Dynamic Clamp System

1. Personal computer running a hard real-time operating system (Real-Time Application Interface (RTAI) patched Linux Kernel) (*see* **Note 5**).
2. A personal computer with the dynamic clamp software, RTXI, installed (*see* **Note 6**).
3. Multipurpose A/D-D/A device installed in the computer (PCI-6259; National Instruments, USA) and shielded connector block (BNC-2110; National Instruments, USA) (*see* **Note 7**).
4. Bayonet Neill–Concelman (BNC) cables.

3 Methods

Subheading 3.1 describes the setup of a typical dynamic clamp system running RTXI, which revolves around adequate communication between RTXI and the patch clamp amplifier. The simplest setup uses a personal computer, running RTXI, interfacing with a patch clamp amplifier using an analog voltage input and output channel through a data acquisition (DAQ) device (Fig. 4). Most patch clamp amplifiers possess the necessary analog output and analog external input to interface with RTXI. Depending on the capabilities of the amplifier, greater integration can be achieved, such as automatic detection of amplifier mode or compensation for changes in output gain (*see* **Note 8**). Also, RTXI modules designed specifically for controlling several of the most common models of patch clamp amplifiers are available (*see* **Note 9**).

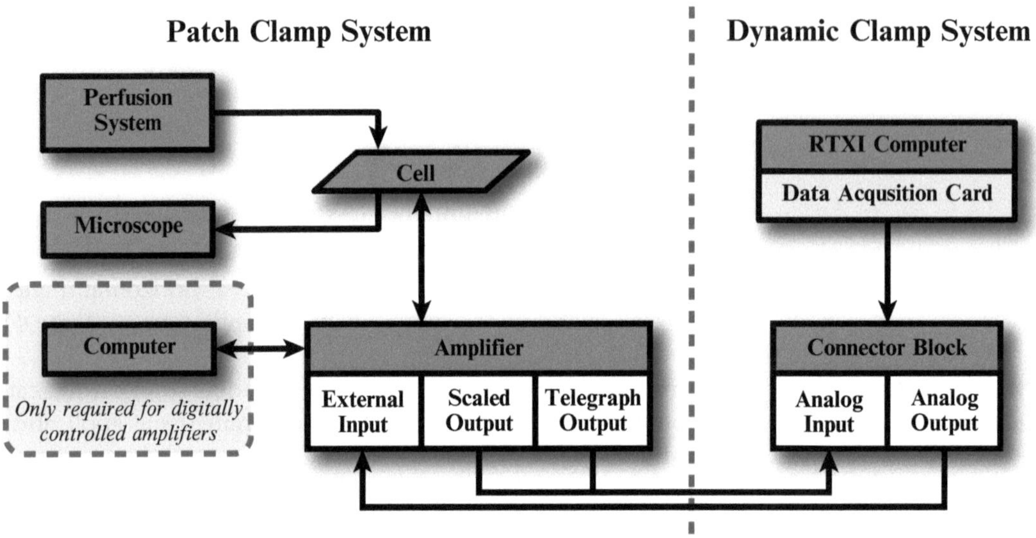

Fig. 4 Diagram of a typical dynamic patch clamp setup. *Arrows* indicate flow and direction of information. The traditional patch clamp setup is shown on the left, including a computer required for digitally controlled amplifiers (*see* **Note 7**). The dynamic clamp system is shown on the *right*. The RTXI computer interfaces with the patch clamp amplifier through a connector block linked to a data acquisition card. BNC cables from the connector block to the amplifier connect the two devices

In subsequent sections, descriptions of two dynamic clamp experiments from the field of cardiac electrophysiology and neural electrophysiology are presented. These two experiments are accomplished using different approaches in development and execution of the experiment protocol. Subheading 3.2 describes the injection of an artificial potassium conductance, specifically IK1, into a cardiomyocyte. Focus is on development of a custom RTXI module written using the programming language C++. As shown in Fig. 3a,this module will sample membrane voltage, calculate a current based on equations describing the IK1 current, and inject this current into a patched cardiomyocyte. Subheading 3.3 describes the construction of a neuronal hybrid network with a real and virtual neuron through an alpha synapse (*see* Fig. 3b). Focus is on using already available RTXI modules, either built into RTXI or shared by the community. Each of these sections contains brief descriptions of cell/tissue preparation and patch clamping. More thorough descriptions can be found throughout this book.

3.1 Dynamic Clamp Setup

1. RTXI Installation:
   RTXI requires a real-time Linux operating system (*see* **Note 5**). Typically, these operating systems are patched, compiled manually, and then installed on a personal computer. Afterwards, RTXI and other required software are installed on the real-time operating system. This process has been simplified through the

use of a "Live USB." The Live USB is a complete bootable real-time Linux operating system with RTXI pre-installed packaged onto a USB drive. Normally the Live USB is used to install the package onto a hard drive like any other operating system. However, it also can be run from the USB directly, resulting in a portable real-time software system. For more details on the installation of RTXI onto a personal computer, visit the installation Web page (www.rtxi.org/install).

2. Dynamic Clamp Hardware Setup:
   The basic dynamic clamp setup is outlined in Fig. 4. Information is passed between RTXI and the amplifier through a data acquisition device. Several patch-clamp amplifier models are digitally controlled, in contrast to the traditional knob and button interface. The controlling software is usually only available on both Windows or Mac operating systems, and not Linux. In these situations, a second computer is required to control the amplifier.
   (a) Install the data acquisition card into the personal computer with RTXI installed, and attach the connector block.
   (b) Using a BNC cable, connect the labeled AO 0 (Analog Output 0) of the connector block to the external input of the amplifier.
   (c) Connect the labeled AI 0 (Analog Input 0) to the Output of the amplifier.
3. Interfacing RTXI and the Patch-Clamp Amplifier:
   RTXI receives information from the amplifier in the form of a voltage signal read through a data acquisition device. Amplifiers apply various gains to these signals to achieve adequate signal to noise ratios. In order for RTXI to accurately read and send signals to the amplifier, these gains must be compensated for. For example, if a gain of 10 is applied to an amplifier output, the corresponding input must be reduced by a factor of 10 by RTXI. Unit mismatches are a frequent source of error. This section explains how to manually change the scaling values for the input/output signals in RTXI. As previously discussed, there are modules available that simplify or automate this process (*see* **Note 9**). In this and subsequent sections, various commands will need to be entered by the reader using the standard Linux terminal. These commands will be preceded by "$".
   (a) Open RTXI with the terminal.
      - Open the terminal (System → Terminal in Ubuntu 9.10).
      - $ rtxi
   (b) Navigate to the system tab, and open the control panel.
      - System → Control Panel.

(c) Activate the 0 channel for analog input and the 0 channel for analog output.
- Set channel: Input|0.
- Toggle the Active button.
- Click the Apply button.
- Repeat for channel: Output|0.

(d) Adjust the scaling values accordingly (*see* **Note 10**).

(e) Set thread rate of RTXI to 10 kHz (*see* **Note 11**).
- Switch to Thread Tab.
- Set Frequency: 10,000|Hz.
- Click the Apply button.

3.2 Artificial Blocking of the Inward Rectifier K Current in Cardiomyocytes Through Dynamic Clamp

A dynamic clamp experiment to artificially block a current requires a computational model that adequately simulates the current in the target cell (Fig. 3a). In this example, the target cell is an isolated guinea pig ventricular myocyte. In this section, we use equations of the IK1 current from the Livsthitz Rudy 2009 guinea pig ventricular myocyte model [7]. From these equations, we describe a basic workflow to construct a RTXI module that can calculate the IK1 current in real-time (*see* Subheading 3.2.1). In Subheading 3.2.2 we then describe the isolation of ventricular myocytes from a guinea pig heart [8]. Finally, we implement the module in an IK1 conductance block experiment (*see* Subheading 3.2.3). This section can be used as a guideline to develop a similar dynamic clamp experiment to introduce, block, or scale any modeled current in other species.

3.2.1 Design and Construction a Custom RTXI Module

The ability for the user to create a custom module is the most powerful feature of RTXI, allowing the development of unique and complex experimental protocols. While some familiarity with C++ is required, by no means does the reader need to be an expert programmer. The custom module will be built using RTXI's default_gui_model library. This library was designed to create modules quickly and easily without in-depth understanding of software's architecture. Starting with a template, this section explains the code necessary for each program part of the IK1 block RTXI module.

1. Open the terminal.
2. Create a directory for a new RTXI module.
   (a) $ mkdir IK1_Block
3. Copy the module program template files from your RTXI installation directory (example directory used).
   (a) $ sudo cp /home/rtxi/doc/my_plugin/* IK1_Block
4. Enter the module directory.
   (a) $ cd IK1_Block

```
#include <IK1_Block.h>
#include <math.h>
#include <qwhatsthis.h>
#include <iostream>

extern "C" Plugin::Object *
createRTXIPlugin(void)
{
    return new IK1_Block();
}

static DefaultGUIModel::variable_t vars[] =
    {
        { "Input Voltage (V)", "Input Voltage (V)", DefaultGUIModel::INPUT, }
        { "Output Current (A)", "Output Current (A)", DefaultGUIModel::OUTPUT, }
        { "BCL", "Basic cycle length for pacing (ms)", DefaultGUIModel::PARAMETER
          | DefaultGUIModel::INTEGER, },
        { "Stim Mag", "Stimulus magnitude (nA)", DefaultGUIModel::PARAMETER
          | DefaultGUIModel::DOUBLE, },
        { "Stim Length", "Stimulus length (ms)", DefaultGUIModel::PARAMETER
          | DefaultGUIModel::DOUBLE, },
        { "Pace", "(0: Pace off) (1: Pace on)", DefaultGUIModel::PARAMETER
          | DefaultGUIModel::INTEGER, },
        { "Cm", "Membrane capacitance (pF)", DefaultGUIModel::PARAMETER
          | DefaultGUIModel::DOUBLE, },t
        { "Scale Factor", "IK1 scaling factor (-1 = block)", DefaultGUIModel::PARAMETER
          | DefaultGUIModel::DOUBLE, },
        { "Experiment", "(0: Experiment off) (1: Experiment on)", DefaultGUIModel::PARAMETER
          | DefaultGUIModel::INTEGER, },
        { "IK1", "Calculated IK1 current", DefaultGUIModel::STATE, },
    };

static size_t num_vars = sizeof(vars) / sizeof(DefaultGUIModel::variable_t);

IK1_Block::IK1_Block(void) :
    DefaultGUIModel("IK1_Block", ::vars, ::num_vars)
{
    QWhatsThis::add(this, "<p><b>IK1_Block:</b><br>Module to block IK1 current in cardiomycoytes.</p>");
    createGUI(vars, num_vars); // this is required to create the GUI
    update(INIT); // this is optional, you may place initialization code directly into the constructor
    refresh(); // this is required to update the GUI with parameter and state values
}

IK1_Block::~IK1_Block(void)
{
}
```

Fig. 5 Header file for IK1 Block RTXI module. The C++ header file contains the class definitions and variable declarations required for the IK1 block module. These variables are model parameters, experimental parameters, or module control flags, and are kept in memory during the life of the module. This is in contrast to variables declared in the source code of module, which are only kept in memory for the duration of the function call

5. Rename the template files
    (a) $ sudo mv my_plugin.cpp IK1_Block.cpp
    (b) $ sudo mv my_plugin.h IK1_Block.h
6. Open the template files in the preferred code editing program (*see* **Note 12**).
    (a) $ emacs IK1_Block.cpp IK1_Block.h
7. Add the necessary code to each part of the template file.
    (a) IK1_Block.h (Fig. 5).
        - Replace all instances of my_plugin with IK1_Block.
        - Declare all variables.
        - Save file.

$$I_{K1} = \overline{G}_{K1} * K1_{\infty} * (V - E_{K1})$$
$$\overline{G}_{K1} = 0.75\sqrt{[K^{+}]_o/5.4}$$
$$K1_{\infty} = \frac{1}{1+0.004e^{0.6987(V-E_K+11.724)}/(1+e^{0.6168(V-E_K+4.872)})}$$

```
void IK1_Block::execute(void)
{
    out = 0;
    V = input(0) * 1e3; // Convert units (V -> mV)

    if( pace == 1 ) { // Pacing
        if( count < stimLengthConverted ) // Pacing stimulus
            out += (stimMag * 1e-9); // Convert units (nA -> A)
        if( count >= bclConverted ) // Reset count at end of beat
            count = 0;
        else
            count++;
    }

    if( experiment == 1 ) {
        K1_ss = 0.004 * (1 + exp(0.6987*(V-EK+11.724))) / (1+exp(0.6168*(V-EK+4.872)));
        IK1 = GK1_ * sqrt(Ko/5.4) * (V-EK) / (1+K1_ss);

        out += (-1 * (scale * IK1)) * (cm * 1e-12); // Sign change due to amplifier and
                                                    // model signs being reversed
    }

    output(0) = out; // Final output in Amps

    return;
}
```

Fig. 6 Source file for IK1 Block RTXI module—vars structure, constructor, and destructor. (*Line 12*) The vars structure defines the inputs, outputs, parameters, and states. An INPUT creates a slot for the module to acquire data from the DAQ device or another module. An OUTPUT is a signal that can be sent to the DAQ device or another module. Using RTXI's signal and slot architecture, modules can share data through connections between inputs and outputs. A STATE is a numeric data-type that cannot be modified by the user. These are typically used to track values of intermediate or computed quantities. A PARAMETER is another numeric data-type, but can be modified on the fly by user input through the GUI. These module data-types are also what the Data Recorder module is capable of recording. (*Line 35*) The module constructor is the function first run upon module loading, and normally does not require any change. (*Line 44*) The module destructor is the function run upon module unloading. This is typically where any garbage collection takes place, such as release of any dynamically allocated memory through the use of the new operator

(b) IK1_Block.cpp
   - Replace all instances of my_plugin with IK1_Block.
   - Add the necessary inputs, outputs, parameters, and states to the "vars" struct (Fig. 6).
   - Add the real-time executed code in the "execute" function (Fig. 7).
   - Add the code executed during module initialization and modification (Fig. 8).
   - Save file.

8. Edit the program Makefile
   (c) Replace all instances of my_plugin with IK1_Block.
   (d) Save file.
9. Compile and install the program.
   (e) $ sudo make install.

3.2.2 IK1 Block: Experiment Setup in RTXI

1. Open RTXI and the newly built module.
   (a) $ rtxi
   (b) Modules → Load User Module → IK1_Block.so.

```
1  void IK1_Block::update(DefaultGUIModel::update_flags_t flag)
2  {
3      switch (flag)
4          {
5  
6          case INIT: // Performed when module is loaded
7              period = RT::System::getInstance()->getPeriod() * 1e-6; // ms
8  
9              // Initialize parameters to defaults
10             bcl = 1000;
11             stimMag = 0;
12             stimLength = 0;
13             pace = 0;
14             cm = 0;
15             scale = 0;
16             experiment = 0;
17 
18             setParameter("BCL", bcl);
19             setParameter("Stim Mag", stimMag);
20             setParameter("Stim Length",stimLength);
21             setParameter("Pace", pace);
22             setParameter("Cm", cm);
23             setParameter("Scale Factor", scale);
24             setParameter("Experiment", experiment);
25             setState("K1_ss", K1_ss);
26             setState("IK1", IK1);
27 
28             IK1 = 0;
29             K1_ss = 0;
30             count = 0;
31             EK = (R*T/F) * log(Ko/Ki);
32 
33             bclConverted = bcl / period; // Use unitless integers to avoid rounding errors
34             stimLengthConverted = stimLength / period;
35             break;
36 
37         case MODIFY: // Parameters are updated when modify button is clicked
38             bcl = getParameter("BCL").toInt();
39             stimMag = getParameter("Stim Mag").toDouble();
40             stimLength = getParameter("Stim Length").toDouble();
41             pace = getParameter("Pace").toInt();
42             cm = getParameter("Cm").toDouble();
43             scale = getParameter("Scale Factor").toDouble();
44             experiment = getParameter("Experiment").toInt();
45 
46             bclConverted = bcl / period; // Use unitless integers to avoid rounding errors
47             stimLengthConverted = stimLength / period;
48             break;
49 
50         case UNPAUSE:
51             break;
52 
53         case PAUSE:
54             break;
55 
56         case PERIOD: // Performed when RTXI thread rate is changed
57             period = RT::System::getInstance()->getPeriod() * 1e-6; // ms
58             bclConverted = bcl / period; // Use unitless integers to avoid rounding errors
59             stimLengthConverted = stimLength / period;
60             break;
61         default:
62             break;
63         }
64 }
```

Fig. 7 Source file for IK1 Block RTXI module—execute function. The execute function represents the real-time code that is run every time step during program execute. It is imperative that the execute code must be as efficient as possible. (*Line 4*) Voltage is first sampled from an input, typically from a DAQ device or a model, and used in the calculation of IK1 current. (*Line 6*) The pace parameter controls the modules periodic stimulus. When changed to 1, stimulation begins based on the BCL (basic cycle length), Stim Mag (stimulation magnitude), and Stim Length (stimulation length) parameters. (*Line 15*) The experiment parameter controls IK1 current injection. (*Line 15*) When the experiment variable is set to 1, voltage is used to calculate the IK1; (*Inset*) original IK1 equations are shown. (*Line 23*) This calculated current is then passed through an output, which is usually the analog output of a DAQ device connected to a patch clamp amplifier

```
1   #include <default_gui_model.h>
2
3   class IK1_Block : public DefaultGUIModel
4   {
5
6   public:
7       IK1_Block(void);
8       virtual
9       ~IK1_Block(void);
10      void execute(void);
11
12  protected:
13      virtual void update(DefaultGUIModel::update_flags_t);
14
15  private:
16      double period; // thread rate of RTXI
17
18      double V; // membrane voltage
19      double K1_ss;
20      double IK1; // total IK1 current (A/F)
21      double EK; // potassium reversal potential
22
23      double out; // total output of model
24      double scale; // scaling factor for IK1 injection
25      int experiment; // run experiment flag
26      int pace; // cell pacing flag
27      int bcl; // basic cycle length (ms)
28      double stimMag; // stimulus magnitude (nA)
29      double stimLength; // stimulus length (ms)
30      double cm; // membrane capacitance (pF)
31      int count;
32
33      // Unitless variables for pacing to eliminate rounding errors
34      int bclConverted;
35      int stimLengthConverted;
36
37      // Constants
38      const static double Ko = 4.5; // extracellular potassium constant (uM)
39      const static double Ki = 142; // intracellular potassium (uM)
40      const static double GK1_ = 0.75;
41      // Physical constants
42      const static double F = 96485; // Faraday's constant, C/mol
43      const static double R = 8314; // gas constant, mJ/K
44      const static double T = 273+37; // absolute temperature, K
45  };
```

Fig. 8 Source file for IK1 Block RTXI module—update function. The update function is called during several module events. (*Line 6*) The INIT case is run during module initialization. (*Line 18*) The setParameter("parameter name," variable) and setState("state name," variable) functions are used to associate the datatypes described in the vars structure with program variables. (*Line 37*) The modify case is called whenever the modify button is clicked. Parameters are set to corresponding values retrieved from the GUI using the getParameter("parameter name") function. Any other variables dependent on the parameters are also updated, such as bclConverted and stimlengthConverted. (*Lines 46* and *47*) These variables are unit-less integers calculated by dividing BCL or stimulation length by RTXI's thread period. Through the use of integers, rounding errors that may occur during the timing of the stimulus are eliminated. (*Lines 50* and *53*) The un-pause and pause states are called during toggling of the pause button. (*Line 56*) The period case is run when the RTXI thread rate is changed through the control panel. All variables dependent on the RTXI thread rate are updated

2. Load additional modules required for dynamic clamp experiment.
   (a) Utilities → Patch Clamp → Membrane Test.
   (b) System → Oscilloscope.
   (c) System → Data Recorder.

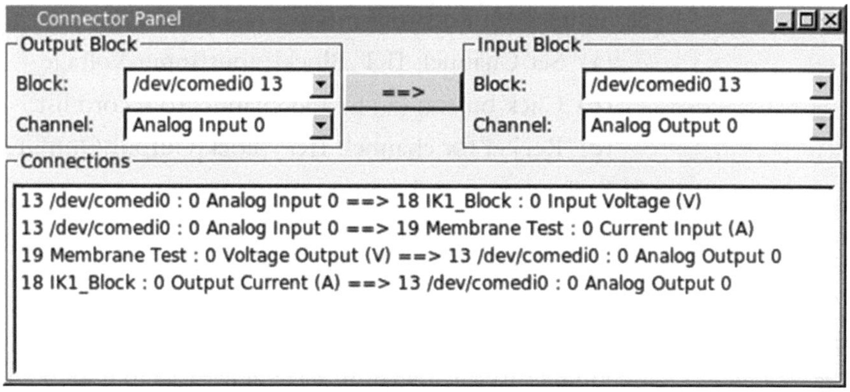

Fig. 9 RTXI connector settings for IK1 block dynamic clamp experiment. The connector panel showcases the signal and slot architecture of RTXI. The analog input 0 channel of a DAQ device (signal) is connected to the input of the IK1_Block module and Membrane Test module (slot). Conversely, the outputs of these two modules (signal) are connected to the analog output 0 channel (slot). The analog output channel is in turn connected to a patch clamp amplifier

3. Connect the modules' input/output (Fig. 9).
   (a) Open connector window (System → Connector).
   (b) Set the Output block: IK1_Block—Output Current.
   (c) Set the Input block: /dev/comedi0—Analog Output 0 (*see* **Note 13**).
   (d) Click Arrow button to complete connection.
   (e) Repeat for the following combinations:
      - /dev/comedi0—Analog Input 0|IK1_Block—Input Voltage.
      - Membrane_Test—Voltage Output|/dev/comedi0—Analog Output 0.
      - /dev/comedi0—Analog Input 0|Membrane_Test—Current Input.
4. Set oscilloscope to display cardiomyocyte membrane voltage (*see* **Note 14**).
   (a) Open oscilloscope properties (right-click on oscilloscope → Properties).
   (b) Set Channel: IK1_Block|Input|Input V.
   (c) Toggle the Active button.
   (d) Set Scale: 20 mV/V.
   (e) Click the Apply button.
   (f) Click Display tab.
   (g) Set Time Scale: 100 ms/div.
   (h) Set Screen Refresh: 50.

5. Setup Data Recorder module (*see* **Note 15**).
   (a) Set Channel: IK1_Block|Input|Input Voltage.
   (b) Click button (>) to add channel to record list.
   (c) Repeat for channel: IK1_Block|Output|Output Current.
6. Save workspace for future experiments.
   (a) File → Save Workspace.
   (b) Choose filename and click Save button.

3.2.3 Single Cardiomyocyte Isolation

1. Inject the guinea pig (400–500 g) with 500 units of heparin by intraperitoneal injection, allow at least 15 min for heparin to act.
2. Anesthetize the guinea pig by intraperitoneal injection of sodium pentobarbital.
3. Excise the heart by thoracotomy.
4. Once extracted, place the heart in oxygenated and 0 °C Common Tyrode's solution. Immediately massage the heart to push blood out of the major chambers.
5. Using fine bent forceps and scissors, remove excess tissue around aorta: lungs, pericardium, and fat.
6. Using a pair of fine bent forceps and surgical suture, cannulate the aorta and begin retrograde perfusion with oxygenated Common Tyrode's solution at 37 °C at 16 ml/min for 5 min at a pressure of at least 80 mm Hg. Heart should begin to beat vigorously and clear blood after 30 s.
7. Perfuse oxygenated calcium-free Tyrode's solution for 4 min. Heart should cease to beat.
8. Perfuse oxygenated 2-enzyme solution for 1–4 min. Length of time is determined by appearance and feel of tissue. Perfusion is stopped once tissue becomes soft, bloated, and the color lightens.
9. While cannulated, cut only the ventricles into 5 ml of 1-enzyme solution using scissors and mince the tissue using a scalpel.
10. Once minced, transfer tissue and solution into a 50 ml conical tube, and add additional 10 ml of 1-enzyme solution.
11. Shake tissue containing solution for 30 s.
12. Filter the now cloudy solution with a 250 μm cell strainer while keeping tissue in the tube. Add 15 ml of 1-enzyme solution to the remaining tissue, and place in shaking water bath at 37 °C for 3 min.
13. Centrifuge the filtered solution for 1 min at 400 rpm and remove supernatant. Resuspend the cell pellet in 5 ml of wash solution.
14. Repeat the previous steps two more times, producing a total of three aliquots of cells in wash solution.
15. Centrifuge all aliquots for 1 min at 400 rpm, remove supernatant and resuspend in wash solution.

16. Inspect the cell aliquots. Dilute if necessary. Healthy cells should be quiescent with visible striations and crisp edges (*see* **Note 16**).

3.2.4 Dynamic Clamp: Blocking of the IK1 Current in Cardiomyocytes

1. Patch cardiomyocyte.
   (a) Add a drop from a cell aliquot into the bath. Fill the bath with extracellular solution, and wait 5 min for cells to adhere to glass.
   (b) Begin perfusion with extracellular solution warmed to 37 °C.
   (c) Fill a pulled patch pipette with appropriate intracellular solution, and attach it to amplifier headstage.
   (d) With the inverted microscope, select a target cardiomyocyte. Cell should not be touching any other cell or debris.
   (e) Set patch clamp amplifier to voltage clamp mode.
   (f) While measuring electrode resistance (*see* **Note 17**), apply light negative pressure to form a gigaohm seal.
   (g) Apply large and quick pulses of negative pressure to rupture the membrane to gain electrical access.
2. Measure membrane capacitance of the cell with Membrane Test module (*see* **Note 18**).
   (a) Click on the radio button corresponding to −80 mV.
   (b) Toggle down Pulse button.
   (c) Switch to the Membrane Properties tab.
   (d) Set pull-down menu: Continuous.
   (e) Toggle down Acquire to begin membrane parameter calculation.
   (f) Input measured Cm into the Cm parameter of IK1_Block module, and click the Modify button.
   (g) Switch to Resistance Tab in Membrane Test module.
   (h) Toggle up Pulse button.
3. Set patch clamp amplifier to current clamp mode.
4. Start stimulation of cardiomyocyte.
   (a) Toggle Pause button up to un-pause IK1_block module.
   (b) Change the Stim parameter of the dynamic clamp module to 1, and click the Modify button.
5. Find appropriate stimulation magnitude by slowly increasing the preset value.
   (a) Add 0.5 to the Stim Mag parameter, and click the Modify button.
   (b) Repeat these steps until an action potential is evoked to find threshold current.
   (c) To ensure reliable action potentials, increase threshold current by 50 %.

6. Stimulate the cell for 5 min at a basic cycle length (BCL) of 1,000 ms to allow cell to reach a steady state (*see* **Note 19**), and balance bridge to compensate for access resistance.
   (a) Change the BCL parameter to 1,000, and click the Modify button.
   (b) Balance bridge to compensate for access resistance in order to remove stimulus artifact (*see* **Note 20**).
7. Record baseline action potentials for 2 min with the Data Recorder module.
   (a) Click Choose File, input filename, and choose save location.
   (b) Click the Start Recording button.
   (c) After 2 min, click the Stop Recording button to end data collection.
8. Start IK1 block dynamic clamp.
   (a) Change the Scale parameter to –1 for full conductance block.
   (b) Change the Experiment parameter to 1 for conductance injection.
   (c) Click the Modify button.
9. Record dynamic clamped action potentials for 2 min with the Data Recorder module.
   (a) Click Choose File, input filename, and choose save location.
   (b) Click the Start Recording button.
   (c) After 2 min, click the Stop Recording button.
10. Observe depolarized resting membrane potential and shortened action potentials (*see* Fig. 10).

3.3 Reciprocal Coupling Between a Virtual Neuron and Live Neuron Through Dynamic Clamp

Using the library of modules built into RTXI and freely available on the RTXI Web site (www.rtxi.org), numerous types of dynamic clamp experiments can be performed without any need of programming. This section describes using a Hodgkin–Huxley model neuron and alpha synapse model to construct a simple neuronal network (*see* Fig. 3b). This network can easily be expanded through additional virtual or real neurons. While no programming is required, the available module source code allows for alteration of the model equations.

3.3.1 Installation of a RTXI Module and Experiment Setup

1. Download the community built modules—Spike Detector and Reciprocal Coupling.
   (a) Visit the RTXI Web site (www.rtxi.org).
   (b) On the Web site, navigate to Modules → Neural Electrophysiology.
   (c) Download the Spike Detector and Reciprocal Coupling modules.

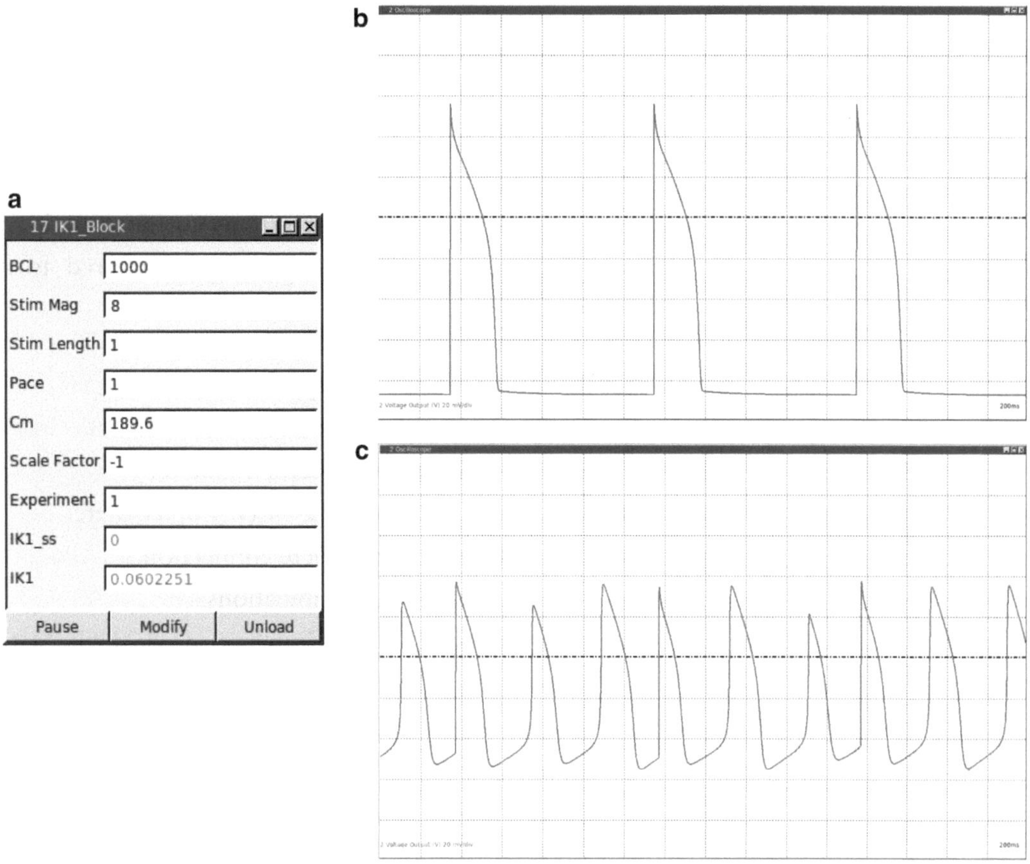

Fig. 10 Results for IK1 conductance block in ventricular guinea pig cardiomyocyte through dynamic clamp. (**a**) Screenshot of the resulting GUI in RTXI of the constructed IK1_Block module during experiment. (**b**) Screenshot of membrane potential of the cardiomyocyte, showing evoked action potentials from a 1,000 BCL pacing protocol using the IK1_Block module using RTXI's Oscilloscope module in real-time. (**c**) Subsequent screenshot after IK1 conductance block dynamic clamp is turned on. Plot now shows pacemaker-like activity of the cardiomyocyte after conductance block is initiated

2. Navigate to the download directory of the modules, unpack the contents, and install the module.
   (a) $ cd ~/Downloads/ (*see* **Note 21**).
   (b) $ tar –xjvf SpikeDetect.tar.gz
   (c) $ cd SpikeDetect
   (d) $ sudo make install
   (e) Repeat for Reciprocal Coupling module.

3.3.2 Reciprocal Coupling: RTXI Experiment Setup

1. Open RTXI and load required modules for dynamic clamp experiment.
   (a) $ rtxi
   (b) Modules → Load User Module → spikedetect.so (referred to as Spike Detector 1) (*see* **Note 22**).

   (c) Repeat to load a second Spike Detector module (referred to as Spike Detector 2).
   (d) Modules → Load User Module → coupling.so.
   (e) Utilities → Model HH Neuron.
   (f) Utilities → Patch Clamp → Membrane Test.
   (g) System → Oscilloscope (referred to as Oscilloscope 1).
   (h) Repeat to load a second Oscilloscope (referred to as Oscilloscope 2).
   (i) System → Data Recorder.
2. Connect the modules' input/output (*see* Fig. 11).
   (a) Open connector window (System → Connector).
   (b) Set Output block: Spike Detector 1—State.
   (c) Set the Input block: Coupling—Cell 1 Spike State.
   (d) Click Arrow button to complete connection.
   (e) Repeat for the following combinations:
      - Spike Detector 2—State|Coupling—Cell 2 Spike State.
      - Neuron—Vm|Coupling—Cell 1 Vm.
      - /dev/comedi0|Coupling—Cell 2 Vm.
      - Neuron—Vm|Spike Detector 1—Vm.
      - /dev/comedi0—Analog Input 0|Spike Detector 2—Vm.
      - Coupling—Isyn 1-2|/dev/comedi0—Analog Output 0.
      - Coupling—Isyn 2-1|Neuron—Iapp.

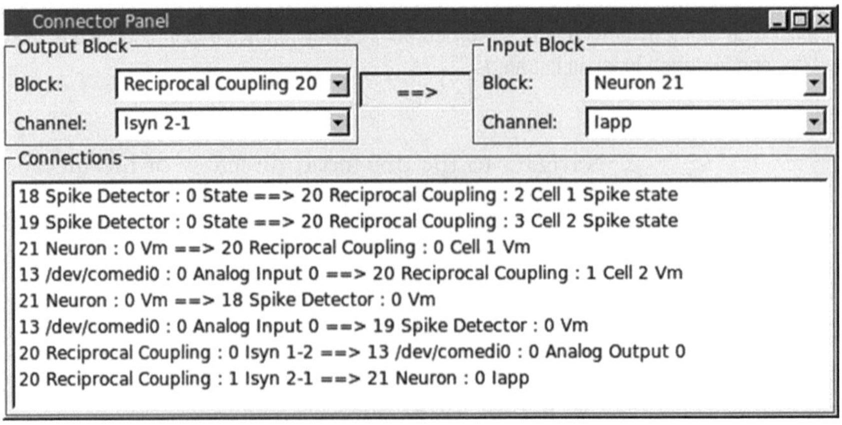

Fig. 11 RTXI connector settings for reciprocal coupling of a live and virtual neuron dynamic clamp experiment. Inputs and outputs are not only used to connect modules to the DAQ device, but also to other modules. One spike detector retrieves input through a DAQ device, while another retrieves its input from the Neuron module. The outputs of both modules are then passed to the Reciprocal Coupling module

3. Set oscilloscope to display virtual neuron membrane potential on Oscilloscope 1 and live neuron membrane potential on Oscilloscope 2.
   (a) Open Oscilloscope 1 properties (Right click on Oscilloscope 1 → Properties).
   (b) Set Channel: Neuron|Output|Vm.
   (c) Toggle the Active button.
   (d) Set Scale: 20 mV/V.
   (e) Click the Apply button.
   (f) Click Display tab.
   (g) Set Time Scale: 100 ms/div.
   (h) Set Screen Refresh: 50.
   (i) Repeat for Oscilloscope 2 for Channel: /dev/comedi0|Output|Analog Input 0.
4. Setup Data Recorder module.
   (a) Set Channel to: Reciprocal Coupling|Input|Cell 1 Vm.
   (b) Click button (>) to add channel to record list.
   (c) Repeat for the following under Reciprocal Coupling:
      - Input|Cell 2 Vm.
      - Input|Cell 1 Spike State.
      - Input|Cell 2 Spike State.
      - Output|Isyn 1-2.
      - Output|Isyn 2-1.
      - State|Phase Diff.
      - State|Time.
5. Save workspace for future experiments.
   (a) File → Save Workspace.
   (b) Choose filename and click Save button.

3.3.3 Brain Tissue Slice Preparation

1. Place rat into sealed chamber with isoflurane vapor until it is anesthetized.
2. Decapitate animal with guillotine. Open skull with scissors.
3. With Scoopula, remove brain and place in 0 °C ACSF.
4. With scalpel, isolate chunk(s) of tissue containing the brain region(s) of interest.
5. Glue tissue to the bottom of a walled slice chamber with cyanoacrylate glue. Fill chamber with 0 °C ACSF.
6. Attach chamber to vibratome. Cut 300–400 μm thick slices. Place tissue into chamber containing ACSF bubbled with 95 % O_2 and 5 % CO_2.
7. Allow tissue to recover for at least 1 h before transferring to microscope stage chamber.

3.3.4 Dynamic Clamp: Reciprocal Neuronal Coupling

1. Patch neuron.
   (a) With a microscope using dual interference contrast microscopy, select a target neuron in slice tissue.
   (b) With micromanipulator, place electrode tip against the cell body.
   (c) Set patch clamp amplifier to voltage clamp mode.
   (d) While measuring electrode resistance, apply light negative pressure to form a gigaohm seal between electrode and neuron.
   (e) Apply negative pressure to pop through the membrane.
   (f) Balance bridge and compensate for capacitance (*see* **Note 23**).
   (g) Observe live neuronal spiking on Oscilloscope 2.
2. Set patch clamp amplifier to current clamp mode.
3. Un-pause Spike Detector 1 and Spike Detector 2.
4. Start Neuron simulation.
   (a) Toggle Pause button up to un-pause Neuron module.
   (b) Set Iapp_offset: 9.
   (c) Click the Modify button.
   (d) Observe virtual neuronal spiking on Oscilloscope 1.
5. Setup Reciprocal Coupling module.
   (a) Set Gmax 2-1: 9e4 (*see* **Note 24**)
   (b) Check the Automate Coupling Delay checkbox.
   (c) Click the Modify button.
   (d) Toggle Pause button up.
6. Start data recording with the Data Recorder module for 2 min.
   (a) Click Choose File, choose save directory, and input an appropriate filename.
   (b) Click the Start Recording button.
7. Begin reciprocal coupling.
   (a) Toggle Couple Neurons button down.
8. Stop data recording after 5 min.
   (a) Click the Stop Recording button of the Data Recorder module.
9. Observe neuronal synchrony (*see* Fig. 12).

Fig. 12 (continued) of the real and virtual neuron, respectively. (**b**) A plot using the Oscilloscope module of the calculated phase difference performed by the Reciprocal Coupling module. Reciprocal coupling dynamic clamp was switched on in the *middle* of the plot, resulting in a constant phase difference. (**c**) Screenshot of Oscilloscope module while plotting the membrane potential of both neurons, live and virtual, using the trigger function. Note the synchrony of the spikes due to the artificial coupling

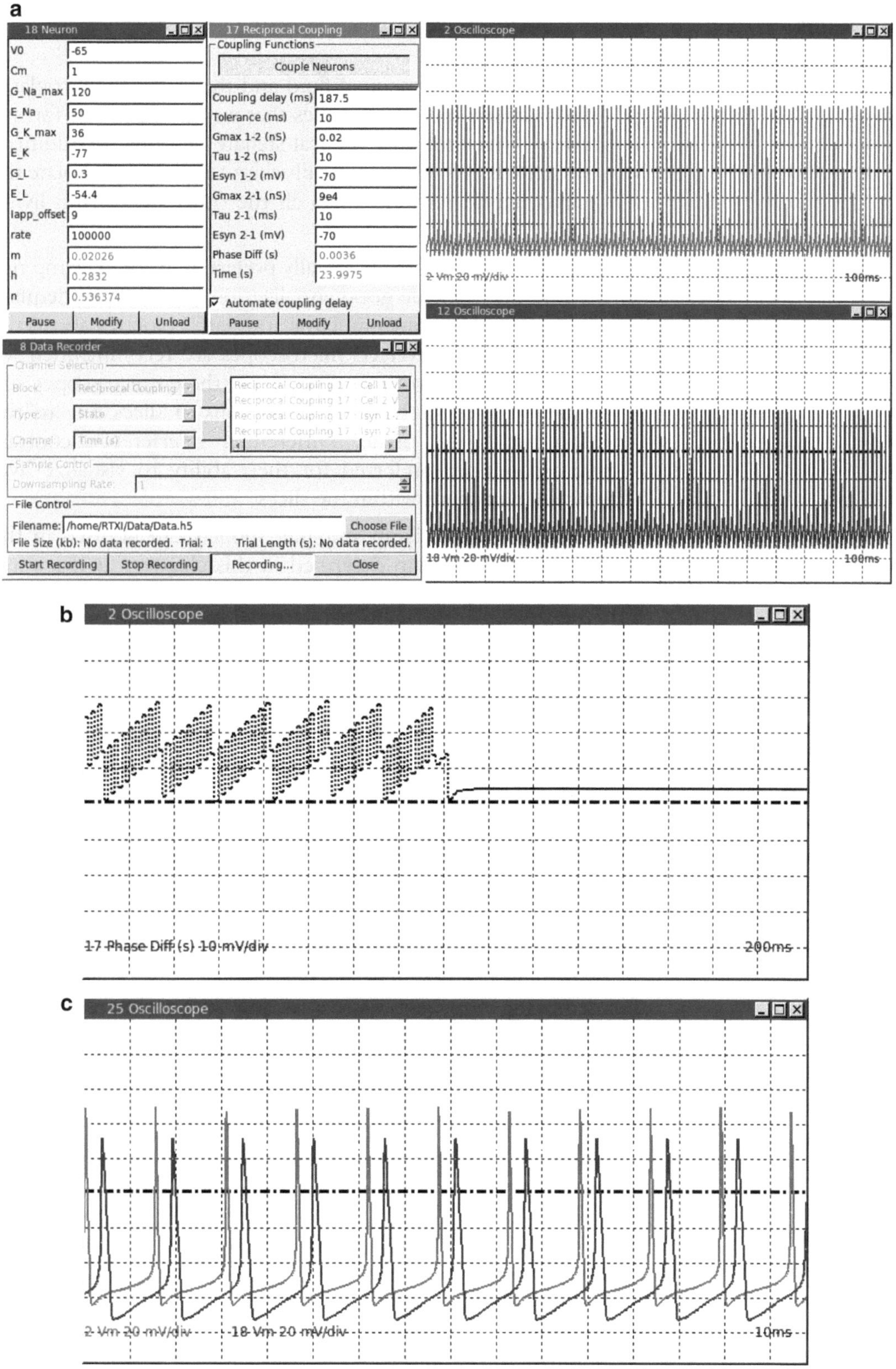

Fig. 12 Results for reciprocal coupling between a live and virtual neuron through dynamic clamp. (**a**) Screenshot of the RTXI workspace during the experiment. *Top* and *bottom* oscilloscopes display the membrane potential

4 Notes

1. The artificial cerebrospinal fluid and the artificial neuroplasm listed here are merely examples that we find sufficient for our studies with entorhinal cortical stellate neurons. In addition, you may wish to add synaptic blockers to chemically isolate the desired hybrid network from all the other neurons in living tissue.
2. Inverted microscopes are usually preferred in patch clamp rigs because they provide good visualization of cells, and adequate access of the micromanipulator and patch pipettes above the cell bath. Thus, inverted microscopes are recommended for patching cardiomyocytes. However, these microscopes are unable to adequately visualize neurons in slices due to the thickness of the tissue. Differential interference contrast microscopes are preferred for their ability to visualize live, unstained neurons within the slices.
3. Almost all patch clamp amplifier models can be used in the dynamic clamp setup described in this chapter. At the minimum, the amplifier must have an external analog output and analog input to interface with the dynamic clamp computer. It is also important to note the maximum current the amplifier can apply, especially for cardiomyocyte recordings. Cardiomyocytes may need upwards of 14 nA of a 1 ms current pulse to evoke an action potential. The amplifier should also be able to bridge balance in order to adequately compensate for access resistance. This is especially important in dynamic current clamp, where the level of current is being changed on the microsecond time scale.
4. Pipette shape and tip size are largely dependent on the cell type. For cardiomyocytes, we recommend a pipette tip resistance of 2–3 MΩ, and to approach the cell as close to perpendicular as possible due to the relative flag surface of the membrane. For neurons, we recommend a pipette tip resistance of 6–7 MΩ, and to approach the cell from a 45° angle.
5. In order to adequately mimic a biological conductance, real-time control is required to execute the task of calculation and injection of the current. The dynamic clamp software must operate on the same time scale as the currents it is trying to mimic. Hard real-time operating systems allows for execution of these tasks at a guaranteed timing. With a standard operating system, multitasking allows a single CPU to switch from multiple tasks to give the appearance all programs are being executed at the same time. This makes the operating system inherently unpredictable, and makes real-time processing unreliable. The software packages, RTAI (www.rtai.org) or

Xenomai (www.xenomai.org), are available to modify the Linux operating system into a hard real-time operating system. These free and open-source extensions are used by RTXI to provide real-time control.

6. Any modern computer system compatible with the Linux operating system can be used in the dynamic clamp setup. This chapter assumes RTXI was installed through its Live USB installation, which uses the Ubuntu distribution of the Linux operating system. It is important to note that modifications to the CPU clock rate can disrupt real-time control. Thus, it is important to prevent the adjustment of the CPU clock rate during experimentation. This can be done by disabling any features, such as "Turbo Mode," through the computer's BIOS.
7. RTXI uses the open-source library of DAQ drivers, Comedi (www.comedi.org), to interface with DAQ devices. A list of supported hardware can be found on the Comedi Web site.
8. Many patch clamp amplifiers have telegraphs, which output analog voltage signals that represent the amplifier's current configuration, i.e., the output gain. Others also allow control of a setting through an analog voltage input. For example, the Axon Multiclamp 700B mode can be changed through a mode external input. These additional amplifier input and outputs can be used by RTXI for better communication with the amplifier.
9. An RTXI module can be designed to take advantage of the various telegraphs of an amplifier, and automatically compensate for any configuration changes. For example, an amplifier telegraph could signal an RTXI module when the output gain was changed. The module could then modify settings within RTXI, signal-scaling values in this case, and compensate accordingly. Several amplifier control modules are bundled with the RTXI installation for common models of patch clamp amplifiers: Axon Multiclamp 700 Series, Axon Axopatch 1D, and the Axon 200 Series. The capabilities of these modules are limited to the amplifier's features, but all are capable of compensating for amplifier gains through input from the amplifier itself or the user.
10. Gains are applied to amplifier inputs and outputs to improve signal to noise ratio. These gains are described in the amplifier's manual, and must be compensated for by RTXI in order to accurately measure a signal. For example, if output gain is set to 10×, RTXI must divide that signal by a factor of 10. This is done by scaling factors that can be set through RTXI's control panel (System → Control Panel). Each channel can have a unique scaling factor set to it. As previously mentioned, RTXI modules for common amplifier models that automatically compensate for these gains are available for download (*see* **Note 8**).

11. The RTXI thread rate controls the real-time period of the system, and can be set in the Thread tab of the Control Panel (System → Control Panel). The real-time period controls how often a module's execute section is run, as well as the sampling rate of active inputs and outputs of the DAQ device. In a dynamic current clamp experiment, thread rate is usually set to the speed necessary to adequately recapitulate the current being injected. However, it is limited by several factors. First, the computation time required to run the model equations. Second, the time required to sample active inputs and outputs from the DAQ device. The sum of these must consistently be below the real-time period. Measurement of the computation time and real-time period can be done through the Real-Time Benchmarks module (System → Real-Time Benchmarks).
12. While any code editor can be used, commands listed use Emacs as an example. Emacs can be installed on an Ubuntu operating system through the following command: "sudo apt-get install emacs".
13. Based on the naming scheme of the Comedi (www.comedi.org) driver, "/dev/comedi0" refers to the first detected DAQ device, i.e., "/dev/comedi1" and "/dev/comedi2" would be the subsequently detected devices. While it is assumed only one DAQ device is being used, RTXI is capable of interfacing with many DAQ devices simultaneously.
14. RTXI's digital oscilloscope module is capable of visualizing any of RTXI's module data-types: INPUT, OUTPUT, STATE, and PARAMETER. The settings window can be accessed by right-clicking a loaded Oscilloscope module to bring up its context menu, and selecting Properties. Multiple channels can be displayed in a single instance.
15. The Data Recorder (System → Data Recorder) is capable of recording any module data-types: INPUT, OUTPUT, STATE, and PARAMETER. Rate of recording is equivalent as RTXI's thread rate, but can be down-sampled. Files are recorded in the HDF5 (www.hdfgroup.org/HDF5/) format and imported directly into MATLAB using the included scripts located in the following directory: /[rtxi install directory]/hdf/. HDF5 Data files can also be converted into ASCII format through the use of the rtxi_hdf_reader program with the following command: rtxi_hdf_reader –r [sample start]:[sample rate]:[sample end] [file name] > [output file], i.e., rtxi_hdf_reader –r 1:1:end test.h5 > test.dat.
16. Health of isolated cardiomyocytes can be negatively affected at many steps of the isolation protocol. Typical issues include the following: inadequate placement of the cannula (past aortic valve will result in inadequate perfusion), use of water containing endotoxins, unclean perfusion equipment, too short or

too long perfusion of the enzyme solution, or too low or too high collagenase concentration in enzyme solution. The length of time and amount of collagenase to use is determined through trial and error, and the times and concentrations listed should be used as starting guidelines.

17. Pipette tip resistance can be measured in a number of ways: with the amplifier's built-in resistance measurement, through the amplifier's included software, or with RTXI's Membrane Test module (Utilities → Membrane Test).
18. Capacitance can be measured with the amplifier's included software, or with RTXI's Membrane Test module (Utilities → Membrane Test). Accurate measurement of cell capacitance is vital to scale the calculated model current. The Membrane Test module calculates cell capacitance through measurement of the capacitive transient evoked through a standard square wave voltage pulse. Measurement should be performed at resting membrane potential to eliminate the effect of the fast sodium current.
19. Determining if a steady state for a cardiomyocyte has been reached can be done by plotting action potential duration (APD) versus time. Over the course of several hundred beats, APD will gradually decrease until it reaches a stable value. It is of note that beat to beat variability will still exist once the cell has stabilized. If APD continues to decrease without stabilizing, the health of the cell should be in question.
20. Stimulus artifact during current injection results from the voltage drop caused by the access resistance of the electrode. Through the use of bridge balance compensation, this voltage drop can be compensated for and eliminated. While the stimulus artifact can be ignored and removed during data analysis in traditional patch clamp experiments, the artifact can have significant effects during dynamic clamp. During artificial conductance injection, the stimulus artifact will be included in the current calculation and result in an inaccurate solution.
21. The "~" represents the current user's home directory. "~/Downloads" is the typical directory where the installed Internet browser will download files.
22. The reciprocal coupling setup uses two instances of the Spike Detector module, one each for the real neuron and virtual neuron. RTXI assigns an ID number to each module listed on the top left of the module's window. Use this ID number to differentiate between multiple instances of an identical module.
23. Depending on amplifier model, bridge balance and capacitance compensation will need to be performed during current-clamp (Axon Axopatch 200 Series) mode or voltage-clamp mode (Axon MultiClamp 700 Series).

24. Both the real and virtual neurons receive synaptic input with a true conductance (nS) from the Reciprocal Coupling module. The Hodgkin–Huxley neuron model bundled with RTXI, however, uses parameters with units that are in conductance densities (mS/cm^2). Thus, while the Gsyn $2>1$ is listed as 9e4 nS (90 μS), the value interpreted by the Hodgkin–Huxley neuron model is 90 $\mu S/cm^2$, to match the units of the Hodgkin–Huxley model.

Acknowledgments

This work was supported by NIH R01 RR020115 (to D.J.C., R.J.B., and J.A.W.). We thank R.J. Lin for invaluable programming assistance, and F.R. Fernandez for helpful comments regarding this chapter.

References

1. Robinson HP, Kawai N (1993) Injection of digitally synthesized synaptic conductance transients to measure the integrative properties of neurons. J Neurosci Meth 49:157–165
2. Sharp AA, O'Neil MB, Marder E (1993) Dynamic clamp: computer-generated conductances in real neurons. J Neurophysiol 69: 992–995
3. Dorval AD, Christini DJ, White JA (2001) Real-Time Linux dynamic clamp: a fast and flexible way to construct virtual ion channels in living cells. Ann Biomed Eng 29:897–907
4. Lin RJ, Bettencourt J, White JA et al (2010) Real-time experiment interface for biological control applications. Conf Proc IEEE Eng Med Biol Soc 1:4160–4163
5. Economo MN, Fernandez FR, White JA (2010) Dynamic clamp: alteration of response properties and creation of virtual realities in neurophysiology. J Neurosci 30:2407–2413
6. Bettencourt JC, Lillis KP, Stupin LR, White JA (2008) Effects of imperfect dynamic clamp: computational and experimental results. J Neurosci Meth 169:282–289
7. Livshitz L, Rudy Y (2009) Uniqueness and stability of action potential models during rest, pacing, and conduction using problem-solving environment. Biophysical J 97:1265–1276
8. Wan X, Laurita KR, Pruvot EJ, Rosenbaum DS (2005) Molecular correlates of repolarization alternans in cardiac myocytes. J Mol Cell Cardiol 12:196–206

Chapter 22

Ion Selectivity of Pore-Forming Peptides and Ion Channels Measured in Xenopus Oocytes

Thierry Cens and Pierre Charnet

Abstract

The *Xenopus laevis* oocyte is a widely used system for heterologous expression of exogenous ion channel proteins. They are easy to obtain, mechanically and electrically stable, have a large size, enabling multiple types of electrophysiological recordings: two-electrode voltage clamp, single cell-attached or cell-free patch-clamp and macropatch recordings. The size of an oocyte (1 mm in diameter) also allows for the use of additional electrodes (from 1 to 3) for injection of diverse materials (Ca^{2+} chelators, peptides, chemicals, antibodies, proteic-partners, etc.) before or during the course of the electrophysiological experiment.

We have successfully used this system to analyze the biophysical properties of pore-forming peptides. Simple extracellular perfusion of these peptides induced the formation of channels in the oocyte plasma membrane; these channels can then be studied and characterized in diverse ionic conditions. The ease of the perfusion and the stability of the voltage-clamped oocyte make it a powerful tool for such analyses. Compared to artificial bilayers, oocytes offer a real animal plasma membrane where biophysical properties and toxicity can be studied in the stable environment.

Key words Voltage clamp, Reversal potential, Calcium channels, Anomalous mole fraction

1 Introduction

The electrophysiological analysis of naturally occurring as well as synthetic channel-forming peptides has been shown to be useful to identify, characterize and understand many aspects of their insertion and their ionotropic, antibacterial, or toxic activities. The basic steps of this approach rely on the incorporation of the peptide in a membrane system that can be voltage-clamped in known ionic conditions for a fine description (in terms of ionic and electric environment) of the conditions of insertion, of the channel conductance and selectivity and, possibly, of the regulation of these properties. In these approaches, the determination of the ionic selectivity, by identifying the nature and the direction of the membrane currents induced by insertion of the peptide in the natural

Marzia Martina and Stefano Taverna (eds.), *Patch-Clamp Methods and Protocols*, Methods in Molecular Biology, vol. 1183,
DOI 10.1007/978-1-4939-1096-0_22, © Springer Science+Business Media New York 2014

membrane, provides important information on their toxicity, and, coupled with mutagenesis, helps to identify the residues involved in channel formation and permeation. All these experiments can be performed in Xenopus oocytes with a simple two-electrode, and/or single-channel recording, setup. The ease with which oocytes can be microinjected and their capacity to express foreign RNA provide an additional technical advantage when pore-forming peptides require intracellular protein partners for the formation of the channel. The following article describes two methods that have been used to analyze ionic selectivity and to determine the multi-ion nature of channel pores by reversal potential and anomalous mole fraction measurements.

2 Materials

2.1 Xenopus laevis Oocyte Preparation

1. Adult *Xenopus laevis* females are from Centre National d'Elevage de Lavalette, Animalerie CRBM, 859 rue Jean-François Breton, 34033 Montpellier, France.
2. Anesthetic solution: 0.2 % MS222 (ethyl 3-aminobenzoate methanesulfonate salt) is dissolved in tap-water.
3. Dissociation solution: collagenase (type IA) is dissolved at 1 mg/ml in the following solution: 82.5 mM NaCl, 2 mM KCl, 1 mM $MgCl_2$, 5 mM HEPES, pH = 7.2 with NaOH (*see* **Note 1**).
4. Washing solution: same as above without collagenase.
5. Survival solution: 96 mM NaCl, 2 mM KCl, 2 mM $CaCl_2$, 1 mM $MgCl_2$, 5 mM HEPES, 2.5 mM pyruvic acid, and 0.05 g/l gentamycin, pH = 7.2 with NaOH.
6. Oocytes shaking. Falcon tubers and/or petri dishes containing oocytes are placed on a shaker (120 rpm) for continuous agitation.
7. RNA injection. Performed using a homemade pneumatic injector which connects either vacuum (position electrode filling) or pressure (1 bar, position RNA injection) to the injection needle via a 3-way micro-electrovalve for the time desired.

2.2 Recording Solutions

1. Monovalent solutions.
   (a) Basic recording solution: 100 mM NaCl, 5 mM HEPES, 2 mM $MgCl_2$, pH = 7.2 with NaOH (Na100 solution).
   (b) For ionic selectivity measurements, NaCl is replaced by an equimolar concentration of either KCl, TEA-Cl, LiCl, CsCl, or choline-Cl and pH is adjusted using KOH, TEAOH, LiOH, CsOH, or NaOH, respectively (K100, TEA100, Li100, Cs100, or choline100 solutions). $MgCl_2$ was found to be necessary for the stability of the recordings.

Removing $MgCl_2$ is possible, but induces the appearance of a leak current that increases with time. For chloride permeability, NaCl is replaced with equimolar Na-acetate, and $MgCl_2$ is removed from both Cl-containing and Cl-free solutions.

2. Solutions for anomalous mole fraction effects.
   (a) Bant10: 10 mM BaOH, 20 mM TEAOH, 50 mM NMDG, 2 mM CsOH, 10 mM HEPES, pH set to 7.2 using methane sulfonic acid.
   (b) Cant10: 10 mM CaOH, 20 mM TEAOH, 50 mM NMDG, 2 mM CsOH, 10 mM HEPES, pH set to 7.2 using methane sulfonic acid.
   (c) The solutions of different Ba^{2+} and Ca^{2+} concentrations (respective Ba^{2+}/Ca^{2+} concentrations in mM 10/0, 9/1, 8/2, 6/4, 4/6, 2/8, and 0/10) are prepared by appropriate mixing of the two Bant10 and Cant10 solutions.
3. BAPTA solution and injection.
   (a) BAPTA solution: 100 mM 1.2-bis (2-Aminophenoxy) ethane-*N*,*N*,*N*′,N′-tetraacetic acid (BAPTA), 10 mM HEPES, 10 mM CsOH, pH = 7.2.
   (b) Injection needles: pulled from borosilicate glass capillaries using a vertical micropipette puller and subsequently broken under the stereomicroscope (×50 magnification) at 5–10 μm. This is done by gently touching the tip of the needle on the bottom of a plastic petri dish.

2.3 Electrophysiological Measurements

1. Two-electrode voltage clamp.
   (a) Agar bridges. Clark capillaries (with filament) are bent at ~120° under flame and subsequently immersed in 60 mm petri dishes filled with almost boiling agar (high gel strength) dissolved at 1 % in 3 M KCl (5–10 capillaries can be placed per dish). When immersed, these bridges are usually filled naturally (by capillarity) by the hot agar solution (*see* **Note 2**). After cooling, two bridges are "dissected" from the agar and placed in the bath-electrode holder previously filled with 3 M KCl.
   (b) Voltage and current electrodes: pulled from borosilicate glass capillaries using a horizontal puller. They have a resistance of 0.5–2 MΩ when filled with 3 M KCl.
   (c) Gravity driven perfusion. The homemade recording chamber (50 μl) is connected to an array of eight reservoirs (50 ml syringes) containing the various solutions. The flow of solution from each syringe can individually be automatically switched on or off by micro-electrovalves.

(d) In both two-electrode voltage clamp and single channel recordings, voltage protocols and current recordings are made using version 9.0 of the pClamp program (Axon Inst., Molecular devices) connected to the GeneClamp 500 amplifier (Axon Instruments, Inc.) for two-electrode voltage clamp or the Axopatch 200B amplifier (Axon Instruments, Inc.) for single channel recordings.

2. Single channel recordings.

(a) Hyperosmotic solution: 200 mM NaCl, 2 mM KCl, 2 mM $MgCl_2$, 10 mM HEPES, pH = 7.2 with NaOH.

(b) Depolarizing solution: 100 mM KCl, 2 mM $MgCl_2$, 10 mM HEPES, 10 mM EGTA, pH = 7.4 with KOH. The oocyte membrane potential measured in this solution was −2.2 ± 1.5 mV.

(c) Patch-clamp pipettes, pulled from borosilicate glasses, treated with Sylgard™ (*see* **Note 3**) and fire-polished using a microforge, have a resistance of 10–15 MΩ when filled with Na100.

3 Methods

3.1 Oocyte Preparation

1. Ovaries are surgically removed from *Xenopus laevis* females anesthetized by 15–20 min immersion into the anesthetic solution. A small abdominal incision through the skin and the gut wall is sufficient. Skin and gut wall are stitched separately after ovarian lobe removal.

2. Pieces (~10 ml) of dilacerated lobe of ovaries containing the oocytes are placed in a 50 ml Falcon tube and washed 3–4 times with 40 ml of washing solution.

3. The Falcon tube is then filled with 30 ml of dissociation solution and placed at 20 °C under gentle agitation using an orbital shaker. The isolation procedure is usually stopped after 2–3 h, once the follicular cells have detached from the oocytes. This can be estimated by observation of a sample under the stereomicroscope.

4. After extensive washing (3–5 times with 40 ml of wash solution), a last wash is made with the survival solution.

5. Isolated stage V–VI oocytes are placed in a 100 mm petri dish filled with the survival medium and batches of 20–100 oocytes are selected under a stereomicroscope and kept in the survival solution under gentle agitation in 35 mm petri dishes at 20 °C.

6. The survival solution is changed every day (and dead oocytes discarded) until use. Oocytes may be used directly for peptide perfusions, or microinjected with either RNA or cDNA and then left 48–72 h before electrophysiological recordings.

3.2 Estimation of the Junction Potential

The junction potential between the different solutions is estimated by two methods (*see* **Note 4**).

1. The first one uses the tool "junction potential calculator" provided by version 9.0 of the pClamp program.
2. The second one simply consists in zeroing the junction potential of both current and voltage electrodes when they are in the bath in presence of the Na100 solution. We then perfuse the different recording solutions and measure the junction potentials on the voltage or current electrode DC meter of the GeneClamp 500 amplifier. Both methods give differences in junction potentials compared to Na100 solution smaller than 2 mV, except for the solution with acetate ions replacing chloride (junction potential = 5 mV).

3.3 BAPTA Solution and Injection

In some experiments using external divalent cations (Ca^{2+}, Ba^{2+}, Sr^{2+}) the contaminating endogenous Ca^{2+}-activated Cl^- current must be suppressed. The available inhibitors of this current (9 AC—anthracene-9-carboxylic acid—, or niflumic acid for example) are usually not sufficiently efficient and specific to completely eliminate this conductance and thus we find it preferable to inject a Ca^{2+} chelator (BAPTA has our preference for its fast chelating capabilities, *see* **Note 5**) into the oocyte during the course of the experiment.

1. Pull an injection needle as described in Subheading 2.
2. Place this needle in the holder of a pneumatic injector connected to vacuum.
3. Backfill this needle by immersing the tip in a drop (1–2 μl) of BAPTA solution placed on the bottom of a petri dish (this is done under a stereomicroscope at 30× magnification).
4. Apply vacuum to the needle (this is most conveniently done with a footswitch connected to the homemade pneumatic injector) and follow the backfilling under the stereomicroscope (*see* **Note 6**).
5. Remove the filled needle and place it on an electrode holder fixed on a micromanipulator on the top of the recording chamber and connected to air pressure (1 bar) via a solenoid valve with time-controlled aperture (*see* **Note 7**; injection lasted from 10 ms to several seconds).
6. The needle is impaled into the oocyte at the start of the recording session and injection of BAPTA is made by repetitive application of a positive pressure (100 ms 2–5 times). The inhibition of the chloride conductance is followed online during injection.
7. The final intra-oocyte BAPTA concentration can be estimated to be around 2–5 mM (*see* **Note 8**).

3.4 Electrophysiological Characterization of Pore-Forming Peptides

1. Reversal potential measurements.
   (a) Macroscopic whole-cell currents are recorded under two-electrode voltage clamp using the GeneClamp 500 amplifier. Voltage-command, sampling, acquisition, and analysis are done using a Digidata 1200 and the pClamp program. All experiments are performed at room temperature (20–25 °C).
   (b) An oocyte is placed in the recording chamber filled with the desired solution (usually Na100).
   (c) Junction potentials (typically less than 3–5 mV) at the two electrodes are first cancelled using the zero button of the amplifier in Na100 solution, and the electrodes resistance is checked before introduction into the oocyte.
   (d) Both electrodes are impaled into the oocyte and the resting membrane potential is measured (typically −30/−50 mV). The oocyte is then voltage-clamped at −80 mV.
   (e) A third electrode is impaled to inject BAPTA as described above. This step is necessary when contaminations by Ca^{2+}-activated Cl^- currents are suspected.
   (f) Then, 50 μl of the tested protein or peptide are applied directly to the bath (with the main perfusion stopped) at the final working concentration (usually 20–50 μM), after proper dilution from a stock solution (500 μM in water) into the desired recording solution. We assume that there is no dilution in the recording chamber which has a volume of around 50 μl.
   (g) Membrane currents are measured during voltage-ramps (from −80 mV to +20 mV, 450 ms long), applied from the holding potential of −80 mV every 3 s, continuously, before and after addition of the peptide. The change in membrane conductance is followed by the modification in the slope of the current–voltage curve (current recorded during the voltage ramp, with the corresponding voltage axis; *see* Fig. 1).

      gbis. Similar recordings can be done at different holding potentials to analyze the voltage dependence of peptide insertion (Fig. 2).
   (h) Washing-out of the peptide is done by a gravity-driven perfusion of the chamber with the same solution without protein (*see* **Note 9**).
   (i) Similar experiments are performed with the different solutions using different oocytes (a new oocyte for each ionic condition).
   (j) Off-line analysis of the changes in the membrane conductance (slope of the current–voltage curve) is made for each membrane potential by a least-square fit of the

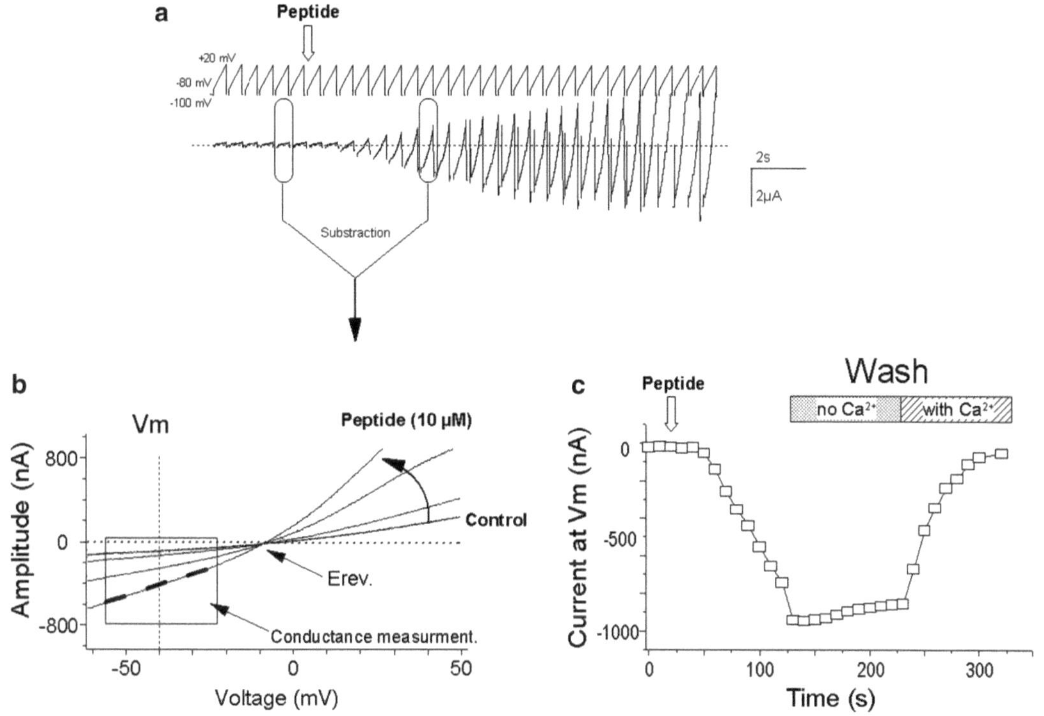

Fig. 1 (**a**) Membrane currents elicited in an oocyte voltage-clamped at −100 mV and submitted to 400 ms-long voltage-ramps from −80 to +20 mV, before and after addition of 50 μl of a peptide (in this case puroindoline at 50 μM). *Top*, voltage protocol; bottom, membrane currents. Ramps are applied every 15 s, and the 400 ms long recording traces have been concatenated for iconographic purposes. Note the increase in membrane current following addition of the peptide. (**b**) Superimposed current–voltage curves recorded during the experiments shown in (**a**) before and after addition of the peptide. Vm, denotes the potential at which the current is measured and the *shaded area* depicts the zone where the linear regression is calculated to measure the slope conductance. Erev, the crossing point of all the currents recorded during peptide application, is the current reversal potential of the peptide-induced current [12]. Erev is the zero-current potential when traces recorded before and after peptide application are subtracted (*see* **a**). (**b**) Time course of current amplitude recorded at the potential Vm before and after application of the peptide and during perfusion of peptide-free solutions without (no Ca^{2+}) and with (with Ca^{2+}) 5 mM calcium. Note that reversibility is only obtained with Ca^{2+}-containing peptide-free solution. (Republished with permission from ref. 13)

current using the pClamp software (ver 9.0, Axon Inst.). These slopes can be plotted as a function of time during peptide application (Fig. 1).

(k) Current Reversal potential (Erev(X)) is measured. This is the point where the current traces cross each other during peptide application (potential "Erev" in Fig. 1b). It can be measured as the zero-current potential after digital subtraction of traces recorded before and after peptide/protein application. This potential is measured for each ionic condition (Na100 K100, TEA100, Li100, Cs100, or choline100 solutions) and eventually corrected for the junction potential (see above).

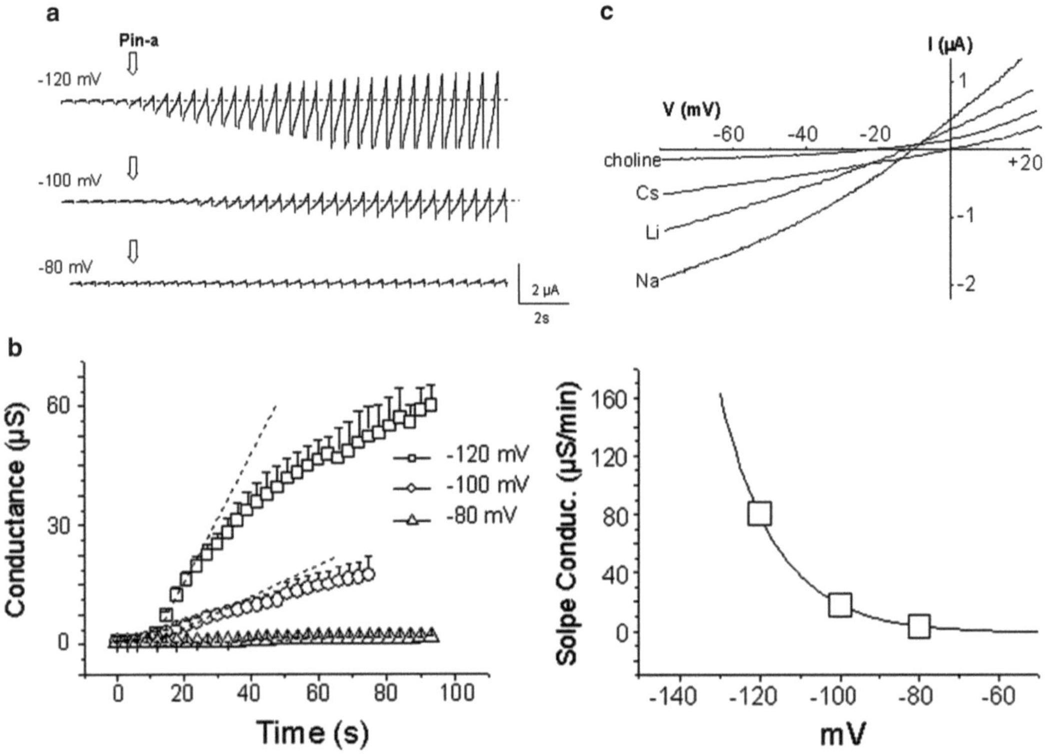

Fig. 2 (**a**) Effect of application of a peptide on an oocyte held at −120, −100, or −80 mV and submitted to 400 ms-long voltage-ramps from −80 to +20 mV. The change in membrane current is null at −80, small at −100 and large at −120 mV [2]. (**b**) *Left*: averaged slope-conductance (calculated as described in Fig. 1) recorded during application of the peptide at −80, −100, or −120 mV. *Right*: voltage-dependence of the increase in membrane current induced by the peptide. In this case, the maximal variation of conductance (depicted as a dashed line in the *left panel*) is calculated by linear regression as the maximal slope variation of the conductance for each potential and plotted as a function of the potential. (**c**) Peptide-induced currents recorded in different ionic conditions. Current traces recorded before and after perfusion of the peptide in different ionic solutions are digitally subtracted and plotted against membrane potential (*see* Subheading 2 for the composition of the different solutions). The potentials at which the current traces cross the abscissa (zero current) are the reversal potentials of the currents in the different solutions, used to calculate the permeability ratios (see text). (Republished with permission from ref. 13)

(1) Ionic permeability ratios of peptide-induced channels (permeability of cation X relative to Na permeability: P_X/P_{Na}) can be deduced for each cation X from the shift in reversal potentials Erev(X) − Erev(Na) using the Goldman–Hodgkin–Katz (GHK) equation [1]:

$$\mathrm{Erev(X)} - \mathrm{Erev(Na)} = RT/zF\ \mathrm{Ln}(P_X\,[X]_o/P_{Na}\,[\mathrm{Na}]_o)$$

$$P_X/P_{Na} = [\mathrm{Na}]_o/[X]_o\ e^{[(\mathrm{Erev(X)}\,-\,\mathrm{Erev(Na)})zF/RT]}$$

where $[X]_o$ is the extracellular concentration of ion X and R, T, z, and F have their usual meanings ($RT/F \sim 25.3$ mV at 20 °C).

(m) Washout of the peptide as well as effect of regulators of permeability (Ca^{2+} for example, *see* ref. 2) can be assayed using the perfusion system.

2. Determination of single channel conductance.

(a) The vitelline envelope is removed manually under a stereomicroscope (magnification 50×) using tweezers style #5 after placing the oocytes for 10–20 min into the hyperosmotic solution.

(b) Carefully place this "denuded" oocyte in the recording chamber filled with the depolarizing solution that nulls the oocyte membrane potential (in fact around -2 mV).

(c) Patch-Pipettes are filled with Na100 with the desired concentration of peptide (usually, for the ones we have tested, 5–10 μM were necessary, *see* **Note 10**).

(d) They are placed on a holder connected to the 100G-CV5 head stage of a GeneClamp 500 amplifier. For stability reasons, the single-channel recordings are performed preferentially on cell-attached patches.

(e) After establishing a gigaOhm seal (usually >10 GΩ), the patch potential is set at the desired value and recordings are made using pClamp 9 in event-driven mode and digitized by a Digidata 1200. The time necessary for insertion of the peptide is function of the sequence, the solution and the applied voltage. Insertion typically takes several minutes.

(f) Once single channel events start to appear, the patch potential can be set at different values to construct single channel current–voltage curves. It is very important to start recording immediately since the continuous insertion of peptides in the membrane will give rise, with time, to an increased probability of having multiple openings [3], complicating the analysis of the results.

(g) Amplitude histograms (pClamp 9.0) of these single channel events recorded at different patch potentials are made and single-channel current amplitudes determined from the fit of such histograms to Gaussian functions. The slope of the linear regression of the single channel amplitude–voltage curve gives the single channel conductance.

(h) The continuous insertion of the peptide in the membrane patch can make the recordings very unstable after several minutes. Since the insertion is often voltage-dependent, it can be useful to decrease the patch potential as soon as the peptide starts to form channels. In this case the patch potential can be set at different values in an episodic mode just during the recording time (episodic stimulation mode of pClamp 9.0).

3.5 Anomalous Mole Fraction for Ca^{2+} Channels

For ionic channels with multiple ion binding-sites in the pore, the concentration-dependent permeability ratios between different ions display the so-called anomalous mole-fraction effect. This phenomenon is characterized by a channel conductance of smaller amplitude in the presence of a mixture of two ions than in the presence of a pure solution of either of these ions (at the same total concentration, *see* Fig. 3a, c, and ref. 1). In the example described in Fig. 3, Ca^{2+} channels display a higher permeability for Ba^{2+} than for Ca^{2+}. But in the presence of a mixture of Ba^{2+} and Ca^{2+}, the measured conductance is even smaller (Fig. 3a, c). In these recordings, the conductance measured at the peak of the current–voltage curve or the reversal potential can be used [1]. The profile of the anomalous mole fraction curve (Fig. 3c) is determined by (1) the number and height of the ion binding sites, and energy barriers between these sites, present in the pore of the channel and (2) the type and concentration of the ions used. Such experiments can be useful to interpret the effects of mutations on channel pore that affect channel

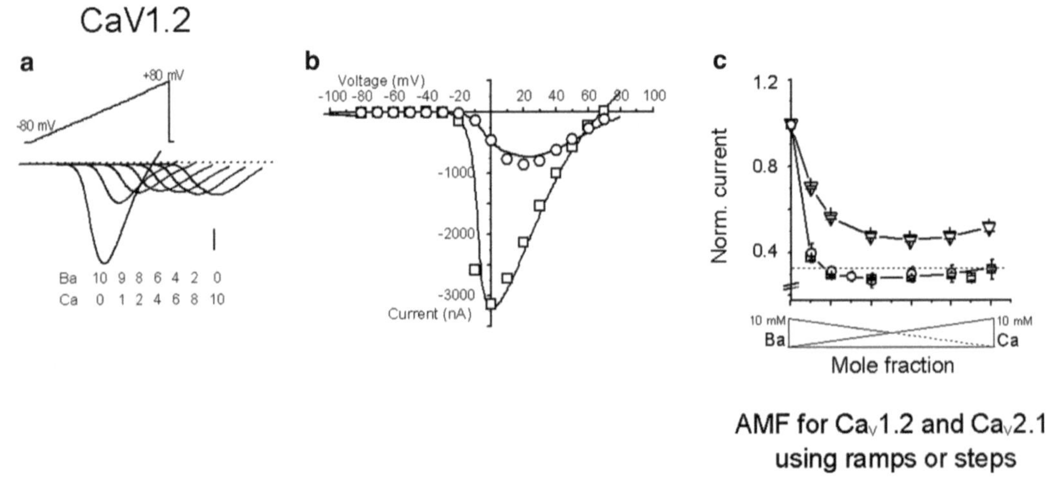

Fig. 3 (**a**) Membrane currents elicited in an oocyte voltage-clamped at −80 mV and submitted to 400 ms-long voltage-ramps from −80 to +80 mV. Currents were recorded from oocytes expressing the $Ca_V1.2$ voltage-gated Ca^{2+} channels (with $\alpha_2-\delta$ and β_2 auxiliary subunits) in solutions of different Ca^{2+} mole fractions (Ba 10 mM, Ba/Ca 9/1, 8/2, 6/4, 4/6, 2/8, and Ca 10 mM). Current traces are superimposed for iconographic purposes. (**b**) Current–voltage curves recorded in 10 mM Ba^{2+} (*open squares*) or Ca^{2+} (*open circles*) using 40 ms voltage steps form −80 mV to 50 mV in 10 mV increments. The superimposed continuous lines are the current–voltage curves recorded with 400 ms-long voltage ramps of the same voltage amplitude. Similar curves are obtained by the two methods. (**c**) Anomalous mole-fraction curves obtained from currents recorded in oocytes expressing the $Ca_V1.2$ voltage-gated Ca^{2+} channels (with $\alpha_2-\delta$ and β_2 auxiliary subunits). Each point represents the peak of the current–voltage curve obtained in a given Ca^{2+} mole fraction (Ba 10 mM, Ba/Ca 9/1, 8/2, 6/4, 4/6, 2/8, and Ca 10 mM). The same curves were obtained whether the current–voltage curves were obtained with voltage-steps (*open circles*) or voltage-ramps (*open squares*). The anomalous mole-fraction curve for $Ca_V2.1$ (with the same auxiliary subunits) is also shown (*open inverted triangle*) and displays a specific shape. (Republished with permission from ref. 13)

selectivity or conductance. We describe here our method to measure anomalous mole fraction in Xenopus oocytes expressing voltage-gated Ca^{2+} channels (either $Ca_V1.2$ or $Ca_V2.1$ [4]).

1. After isolation, the Xenopus oocytes have to be injected with 40 nl of a mixture of RNA encoding the different Ca^{2+} channel subunits ($Ca_V\alpha$, $Ca_V\beta$, and $Ca_V\alpha2-\delta$ subunits at a concentration of 1 ng/nl each) and left 3–4 days for proper expression of these proteins. Alternatively 10 nl of the DNA encoding the same subunits can be injected in the nucleus. In this case expression vectors with eukaryotic promoters have to be used (pcDNA3.1 from Invitrogen for example), and nuclear injection is simply performed by choosing the center of the animal (dark) pole for injection.
2. Put one of these oocytes in the recording chamber, and impale both the recording electrodes, for two-electrode voltage clamp, and the BAPTA injection needle.
3. Voltage-clamp the oocyte at −80 mV and perfuse the pure 10 mM Ba solution (Bant10).
4. After injection of BAPTA and stabilization of the Ba^{2+} currents observed with repetitive depolarization from −80 to +10 mV (400 ms long), record current–voltage curves using either a voltage-ramp protocol (from −60 to +60 mV in 400 ms, *see* **Note 11**) or a series of conventional voltage-steps of increasing amplitude (from −60 to +60 mV in 5 or 10 mV increments). These two methods should give similar results if the speed of the voltage ramp is chosen carefully (Fig. 3b, c).
5. Repeat these recordings with the different solutions of increasing Ca^{2+} molar fraction (with a total concentration $[Ba^{2+}]+[Ca^{2+}]$ kept constant at 10 mM, i.e., 9/1, 8/2, 6/4, 4/6, 2/8, and 0/10 *see* Subheading 2 and Fig. 3c).
6. Often check the stability of the recordings and the lack of current rundown by switching back to the original solution at the end of the experiments.
7. Plot the peak of the current–voltage curves recorded in the presence of each solution of increasing Ca^{2+} molar fraction.
8. Repeat the same experiment with each Ca^{2+} channel type or mutant.

In a typical experiment (*see* Fig. 3) such mole fraction curves are measured for different Ca^{2+} channels ($Ca_V1.2$ and $Ca_V2.1$). In this particular case, the recording of different mole fraction curves with the two types of Ca^{2+} channels suggests that the permeability profiles of these channels are different. Mutation of negatively charged amino-acids that are suspected to form a divalent cation binding sites within the pore can lead to more drastic changes with a complete suppression of the anomalous mole fraction (not shown).

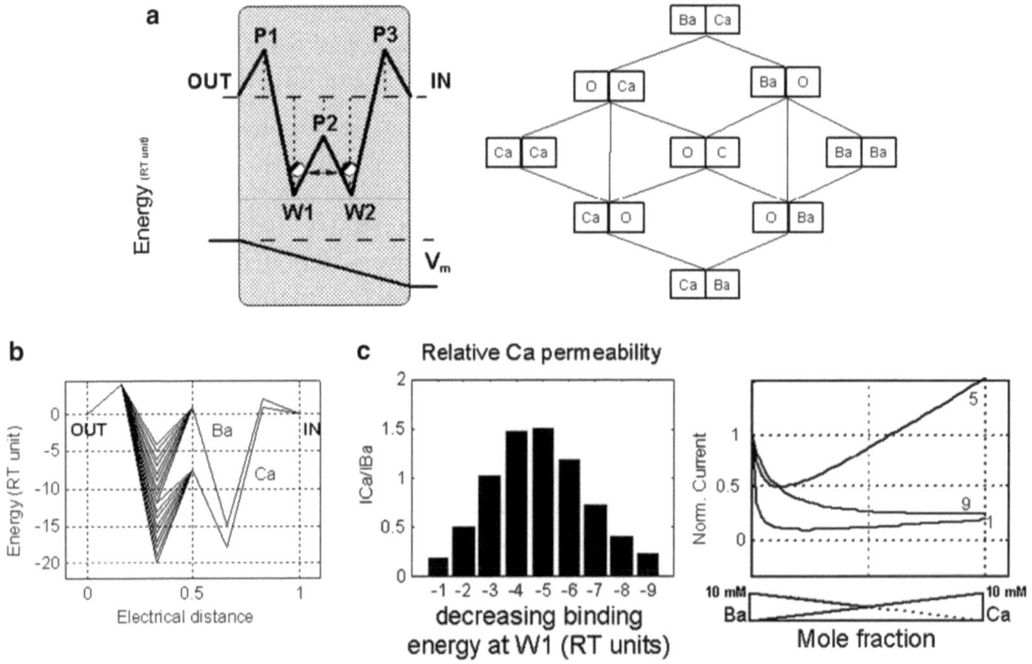

Fig. 4 (**a**) *Left*: Schematic of the free energy profiles for Ca^{2+} and Ba^{2+} of a Ca^{2+} channel with its two Ca^{2+} binding sites (W1 and W2) and three energy barriers (P1, P2 and P3). Vm denotes the membrane potential and schematizes the applied electric field. *Right*: state diagram of the different occupied states of the channel in presence of two different cations (Ba^{2+} and Ca^{2+}) assuming single filling (only one ion per site). (**b**) Different energy profiles for Ba^{2+} and Ca^{2+} ions used in the following numerical simulation. The binding energy of the extracellular Ca^{2+}/Ba^{2+} binding site is decreased progressively. (**c**) Numerical simulation of current–mole fraction curves of a two binding-site Ca^{2+} channel with the energy profiles displayed in (**b**). *Left*: relative Ca^{2+} permeability (current at 10 mM Ca^{2+} divided by the current at 10 mM Ba^{2+}) for the channels with the different profiles. *Right*: current–mole fraction curves for channels with an external site of high (1) intermediate (5) or low (9) affinity for divalent cations. High- and low-binding energies gave the same relative permeability, but anomalous mole fraction is only recorded for high energy binding. (Republished with permission from ref. 13)

Using Eyring rate theory and modeling a channel with two Ca^{2+} binding sites (W1 and W2) separated by three energy barriers (P1, P2, and P3) allows to build a state diagram of all possible occupancies of these two sites by Ba^{2+} and/or Ca^{2+} and to assign rate constants to the transitions between these states (Fig. 4a). The rate constants depend on the channel-dependent energy profiles of the pore for both Ba^{2+} and Ca^{2+}, on an electrostatic repulsion factor arising when the pore is doubly occupied and on the electric field generated by the membrane potential (*see* **Note 12**). We have numerically solved these differential rate equations and calculated the theoretical mole-fraction current for channel with an external cation binding site of decreasing energy (Fig. 4b). It can be seen that this decrease in the binding energy induces a bell-shaped curve of the relative Ca/Ba current, with large Ca current for

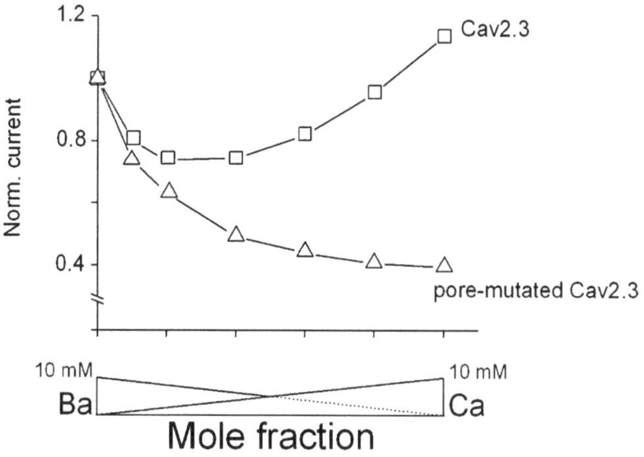

Fig. 5 Anomalous mole-fraction curves obtained from currents recorded in oocytes expressing the $Ca_V2.3$ voltage-gated Ca^{2+} channels (*square*), or a mutated version of $Ca_V2.3$ in which amino-acids in the pore have been mutated (*triangle*) both with $\alpha_2-\delta$ and β_2 auxiliary subunits. Each point represents the peak of the current–voltage curve obtained in a given Ca^{2+} mole fraction (Ba 10 mM, Ba/Ca 9/1, 8/2, 6/4, 4/6, 2/8, and Ca 10 mM). The anomalous mole-fraction for the mutated version of the channels is monotically decreasing, suggesting changes in the pore properties [5]

intermediate binding energies (labeled 5 on part C of Fig. 4) and small Ca currents for either strong (labeled 1) or weak (labeled 9) binding energies on the external site. Interestingly, in these later cases, the mole-fraction curve is only "anomalous" (i.e., displaying a minimum) for strong binding energies (trace 1), while it decreases monotically for weak binding energies (trace 9). These data suggest that decreasing the binding energy of an external cation binding site (by removal of the negative charges) can modify the anomalous mole fraction curve while the relative Ca^{2+} current is only slightly modified. This can be verified experimentally by mutations within the pore, *see* for example Fig. 5 [5].

4 Notes

1. All solutions are prepared with deionized water. After equilibration and pH adjustment they are filtered at 0.22 μm using disposable Nalgen sterilization filter units. They can be stored at 4 °C for several months.
2. After cooling, these petri dishes are sealed with Parafilm™ and can be stored at 4 °C for several months.
3. Once pipettes have been pulled, the tip is painted with Sylgard under a stereomicroscope up to 200–500 μl of the aperture,

and rapidly baked at moderate temperature (for Sylgard polymerization) by insertion into the center of a hot coiled filament. Classical fire-polishing is then performed.

4. For a complete description of the liquid junction potential correction the following reference should be consulted [6].
5. We have also used with success DM nitrophen and derivatives of BAPTA. EGTA and EDTA work less efficiently probably because of their slower chelating kinetics and their inability to chelate Ca^{2+} in the close vicinity of the membrane due to the lack of interaction with phospholipids [7].
6. Be careful to stop backfilling before any air bubbles enters the needle.
7. When the needle is not connected to the pressure, it should be kept open to air pressure and not closed, to avoid continuous injection by the pressure established during the previous injection.
8. For experiments using oocytes expressing voltage-gated Ca^{2+} channels, the BAPTA solution should be injected in the animal (dark) pole of the oocyte where the translation machinery and the Cl^- channels are present in higher concentration. Injection in the other pole will significantly delay the inhibition of the Cl^- conductance. The same injection needle can be used for several oocytes provided that it is not clogged-up.
9. From our personal experience, reversibility of peptide insertion is difficult to obtain. The determination of the reversal potential in different ionic conditions thus requires a new oocyte for each condition. For some peptides, a block of the current can be recorded by perfusing Ca^{2+}-containing (10 mM) solution [2]. This block disappears when Ca^{2+} is removed (even in the absence of continuous perfusion of the peptide).
10. The presence of the peptide in the pipette solution can sometimes prevent the seal formation. In these conditions it can be advantageous to backfill the tip of the pipette with a peptide-free solution.
11. The voltage ramp protocol has to be adjusted for the type of Ca^{2+} channel used considering the specific kinetics of inactivation of the channel. Great care should be taken to avoid any contamination of the shape of the current–voltage curves generated using these ramps, due to the inactivation kinetics of the channel. A comparison of the current–voltage curves recorded during voltage ramps and measured at the peak of voltage steps of increasing amplitude should confirm the lack of contamination by inactivation (*see* Fig. 3b). It is often preferable to use voltage-ramps because the determination of an anomalous mole fraction curve requires the recording of

current–voltage curves in different ionic conditions. These recording are obtained relatively rapidly with ramps but can last more than 30–40 min if voltage steps are used, thus increasing the likelihood of rundown of the current. Moreover the peak of the current–voltage curve should be taken for these curves and not the current at a fixed potential, to take into account the modification in the surface charges induced by the modification of the ionic composition of the extracellular solutions [8].

12. A detailed description of this numerical simulation is beyond the scope of this article, but can be found elsewhere [9–11].

Acknowledgements

This work was supported by CNRS, INSERM, and ANR. The authors would like to thank Dr. I. Lefevre for critical reading of the manuscript and Jean-Marc Donnay for oocyte preparation.

References

1. Hille B (2001) Ion channels of excitable membranes, 3rd edn. Sinauer Associates Inc., Sunderland, MA
2. Charnet P, Molle G, Marion D, Rousset M, Lullien-Pellerin V (2003) Puroindolines form ion channels in biological membranes. Biophys J 84:2416–2426
3. Chaloin L, De E, Charnet P, Molle G, Heitz F (1998) Ionic channels formed by a primary amphipathic peptide containing a signal peptide and a nuclear localization sequence. Biochim Biophys Acta 1375:52–60
4. Mangoni ME, Cens T, Dalle C, Nargeot J, Charnet P (1997) Characterisation of alpha1A Ba^{2+}, Sr^{2+} and Ca^{2+} currents recorded with ancillary beta 1–4 subunits. Receptor Channel 5:1–14
5. Cens T, Rousset M, Kajava A, Charnet P (2007) Molecular determinant for specific Ca/Ba selectivity profiles of low and high threshold Ca^{2+} channels. J Gen Physiol 130:415–425
6. Neher E (1992) Correction for liquid junction potentials in patch clamp experiments. Methods Enzymol 207:123–131
7. Rousset M, Cens T, Vanmau N, Charnet P (2004) Ca^{2+}-dependent interaction of BAPTA with phospholipids. FEBS Lett 576:41–45
8. Green WN, Andersen OS (1991) Surface charges and ion channel function. Annu Rev Physiol 53:341–359
9. Begenisich TB, Cahalan MD (1980) Sodium channel permeation in squid axons. I: reversal potential experiments. J Physiol 307:217–242
10. Campbell DL, Rasmusson RL, Strauss HC (1988) Theoretical study of the voltage and concentration dependence of the anomalous mole fraction effect in single calcium channels. New insights into the characterization of multi-ion channels. Biophys J 54:945–954
11. McCleskey EW (1999) Calcium channel permeation: a field in flux. J Gen Physiol 113: 765–772
12. Pellegrin P, Menard C, Mery J, Lory P, Charnet P, Bennes R (1997) Cell cycle dependent toxicity of an amphiphilic synthetic peptide. FEBS Lett 418:101–105
13. Cens T, Charnet P (2007) Use of Xenopus oocytes to measure ionic selectivity of pore-forming peptides and ion channels. In: Molnar P, Hickman JJ (eds) Patch-clamp methods and protocols. Series: methods in molecular biology, 1st edn. Humana, Totowa, NJ, pp 195–209

Chapter 23

Principles of Single-Channel Kinetic Analysis

Feng Qin

Abstract

Single-channel recording provides high resolution information on gating mechanisms of ion channels that are generally difficult to obtain from macroscopic measurements. Analysis of the data, however, has proven to be challenging. Early approaches rely on half-amplitude threshold detection to idealize the record into dwell-times, followed by fitting duration histograms to resolve kinetics. More recent analyses exploit explicit modeling of the data to improve the idealization accuracy. The dwell-time fitting has also evolved into direct fitting of dwell-time sequences using the maximum likelihood approach while taking account of effects of missed events. Finally, hidden Markov modeling provides an ultimate approach by which both single channel amplitudes and kinetics are analyzed simultaneously without the need of idealization. The progress in theory, along with the advance in computing power as well as the development of user-friendly software, has transformed single-channel analysis, once a specialty task, now readily accessible to a broader community of scientists.

Key words Ion channel, Single channel, Single molecule, Gating, Kinetics, Dwell-time, Idealization, Duration histogram, Maximum likelihood, Global fitting, Missed events, Segmental k-means, Hidden Markov modeling

1 Introduction

Patch-clamp recording is a primary tool for studying structural functions of ion channels. It is among the few techniques that can allow proteins to be detected at the single-molecule level. The measurement contains much information on the heterogeneity of channel conformations and their dynamic properties. Analysis of the single-channel recording provides a unique means for deciphering the molecular mechanisms underlying channel functions. However, extraction of the information requires sophisticated statistical interpretation of the data. Here we review the basic principles for analysis of single-channel measurements. The theory is also applicable to modeling of other types of single-molecule measurements.

Marzia Martina and Stefano Taverna (eds.), *Patch-Clamp Methods and Protocols*, Methods in Molecular Biology, vol. 1183, DOI 10.1007/978-1-4939-1096-0_23, © Springer Science+Business Media New York 2014

1.1 Markov Model for Single-Channel Kinetics

A common feature of single-molecule activity is that it occurs at discrete steps in a seemingly random manner. In the case of ion channels, the current alternates between "on" and "off" like a random telegraph signal. A statistical model that can capture such behaviors is the Markov model [1], which assumes that the molecule exists in a finite number of discrete states and the transitions between states are governed by the first-order rate constants. With the additional assumption that the rate constants are independent of time the kinetics further become a time-homogeneous Markov process.

The assumption of the first-order transition rates implies the system to be stochastic in nature. It says that within an infinitesimal small time interval Δt, the transition of the system from one state to another follows a probability

$$\Pr\left[s(t+\Delta t)=j \mid s(t)=i\right]=k_{ij}\Delta t \tag{1}$$

where $s(t)$ designates the state of the channel at time t. As we will see in the next section, this simple rule completely determines how the system evolves in time.

The Markov formalism of single-molecule kinetics is essentially an extension of the classical chemical reaction theory, where a state corresponds to an energetically stable conformation of the molecule while a rate constant of a transition between two states is determined by the activation energy or the energy barrier separating the different conformations. Other schemes such as diffusion models and fractal models are also available for describing ion channel gating, but they have not been as useful as the Marov model in general [2–5].

The problem of single-channel analysis is to solve the inverse problem, i.e., to identify an appropriate model to explain experimental data. A major challenge to the problem is the limited observability of the system offered by the patch-clamp technique. That is, multiple conformational states may share the same conductance. Transitions between these states are not directly observable. Instead, they can only be inferred from the distinct statistical distributions of their durations. This aggregation of multiple states into a single observation class incurs ambiguity in resolving detailed kinetics of gating of a multi-state channel [6].

1.2 Properties of Markov Model

Two probabilities of a Markov model are particularly useful for modeling patch-clamp data. One is the transition probability between all individual states, and the other is the probability of transitions restricted within a subset of states. Both probabilities can be derived directly from the first-order assumption of the Markov model (*see* Box 1). The unrestricted transition probability is determined by

$$\mathbf{P}(t)=\exp(\mathbf{Q}t) \tag{2}$$

Box 1
Derivation of Transition Probability

Consider a two-state system with a starting probability $\boldsymbol{\pi} = [\pi_1, \pi_2]$. The problem is to calculate the probability of the two states at an arbitrary time t. Since the Markov assumptions only define

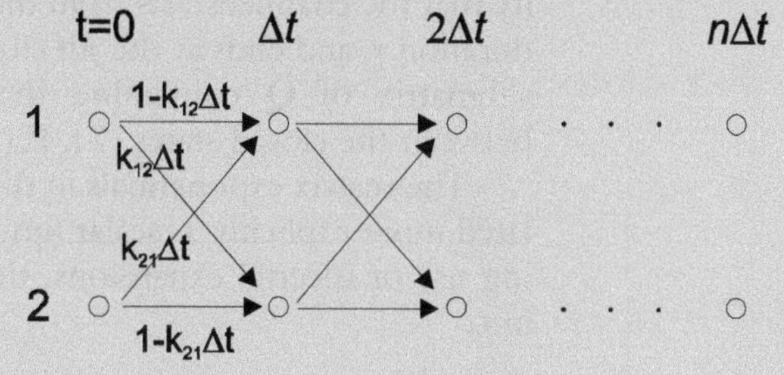

transitions in an infinitesimal time, one can divide the interval $[0, t]$ into n infinitesimal ones, each with a duration Δt. The probability that the system is in state 1 after Δt is equal to $\pi_1(1 - k_{12}\Delta t) + \pi_2 \cdot k_{21}\Delta t$, which is the probability to remain in state 1 plus the probability to leave state 2 for state 1. Similarly, the probability being in state 2 after Δt is $\pi_1 \cdot k_{12}\Delta t + \pi_2 \cdot (1 - k_{21}\Delta t)$. In the matrix form, they can be written as

$$(\pi_1, \pi_2) \cdot \begin{pmatrix} 1 - k_{12}\Delta t & k_{12} \\ k_{21} & 1 - k_{21}\Delta t \end{pmatrix}.$$

Repeating the procedure for subsequent Δt's leads to the probability at time $t = n\Delta t$:

$$(\pi_1, \pi_2) \cdot \begin{pmatrix} 1 - k_{12}\Delta t & k_{12} \\ k_{21} & 1 - k_{21}\Delta t \end{pmatrix}^n.$$

Recall the relationship $(1 + t/n)^n \to e^t$ as $n \to \infty$. The probability at time t can be reduced to $\boldsymbol{\pi} \cdot \exp(\mathbf{Q}t)$ where $\mathbf{Q}$ is the matrix of rate constants defined in the text. The matrix exponential $\exp(\mathbf{Q}t)$ corresponds to the transition probability between states for time t

where $\mathbf{P}(t)$ is a matrix whose (i,j)th element represents the probability that the channel is in state j at time t given that it starts from state i at the beginning, and $\mathbf{Q}$ is the rate constant matrix whose (i,j)th element, k_{ij}, corresponds to the transition rate from state i to state j [7]. The diagonal elements of the matrix are defined by $k_{ii} = -\sum_{j \neq i} k_{ij}$, which is the total rate of the probability flux leaving state i.

The transition probability within a subset of states defines the probability that the channel stays in the subset of states for a period of time without leaving it during the time. For patch-clamp data, such a subset corresponds to a class of states with a common conductance level. For simplicity of discussion, we will consider only binary channels. In this case, the transition probability is determined by

$$\begin{aligned} \mathbf{P}_c(t) &= \exp(\mathbf{Q}_{cc}t) \\ \mathbf{P}_o(t) &= \exp(\mathbf{Q}_{oo}t) \end{aligned} \tag{3}$$

where $\mathbf{P}_c(t)$ is a matrix whose (i,j)th element defines the probability that the channel starts from the ith closed state, stays closed for duration t, and ends at the jth closed state at time t, and $\mathbf{Q}_{cc}$ is the submatrix of $\mathbf{Q}$ containing the rate constants for transitions between the closed states [7]. $\mathbf{P}_o(t)$ and $\mathbf{Q}_{oo}$ are defined similarly.

The matrix exponentials in the above equations can be formulated more explicitly as scalar functions of t. For example, by making use of spectral expansions, the $\mathbf{Q}$ matrix can be decomposed into

$$\mathbf{Q} = \lambda_1 \mathbf{A}_1 + \cdots \lambda_N \mathbf{A}_N$$

where N is the number of states, λ_i's are eigenvalues of $\mathbf{Q}$, and $\mathbf{A}_i$'s are matrices determined from eigenvectors. Then the matrix exponential can be calculated by

$$\mathbf{P}(t) = e^{\lambda_1 t}\mathbf{A}_1 + \cdots e^{\lambda_N t}\mathbf{A}_N. \tag{4}$$

The equation indicates that each element of the transition probability matrix is a sum of exponentials with as many components as the number of states. The spectral expansion generally requires the matrix to be diagonable. Fortunately, the rate constant matrix of a physical process usually satisfies the condition given the microscopic reversibility [8].

1.3 Ensemble Currents

Equations 2 and 3 form the basis for computing the probability of various measurable quantities. The occupancy probability of a channel, for example, determines the ensemble current recorded from a macro-patch or a whole cell. It can be formulated from the transition probability matrix $\mathbf{P}(t)$ as

$$\mathbf{p}(t) = \pi \cdot \exp(\mathbf{Q}t) \tag{5}$$

where $\boldsymbol{\pi}$ is the row vector of the initial probability of each state and $\mathbf{p}(t)$ is the vector of the occupancy probability at time t. In other words, the occupancy probability of the channel at time t is its initial probability multiplied by the transition probability over the time t.

From the state occupancy probability, the current resulting from an ensemble of n channels can be predicted by

$$i(t) = n \cdot \mathbf{p}(t) \cdot \mathbf{I}^{\tau} \tag{6}$$

where $\mathbf{I} = [i_1,\ldots,i_N]$ is the vector of the unitary current of each state of the channel.

Equation 6 is the basis for fitting the time course of whole-cell recording. In practice the starting probability $\boldsymbol{\pi}$ can often be deduced from the equilibrium probability of the channel at the holding condition, which is determined by

$$\pi \cdot \mathbf{Q}_h = \mathbf{0} \quad (7)$$

where $\mathbf{Q}_h$ is the rate constant matrix at the holding condition. Since the transition probability matrix has its elements in the exponential forms, the ensemble current $i(t)$ follows sums of exponentials too. The time constants of the current correspond to the reciprocals of the eigenvalues of the rate constant matrix $\mathbf{Q}$ at the test condition.

When the number of channels is limited, the random openings and closures of the individual channels produce measurable gating noise. At any time t, the variance of the current of a single channel is given by

$$\sum_{k=1}^{N} p_k(t)\cdot\left[i_k - \bar{i}(t)\right]^2$$

where i_k is the unitary current of the kth state and $\bar{i}(t) = \mathbf{p}(t)\cdot\mathbf{I}^{\tau}$ is the mean current of the channel at time t. The variance of an ensemble of n channels is then equal to

$$\sigma^2(t) = n\cdot\mathbf{p}(t)\cdot\mathbf{I}_2^{\tau} - i^2(t)/n \quad (8)$$

where $\mathbf{I}_2 = [i_1^2,\ldots,i_N^2]$. When the channels are either all closed or all open, the variance vanishes. Therefore, if the channels are held at the resting state and then fully stimulated, the variance of the ensemble current will first increase with time and then decrease. The variance contains additional information on the kinetics of the channel and can be combined with the ensemble mean measurement to separate the unitary current and the number of channels in the patch.

In the special case where the channel has two states (one closed and one open), Eq. 8 is reduced to the familiar parabolic form

$$\sigma^2 = i < I > - < I >^2 / n \quad (8)$$

where i is the unitary current of the opening and $<I>$ is the mean measurement. The variance has two zero-crossing points at $<I> = 0$ and $<I> = i\cdot n$ and a peak at $i\cdot n/2$. The slope of the curve at the first zero-crossing point determines the unitary current, whereas the peak position can be used in conjunction with the unitary current to give the number of channels.

In theory, the ensemble current contains information about the kinetics of a channel. In practice, the ensemble averaging reduces the heterogeneity of the kinetics of single channels. The number of exponentials that is necessary to fit the time course of an ensemble recording is generally far less than the number of the kinetic components present in single-channel recordings. Thus the ensemble measurements have a limited resolution on determining the kinetic details of a channel.

1.4 Dwell-Time Distributions (One-Dimensional, 1D)

Single-channel recording reports directly the individual openings and closures of a channel. The duration of a dwell-time at each conductance is a stochastic variable. It consists of three events: the channel enters the conductance at the beginning of the dwell-time, remains in the conductance for a duration t, and finally exits the conductance after time t. The probability of observing a dwell-time is thus the product of the probability of the three events, leading to

$$
\begin{aligned}
f_c(t) &= \pi_c \mathbf{P}_{cc}(t)\mathbf{Q}_{co}\mathbf{1} \\
f_o(t) &= \pi_o \mathbf{P}_{oo}(t)\mathbf{Q}_{oc}\mathbf{1}
\end{aligned}
\tag{9}
$$

where $\boldsymbol{\pi}$'s collect the entry probability of the states in the corresponding conductance, $\mathbf{P}(t)$'s are the transition probability matrix as previously defined, $\mathbf{Q}_{co}$ is the off-diagonal submatrix of $\mathbf{Q}$ containing transitions from closed states to open states, $\mathbf{Q}_{oc}$ from open states to closed states. The $\mathbf{1}$ designates a column vector of ones of appropriate length. The multiplication by the unit vector at the end accounts for the uncertainty of the destination state the channel is leaving for. Strictly speaking, $f(t)$'s are probability distributions because $\mathbf{Q}_{co}$ or $\mathbf{Q}_{oc}$ represents rates rather than transition probability. The latter involves the product of a rate with an infinitesimal time Δt during which the transition occurs.

The entry probability of a state in a conductance can be determined from the equilibrium probability of the channel as [1]

$$
\begin{aligned}
\pi_c &= \frac{w_o Q_{oc}}{w_o Q_{oc}\mathbf{1}} \\
\pi_o &= \frac{w_c Q_{co}}{w_c Q_{co}\mathbf{1}}
\end{aligned}
\tag{10}
$$

where w's are the vectors of the equilibrium probabilities of the states at the corresponding conductance class. The denominators are the normalization factor so that the total probability is equal to 1. Alternatively, the entry probability can be determined by [9]

$$
\begin{aligned}
\pi_c &= \pi_o \mathbf{Q}_{oo}^{-1}\mathbf{Q}_{oc} \\
\pi_o &= \pi_c \mathbf{Q}_{cc}^{-1}\mathbf{Q}_{co}
\end{aligned}
\tag{11}
$$

where the inverse matrices on the right-hand sides can be considered as the integral of the corresponding transition probability matrix over time. Thus the product in the right-hand side is the probability entering the conductance of interest irrespective of the duration of its previous stay in the other conductance. The probability in Eq. 12 defines a set of homogeneous equations, which can be solved in combination with the probability totality constraint using matrix singular value decomposition techniques.

The formalism in Eq. 10 predicts the dwell-time distribution of a channel if its gating follows a Markov process. By representing the matrix exponential with spectral expansions, the distributions may be put in the form of sums of exponentials,

$$\begin{aligned} f_c(t) &= \sum_{i=1}^{N_c} \left(\pi_c \mathbf{A}_i^{(c)} \mathbf{Q}_{co} \mathbf{1} \right) e^{\lambda_i t} \\ f_o(s) &= \sum_{j=1}^{N_o} \left(\pi_o \mathbf{A}_j^{(o)} \mathbf{Q}_{oc} \mathbf{1} \right) e^{\mu_j s} \end{aligned} \tag{12}$$

where N_c is the number of closed states, N_o the number of open states, λ_i's the eigenvalues of $\mathbf{Q}_{cc}$, μ_i's the eigenvalues of $\mathbf{Q}_{oo}$, and $\mathbf{A}_i$'s the spectral expansions of the submatrices. The products in the parenthesizes are the coefficients of the exponentials. These equations allow one to analyze the experimental dwell-time histograms in linear combinations of exponentials. The information available from the analysis includes the number of open and closed states of the channel and the time constants and the populations of individual components.

Direct fitting of the dwell-time distributions with sums of exponentials has some limitations. First, the time constants and the coefficients of the exponentials are complicated functions of rate constants. While they give rise to information on the components of the dwell-times, their relation to the model parameters are difficult to interpret. Second, the data obtained under different experimental conditions are difficult to analyze in combination. An approach that may alleviate these issues is to fit the histograms with a model explicitly [10]. In this approach, the time constants and the coefficients of the exponentials are evaluated from the model parameters and the experimental variables. The rate constants of the model are directly optimized to best fit the histograms.

The dwell-time distribution functions in Eq. 12 have a limited number of degrees of freedom. This places an upper limit on the complexity of a model that may be resolved by histogram fitting. The total number of free parameters involved in the two distributions is $(2N_c - 1) + (2N_o - 1) = 2(N - 1)$. This is exactly the complexity of a linear sequential model or a model with linear branches. Even so, not all models with this complexity can be resolved by histogram fitting. One example is illustrated in Scheme I, which is a model for the Ca^{2+}-activated K^+ channel [11]. The two open states have different lifetimes, one 0.35 ms and the other 3 ms. Regardless of whether the short open state is connected to C_2 or C_3, the resultant model produces identical dwell-time distributions for both openings and closures [10, 12]. In the other words, the histogram fitting cannot uniquely determine such a branched model even though its complexity matches the number of degrees of freedom in the distribution function.

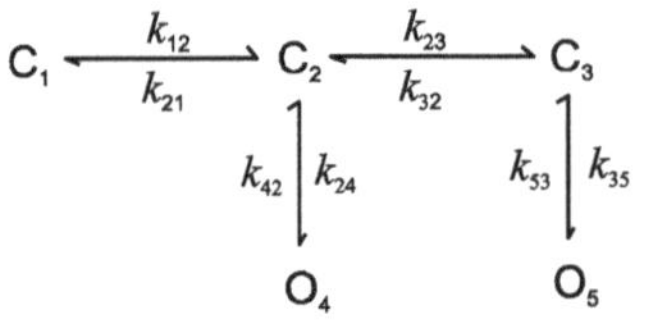

(Scheme I)

1.5 Dwell-Time Distributions (Two-Dimensional, 2D)

For models containing branched openings, correlations arise between specific open and closed states. This correlation information is necessary to discriminate the connections between these states. Unfortunately, such information is not preserved in the binned one-dimensional (1D) dwell-time histograms, which treat the openings and closures independently. In order to exploit the information, a high-order dwell-time distribution has to be used. For binary channels, it has been shown that the two-dimensional (2D) distributions of the successive opening and closure pairs contain adequate information on the correlation between open and closed states [8]. Thus fitting the 2D dwell-time distributions provides a complete approach to make use of all information of data [10, 12].

Similar to the derivation of the 1D dwell-time distribution, the 2D dwell-time distribution can be obtained as

$$\begin{aligned} f_{\mathrm{co}}(t,s) &= \pi_{\mathrm{c}}\mathbf{P}_{\mathrm{cc}}(t)\mathbf{Q}_{\mathrm{co}}\mathbf{P}_{\mathrm{oo}}(s)\mathbf{Q}_{\mathrm{oc}}\mathbf{1} \\ f_{\mathrm{oc}}(s,t) &= \pi_{\mathrm{o}}\mathbf{P}_{\mathrm{oo}}(s)\mathbf{Q}_{\mathrm{oc}}\mathbf{P}_{\mathrm{cc}}(t)\mathbf{Q}_{\mathrm{co}}\mathbf{1} \end{aligned} \tag{13}$$

where $f_{\mathrm{co}}(t,s)$ represents the probability distribution of observing a closed dwell-time t followed by an open dwell-time s, and $f_{\mathrm{oc}}(s,t)$ represents the distribution of an open duration s succeeded by a closed duration t. According to the equations, the distributions are determined by the product of the probability of entering the preceding conductance, the probability of the transitions between the states within the conductance, the rates of the transitions to the following conductance, the probability of the transitions within that conductance, and finally the rates leaving the conductance.

Substituting the matrix exponentials using spectral expansions, one can expand the 2D dwell-time distributions into

$$\begin{aligned} f_{\mathrm{co}}(t,s) &= \sum_{i=1}^{N_{\mathrm{c}}}\sum_{j=1}^{N_{\mathrm{o}}}\left(\pi_{\mathrm{c}}\mathbf{A}_i^{(c)}\mathbf{Q}_{\mathrm{co}}\mathbf{A}_j^{(o)}\mathbf{Q}_{\mathrm{oc}}\mathbf{1}\right)e^{\lambda_i t+\mu_j s} \\ f_{\mathrm{oc}}(s,t) &= \sum_{i=1}^{N_{\mathrm{c}}}\sum_{j=1}^{N_{\mathrm{o}}}\left(\pi_{\mathrm{o}}\mathbf{A}_j^{(o)}\mathbf{Q}_{\mathrm{oc}}\mathbf{A}_i^{(c)}\mathbf{Q}_{co}\mathbf{1}\right)e^{\lambda_i t+\mu_j s} \end{aligned} \tag{14}$$

The 2D dwell-time distributions are also combinations of exponential functions. Different from the 1D dwell-time distribution, however, the exponentials are now two-dimensional, with one time constant for closures and the other for openings. These time constants are the same as those of the 1D dwell-time

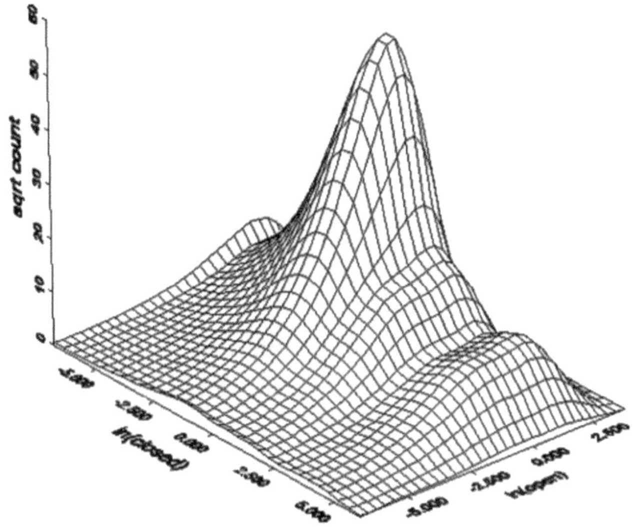

Fig. 1 The 2D dwell-time distribution predicted from Scheme I for Ca^{2+}-activated potassium channels. The existence of coupling between closed and open dwell-times is evident from the different appearances of the 1D open dwell-time distributions in adjacency to different closed durations. Republished with permission from ref. 39

distributions. The coefficient of each 2D component in the distributions measures the volume occupied by the exponential. It is indicative of the strength of the coupling between the closed and open dwell-times associated with the component.

Figure 1 illustrates the 2D dwell-time distribution predicted by the model in Scheme I [13]. When a channel contains a single gateway state for opening, the 2D dwell-time distribution is simply the product of the two 1D dwell-time distributions of openings and closures. It has the same appearance on all sections cut at different values at each axis. To the contrary, the open dwell-time distribution adjacent to the long closures in Fig. 1 exhibits double humps, whereas those adjacent to the short closures show a single sharp peak. Such a dependence of dwell-time distribution on the adjacent dwell-time is indicative of existence of multiple opening pathways of a channel. The analysis of the 2D dwell-time histograms can therefore reveal, for example, whether a channel is able to open when it is partially liganded. As expected, the fitting of the 2D histograms is able to discriminate the models in Scheme I, which the 1D histogram fitting [12] was not able to resolve. Nevertheless, the false solution, i.e., the model with the short opening connected to C_2, remains to be a local maximum [10] of the fitting.

The 2D dwell-time distribution contains a total of $2N_cN_o$ degrees of freedom [9]. This implies that the 2D histogram fitting can resolve models with up to N_cN_o connections between states. Because the 2D dwell-time distribution contains all information of data, this is also the limit of any other approach under a single experimental condition at equilibrium. The limitation can be ascribed to the inability to observe transitions between states at the same conductance. It can be shown that this number is equal to the degrees of the freedom of the 1D distribution if and only if $N_c = 1$ or $N_o = 1$. In the other words, the 1D dwell-time distribution contains all information of data only when the channel has a single open or closed state. Accordingly, the 1D histogram fitting is applicable only to models with a single open or closed gateway state.

1.6 Maximum Likelihood Fitting of Dwell-Time Sequence

A problem inherent to the dwell-time histogram fitting is that the approach is applicable only to data obtained at equilibrium. The construction of 2D dwell-time histograms also requires a large number of events. An alternative approach is to fit the dwell-time sequence directly using the maximum likelihood estimation [14–17]. The approach is applicable to both stationary and nonstationary data and allows for multiple conductance levels. The maximum likelihood estimates also have theoretical advantages of being asymptotically unbiased and having the minimum variance and approximate normal distributions as the sample size increases. These desirable properties make the maximum likelihood estimation the most robust parameter estimation technique in many applications.

Consider a dwell-time sequence $t = \{t_1 t_2 \ldots t_L\}$ with the corresponding conductance sequence $a = \{a_1 a_2 \ldots a_L\}$ where a_i is either closed or open. The probability density of the entire dwell-time sequence can be obtained by multiplying the entry probability, the transition probability within a dwell-time and the transition rates between dwell-times. This leads to

$$f(t,a) = \pi_{a_1} \left\{ \prod_{i=1}^{L} \left[P_{a_j a_i}(t_i) Q_{a_j a_{j+1}} \right] \right\} 1 \tag{15}$$

where the last conductance class a_{L+1} is defined as the collection of all states not in a_L. When viewed as a function of the model, the probability distribution becomes the likelihood of the model, which is often designated as $L(\boldsymbol{\theta})$ where $\boldsymbol{\theta}$ represents the model parameters. The likelihood value provides a natural measure on the goodness of a fit. The maximum likelihood estimation is then to seek the maxima of the likelihood function, which corresponds to the most likely estimates of the parameters.

The maximum likelihood concept can be illustrated using a simple two-state model. The likelihood function for a dwell-time sequence can be written as

$$L(k_{co}, k_{oc}) = \exp(-k_{oc} t_1) \cdot k_{oc} \cdot \exp(-k_{co} t_2) \cdot k_{co} \cdots \exp(-k_{oc} t_L) \cdot k_{oc}$$

where the first and the last dwell-times are assumed to be open, and k_{co} and k_{oc} are the opening and closing rate constants respectively. Suppose the data are generated by a channel with comparable lifetimes on the closed and open states. Then, for a model with very different k_{co} and k_{oc}, its likelihood will be small. This occurs because of the mismatch between the model rate constants and the observed dwell-time durations in the exponents. The likelihood can be maximized by calculating the logarithm of the likelihood and setting its derivatives to zero. The maximum occurs at

$$k_{co}^{-1} = 2\sum t_{2i} / L$$
$$k_{oc}^{-1} = 2\sum t_{2i-1} / L$$

which are the same as the mean estimates of durations of the observed closed and open dwell-times respectively.

For more complicated models, the likelihood function has to be calculated numerically. One approach is to evaluate it recursively by taking advantage of the matrix product structure of the likelihood function [18]. Starting from the entry probability $\alpha_0^{\tau} = \pi_{a_1}$, one then calculates

$$\alpha_k^{\tau} = \alpha_{k-1}^{\tau}\mathbf{P}_{a_k}(t_k)\mathbf{Q}_{a_k a_{k+1}} \tag{16}$$

for $k = 1,2,\ldots,L$. As the procedure progresses through each dwell-time of the sequence, the probability of all partial dwell-time sequences, $\{t_1 \ldots t_k\}$, $k = 1 \ldots L$, are computed. The likelihood for the entire sequence is simply the sum of the components of the final probability vector, i.e., $L(\theta) = \alpha_L^{\tau} \cdot 1$.

One potential problem with the above procedure is numeric overflow. For long data sequence or poor starting values of parameters, the likelihood value could become extremely small and exceed the machine precision range. The problem can be alleviated by adaptively scaling $\boldsymbol{\alpha}$'s into proper numeric ranges and calculating the log likelihood instead of the likelihood itself. The procedure has an overall computational complexity on the order of N^2L where N is the number of states and L is the number of dwell-times. For models of moderate sizes, the procedure is generally fast, taking on the order of seconds. This efficiency renders the histogram fitting techniques largely obsolete. The latter become advantageous only when the full likelihood approach becomes too slow, in which case the binning of the dwell-times can significantly speed up the calculation.

1.7 Missed Events Correction

The theory described above assumes a perfect signal from a Markov model. In practice, the recording apparatus has a limited time response, which causes some short events to inevitably go undetected. The missed events could introduce large errors in the detection of dwell-time durations. For example, if a channel contains long openings interrupted by brief closures, omissions of the short closures will prolong the apparent openings.

Without correction for the effects of the missed events, the results from dwell-time analysis will be inaccurate.

Because of their stochastic nature, the missed events cannot be recovered deterministically. Instead, they can only be taken into account statistically by correcting the probabilistic distribution for the apparent dwell-times. The missed events are usually characterized by a fixed dead time, so that all events longer than the dead time are detected whereas all shorter ones are missed. By this criterion, an apparent dwell-time t may result from any dwell-time sequence $\{\tau_i, i=1\ldots n\}$ which satisfies the following conditions:

1. The first dwell-time has the same conductance as the apparent dwell-time, and its duration $\tau_1 \geq t_d$;
2. The departures to the other conductance are shorter than the dead time, i.e., $\tau_{2i} \leq t_d$ for all i;
3. The total duration of all dwell-times of the sequence is equal to the duration of the apparent dwell-time, i.e., $\sum \tau_i = t$.

The transition probability matrix of the apparent dwell-time is then the sum of the probability of all such dwell-time sequences. The final solution can be written analytically in the Fourier domain as [19]

$$
\begin{aligned}
{}^{e}P_c(t) &= -\frac{1}{2\pi} e^{Q_{cc}t_d} \int_{-\infty}^{\infty} e^{-j\omega(t-t_d)} \left[j\omega I + Q_{cc} - Q_{co}\left(I - e^{(j\omega I + Q_{oo})t_d}\right)\left(j\omega I + Q_{oo}\right)^{-1} Q_{oc} \right]^{-1} d\omega \\
{}^{e}P_o(t) &= -\frac{1}{2\pi} e^{Q_{cc}t_d} \int_{-\infty}^{\infty} e^{-j\omega(t-t_d)} \left[j\omega I + Q_{cc} - Q_{co}\left(I - e^{(j\omega I + Q_{oo})t_d}\right)\left(j\omega I + Q_{oo}\right)^{-1} Q_{oc} \right]^{-1} d\omega
\end{aligned} \quad (17)
$$

These are the equivalents of the transition probability matrix defined in Eq. 3 in the absence of missed events. Substituting them into the various dwell-time distributions results in correction of these distributions for the effect of missed events.

The exact solution in Eq. 17 is difficult to evaluate because of appearance of inversions of matrix functions in the integrand. Various approximate solutions have been proposed [16, 19–22]. One simple solution is the first-order approximation:

$$
\begin{aligned}
{}^{e}P_c(t) &= \exp\left({}^{e}Q_{cc}t\right) \\
{}^{e}P_o(t) &= \exp\left({}^{e}Q_{oo}t\right)
\end{aligned} \quad (18)
$$

where ${}^{e}Q_{cc}$ and ${}^{e}Q_{oo}$ are defined by

$$
\begin{aligned}
{}^{e}Q_{cc} &= Q_{cc} - Q_{co}\left(I - e^{Q_{oo}t_d}\right)Q_{oo}^{-1}Q_{oc} + t_d Q_{co}\left(I - e^{Q_{oo}t_d}\right)Q_{oo}^{-1}Q_{oc} \\
{}^{e}Q_{oo} &= Q_{oo} - Q_{oc}\left(I - e^{Q_{cc}t_d}\right)Q_{cc}^{-1}Q_{co} + t_d Q_{oc}\left(I - e^{Q_{cc}t_d}\right)Q_{cc}^{-1}Q_{co}
\end{aligned} \quad (19)
$$

The corrected transition probability matrix has the same form as in the ideal case. The only change is the replacement of the original **Q** matrix with the corrected one given in Eq. 19. The first-order approximation is obtained under the assumption that the total missed event duration is negligible compared to the apparent duration. There are other solutions which are less sensitive to the assumption [21, 22]. It has also been shown that the exact solution could be evaluated piecewise by dividing an apparent dwell-time into multiples of the dead time [22]. The advantage of the first-order approximation is the simplicity of its evaluation and the availability of its analytical derivatives to rate constants, which are useful for the optimization of fitting. It is also generalizable to channels with multiple conductance levels [16].

1.8 Practical Issues of Analysis

The theory described above explains the basic principles of analysis of single-channel currents. There are many issues concerning the use of the theory in practice. Some typical ones include, for example, how to optimize the likelihood, how to allow for simultaneous fitting of data from multiple conditions, how to impose constraints on rate constants, and how to test whether a model is indeed optimal.

Global fitting. The rate constants of a biological channel are often scaled by experimental conditions. To fit data across multiple conditions, it is necessary to represent the rate constants in terms of some intrinsic parameters that do not vary with the conditions. For voltage-gated channels, the rates typically depend on voltage in an exponential manner. The same is true for force in mechanosensitive channels. For ligand-gated channels, the rates change linearly with ligand concentration. A unified representation of these dependencies may be written as

$$q_{ij} = C_{ij} \exp\left(\mu_{ij} + \nu_{ij} V\right) \tag{20}$$

where C_{ij} is the concentration of the drug that the rate q_{ij} is sensitive to and V is the membrane potential or other global variables such as force. Each rate constant is represented in two variables: μ_{ij}, ν_{ij}. For the rates that are independent of concentration, one can set $C_{ij} = 1$. Similarly, for the rates that are independent of voltage, $\nu_{ij} = 0$. The parameters μ_{ij} and ν_{ij} do not vary with the experimental conditions. Thus they can be chosen as intrinsic parameters of a channel for fitting.

The parameterization of a rate constant in terms of μ_{ij} and ν_{ij} automatically restricts the rate constant to be non-negative. It also has the advantage to facilitate the handling of the detailed balance constraint. These constraints are highly nonlinear in q_{ij}, but become linear with μ_{ij} and ν_{ij}. As seen below, linear constraints can be taken into account explicitly during optimization.

Constraints. Typical constraints imposed on rate constants include: holding some rate constants at fixed values, linear scaling by other rate constants, and detailed balance conditions. All these constraints can be represented as linear equations in terms of variables μ_{ij} and ν_{ij}, leading to

$$\Gamma\theta = \xi \tag{21}$$

where $\boldsymbol{\theta} = (\ldots, \mu_{ij}, \nu_{ij}, \ldots)^{\tau}$ is the vector of variables, $\boldsymbol{\Gamma}$ is the coefficient matrix, and $\boldsymbol{\xi}$ is the constant vector. If the imposed constraints are mutually independent, $\boldsymbol{\Gamma}$ has a full rank.

The inclusion of constraints into optimization leads to constrained optimization, which is generally difficult to solve. For linear equality constraints, however, it is possible to eliminate them by reducing the degree of freedom of the problem. That is, one can formulate the linearly constrained variables into a linear combination of a set of unconstrained variables,

$$\theta = \mathbf{A} \times \mathbf{x} + \mathbf{b} \tag{22}$$

where $\mathbf{A}$ is a constant matrix and $\mathbf{b}$ is a constant vector, both of which can be readily determined from the matrix $\boldsymbol{\Gamma}$ and the vector $\boldsymbol{\xi}$ in Eq. 21 using standard matrix decomposition techniques [16].

For any value of $\mathbf{x}$, the $\boldsymbol{\theta}$ given by Eq. 22 automatically satisfies the constraints in Eq. 21. Thus one can choose $\mathbf{x}$ as the free variables to optimize. By doing so, the optimization subject to the constraints (Eq. 21) is reduced to an equivalent unconstrained optimization problem, which can be solved by conventional optimization techniques.

Derivatives. For efficient optimization and accurate estimation of model parameters, it is necessary to have gradient information of objective functions. There are optimization algorithms that require only function evaluations, such as the downhill simplex method and the Powell's direction set method. These derivative-free approaches are usually not efficient for complex functions such as the likelihood of dwell-time sequences. They are also limited to applications with fewer parameters. Analytical derivatives have been derived for the 1D and 2D dwell-time distributions [10] and the full likelihood function [16, 18]. One central step of the derivation is the calculation of the derivatives of a matrix exponential, which follows [15]

$$\frac{\partial}{\partial\theta}\mathbf{P}_{cc}(t) = \sum_{i=1}^{n_c}\sum_{j=1}^{n_c}\left(\mathbf{A}_i^{(c)}\frac{\partial\mathbf{Q}_{cc}}{\partial\theta}\mathbf{A}_j^{(c)}\right)\gamma\left(\lambda_i;\lambda_j;t\right) \tag{23}$$

where γ is a scalar function determined by the eigenvalues. The derivatives of the dwell-time distributions can then be determined by applying (Eq. 23) to the transition probability matrices of the individual dwell-times using the chain rule. In practice, the calculation of the derivatives of an objective function involves the following steps: the derivatives of the parameters $\boldsymbol{\theta}$ with respect to the free variables $\mathbf{x}$, the derivatives of the $\mathbf{Q}$ matrix to the parameters $\boldsymbol{\theta}$, the derivatives of the corrected $\mathbf{Q}^e$ matrix to the elements of $\mathbf{Q}$, and the derivatives of the dwell-time distributions or the likelihood function with respect to the elements of $\mathbf{Q}^e$. The final derivatives of the objective function with respect to the free variables $\mathbf{x}$ are determined by combining all intermediate derivatives with the chain rule.

Optimization. There are three types of gradient-based approaches for unconstrained optimization: the steepest descent method, the conjugate directions method, and the variable metric method [23]. Among them, the variable metric method generally performs well for fitting dwell-time distributions or the likelihood function. The steepest descent method has a poor performance on the convergence near the optimal solution. The conjugate directions method requires an exact line search at every iteration, which generally takes many function evaluations. The variable metric method, one the other hand, has a quadratic convergence around the optimal point. It also exploits approximate line search for step size determination. The search direction resulting from the parabola interpolation is often sufficiently accurate so that only a few functional evaluations are needed for each line search. Thus the method is particularly suitable to functions whose evaluation requires extensive computing. The method does not need the exact Hessian matrix of the objective function. Instead, it builds up the curvature information from successive parabola interpolation of the objective function along with its first-order gradients.

Error estimates. The uncertainties on the fitted parameters can be estimated from the curvature of the likelihood function at its maximum:

$$\mathrm{Cov}(\mathbf{x}) \approx -\mathbf{H}^{-1}(\mathbf{x}) \tag{24}$$

where Cov is the covariance matrix of the parameters, and $\mathbf{H}$ is the Hessian matrix of the second derivatives. The exact Hessian matrix could be calculated numerically, but involves intensive computing. A faster, although less accurate, solution is the approximate inverse Hessian matrix generated by the variable metric optimizer. For a quadratic function of n variables the matrices generated by the method converge to the correct inverse Hessian matrix after n steps. For a nonquadratic function this no longer holds. However, in the vicinity of a maximum point the likelihood function may be well approximated by a quadratic function. Thus it can still be expected to provide information on the local Hessian at the optimal point.

The error estimates on the free variables $\mathbf{x}$ can be converted to the error estimates of the rate constants. The two are related by [24]:

$$\mathrm{Var}(q_{ij}) \approx \sum_k \left(\frac{\partial q_{ij}}{\partial x_k}\right)^2 \cdot \mathrm{Var}(x_k) \tag{25}$$

where Var denotes variance and Var(x_k) correspond to the diagonal elements of the covariance matrix obtained in Eq. 24.

Multiple channels. Often, there will be more than one channel in the recording. If the constituent channels are statistically independent, the data of multiple channels will still be a Markov process. Suppose that the patch contains n channels and the ith individual channel has n_i kinetic states. The patch at any time is then fully described by a vector

$$s_t = \left(s_t^{(1)}, s_t^{(2)}, \ldots, s_t^{(n)}\right) \qquad (26)$$

where $s_t^{(i)}$ specifies which state the ith channel is in at time t. The state space of $\mathbf{s}_t$ has a dimension of $n_1 n_2 \ldots n_n$. The transition rate of $\mathbf{s}_t$ from state $(i_1, \ldots, i_l, \ldots i_n)$ to state $(i_1, \ldots, j_l, \ldots i_n)$ is equal to the rate from i_l to j_l of the lth channel. In other words, the data of multiple channels can be considered as if generated from a single channel in which the states are defined as vectors and the transition rates are constrained to be those of the constituent channels. The theory developed above can still be applied to these data by operating on the induced Markov model for $\mathbf{s}_t$ and by imposing appropriate constraints on its rate constants. All constraints are of the linear scaling type and are therefore accommodated by the linear equality constraints.

The vector state defined in Eq. 26 is applicable to both identical and nonidentical channels. For identical channels, there exists a more efficient definition by exploiting the indistinguishibility of the constituent channels, namely,

$$n_t = \left(n_t^{(1)}, n_t^{(2)}, \ldots, n_t^{(k)}\right) \qquad (27)$$

where $n_t^{(i)}$ represents the number of channels in state i at time t [14]. The state space of $\mathbf{n}_t$ comprise all possible compositions of n into k parts. So the dimension of the state space is given by the binomial coefficient $\begin{pmatrix} n+k-1 \\ n \end{pmatrix}$. The rate constant for the transition from state $(m_1 \ldots m_i \ldots m_j \ldots m_k)$ to state $(m_1 \ldots m_i - 1 \ldots m_j + 1 \ldots m_k)$ is equal to $m_i q_{ij}$, where q_{ij}'s are the rate constant of the constituent channel.

The composition vectors $\mathbf{n}_t$ for given n and k may be generated automatically by a standard combinatorial analysis algorithm [25]. In general, $\begin{pmatrix} n+k-1 \\ n \end{pmatrix}$ is much less than k^n. So the state space of $\mathbf{n}_t$ is much smaller than that of $\mathbf{s}_t$. Considering that the evaluation of a dwell-time distribution has a complexity at least quadratic on the number of states, a considerable reduction in both processing time and memory space may be expected by using $\mathbf{n}_t$ for modeling multiple identical channels.

Optimality. For maximum likelihood fitting, regardless of the model used, the fitting always produces a set of parameters.

The question is whether the resultant model has the right features of the data. There are two possible scenarios: the model is overly complex, or the model is inadequate. When a model is overly complex, the resultant fit often contains rate constants that are either very small or very large. In either case, it signals that the data do not contain sufficient information on the corresponding transitions. On the other hand, when a model is invalid or too simplistic, the predicted distributions may become significantly deviated from the experimental histograms. In this case, the superposition of the fitted dwell-time distributions with the histograms provides a visual clue on whether the fit is inadequate. However, it should be cautious that a good fit of the histograms does not imply the correctness of the model since the 1D dwell-time histogram does not contain all information of the data. A more sensitive test is to calculate the maximum attainable likelihood, i.e., the likelihood that would be obtained when the true model is used. This true maximum likelihood can be obtained by fitting data with a fully connected and uncoupled model as illustrated in Scheme II.

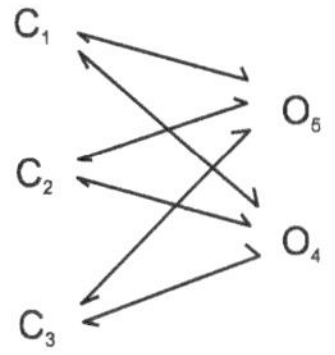

(Scheme II)

The model has $2N_cN_o$ parameters, which coincides with the maximum number of degrees of freedom that can be resolved from the data for the given number of states [8]. Because the open and closed states are uncoupled, the model is fully identifiable [6]. If the likelihood value of the test model is significantly less than that of this uncoupled model, it would suggest that the test model does not have adequate capacity to account for the data. For fitting multiple datasets over different conditions, one may calculate the maximum log likelihood with the uncoupled model for each condition. The maximum log likelihood over the entire datasets can be obtained as the sum of the maximum log likelihood over individual datasets.

1.9 Hidden Markov Model

Traditionally, single-channel currents are analyzed by first idealizing the records into dwell-times, followed by fitting the dwell-times with appropriate models. The idealization is typically performed by threshold detection. The technique relies on heavy filtering, which results in band-limiting distortions. In extreme cases where the current is small and/or the kinetics are fast, a reliable idealization of the data may become impossible. For these problems, the hidden Markov modeling provides an alternative means of analysis. It can be used for both idealization of recording [26] and estimation of model parameters [27–32]. Furthermore, it does not suffer from the missed events problem.

A hidden Markov model refers to a Markov process whose states are not directly observable, for example, a Markov process embedded in noise [33]. When the noise is high, the observation cannot be associated with states unambiguously. Instead, they can only be characterized probabilistically by

$$b_i(y_t) = \Pr(y_t \mid s_t = i)$$

which is the probability of observing a sample y_t given that the underlying Markov process s_t is at state i. In the case of white noise, this observation probability is determined by the Gaussian distribution

$$b_i(y_t) = \frac{1}{\sqrt{2\pi}\sigma_\mu} \exp\left(-\frac{(y_t - I_\mu)^2}{2\sigma_\mu^2}\right) \quad (28)$$

where μ designates the conductance class that state i belongs to, σ_μ^2 is noise variance at the corresponding conductance level, and I_μ is the unitary current of the μth conductance.

Because of noise, the data need to be modeled in discrete samples instead of continuous dwell-times. For a discrete Markov process, the state transition can be more conveniently characterized by the transition probability between adjacent samples, which are collectively represented by a matrix $\mathbf{A} = [a_{ij}]$, where a_{ij} defines the probability of the channel being in state j at the next sampling clock given that it was in state i at the previous sampling clock. The transition probability matrix $\mathbf{A}$ is related to the infinitesimal rate constant matrix $\mathbf{Q}$ by

$$\mathbf{A} = \exp(\mathbf{Q} \quad t) \quad (29)$$

where Δt is the sampling interval. The parameters of a hidden Markov model thus consists of the transition probabilities a_{ij}'s, the current amplitudes I_μ's, and the noise variances σ_μ^2. They are collectively designated as $\lambda = \{\mathbf{A}, I_\mu, \sigma_\mu^2\}$.

Patch-clamp recording contains noise from a variety of sources. In general, the noise is not white. Furthermore, the data is filtered. The filtering introduces correlations in both signal and noise. The standard hidden Markov model can be extended to accommodate these distortions. For the illustration of concepts, we will restrict ourselves to the case of an ideal Markov signal in white noise. For generalization to filtered signal or correlated noise, the readers are referred to 28–30, 32.

1.10 Restoration of Single-Channel Current

One application of hidden Markov modeling for single-channel analysis is the detection of dwell-times. In particular, it allows one to consider an idealization that is optimal in some statistical sense. One such criterion is to seek a state sequence, $\mathbf{s} = s_1 s_2 \ldots s_T$, to maximize the a posteriori probability [34],

$$\Pr(s_1 s_2 \ldots s_T; y_1 y_2 \ldots y_T; \lambda\} = \max \quad (30)$$

where $\mathbf{y} = y_1 y_2 \ldots y_T$ is the sequence of the observation samples and λ designates the underlying hidden Markov model. The solution satisfying the criterion represents the most likely idealization of the data with respect to a model. Accordingly, the probability can be viewed as the likelihood of the idealization.

By Bayes law, the probability in Eq. 30 can be cast into the probability of the state sequence itself, multiplied by the probability of observing the samples given the state sequence, i.e.,

$$\Pr(s_1 s_2 \ldots s_T, y_1 y_2 \ldots y_T; \lambda\} = \Pr(s_1 s_2 \ldots s_T; \lambda\} \cdot \Pr(y_1 y_2 \ldots y_T \mid s_1 s_2 \ldots s_T; \lambda\}$$

The probability of the state sequence is determined by the initial probability of the states (π_i), multiplied by the transition probability through subsequent samples

$$\Pr(s_1 s_2 \ldots s_T; \lambda\} = \Pr(s_1) \cdot \Pr(s_2 \mid s_1) \cdots \Pr(s_T \mid s_{T-1}) = \pi_{s_1} \cdot a_{s_1 s_2} a_{s_2 s_3} \ldots a_{s_{T-1} s_T}$$

The probability of the observed samples given the state sequence is simply the product of the observation probability of individual samples

$$\Pr(y_1 y_2 \ldots y_T \mid s_1 s_2 \ldots s_T) = \Pr(y_1 \mid s_1) \ldots \Pr(y_T \mid s_T) = b_{s_1}(y_1) b_{s2}(y_2) \ldots b_{s2}(y_T)$$

Thus the likelihood function in Eq. 30 can be represented by

$$\Pr(\mathrm{s,;y,;»}) = \pi_{s_1} a_{s_1 s_2} \cdots a_{s_{T-1} s_T} \times b_{s_1}(y_1) b_{s_2}(y_2) \cdots b_{s_2}(y_T).$$

The problem of idealization is then to choose among all possible choices a state sequence $\mathbf{s}$ and a set of model parameters λ so that the probability is maximal.

A straightforward solution to the problem is to enumerate all state sequences and determine the one that gives the maximal probability. Unfortunately, the strategy is unrealistic. Even in the simplest case with two states, there are 2^T sequences for T samples. For a small number of 100 samples, this will result in ~10^{30} state sequences. A simple enumeration of all the sequences, even without calculating their probability, would take >10^{23} years on a computer operating at 1 GHz.

The exhaustive search fails because it evaluates each state at each sample point repetitively for different state sequences. An approach that can avoid the problem is the Viterbi algorithm based on dynamic programming [35]. The algorithm is recursive and proceeds as follows. Let $\phi_1(i) = \pi_i b_i(y_1)$ for $1 \le i \le N$. Then the following recursion for $2 \le t \le T$ and $1 \le j \le N$:

$$\phi_t(j) = \max_{1 \le i \le N} \{\phi_{t-1}(i) a_{ij} b_j(y_t)\} \tag{31a}$$

and

$$\psi_t(j) = i^* \tag{31b}$$

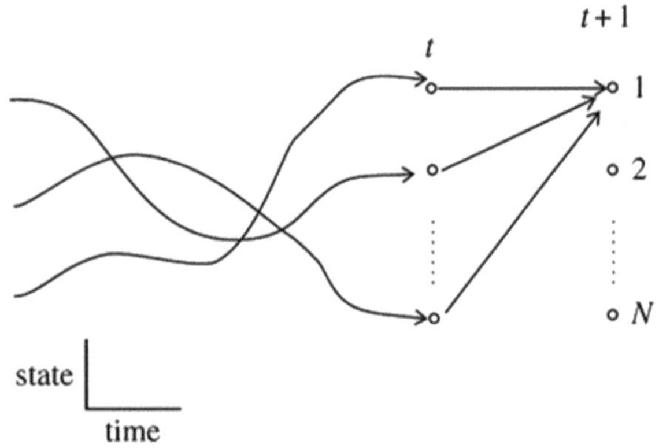

Fig. 2 Illustration of the Viterbi algorithm. The optimal sequence leading to a given state at time $t+1$ can be constructed from the sequences up to time t combined with a single transition from the previous ending state at time t to the given state at time $t+1$. Republished with permission from ref. 39

where i^* is a choice of the state that maximizes $\phi_t(i)$. Upon termination, the likelihood is given by $\Pr\left(s,;y,;»\right) = \max_{1\le i\le N} \phi_T(i)$. The most likely state sequence can be recovered from ψ as follows. Let $s_T = i^*$, which maximizes $\phi_T(i)$. Then for $T \geq t \geq 2$, $s_{t-1} = \psi_t(s_t)$.

The basic idea of the Viterbi algorithm is schematically illustrated in Fig. 2. It performs the idealization through time successively. At each time t, it keeps track of the optimal state sequences (pathways) leading to all possible states at that point. Then, an optimal sequence up to the next time $t+1$ is constructed by examining all existing N sequences up to time t in combination with an appropriate transition from time t to $t+1$. Because the probability of each state sequence up to time t is remembered, the construction of the new extended sequence requires only N^2 computations, as implied by (Eq. 31a). The idealization of the entire dataset then takes on the order of N^2T operations, which is quadratic on the number of states and linear on the number of samples, as opposed to the exponential dependence required by an exhaustive search.

The result of the Viterbi detection is optimal relative to the model used. In practice, the model is unknown priori to analysis. As a result, the model parameters need to be estimated. Given an idealization, the estimation can be done empirically. To estimate the current amplitudes and noise variances, one can classify the samples into clusters according to their conductance. The current amplitude and noise variance can then be estimated as the means and variance of samples within each cluster, respectively,

$$\hat{I}_\mu = \sum_{s_t \in C_\mu} y_t \Big/ \sum_{s_t \in C_\mu} 1 \qquad (31c)$$

$$\hat{\sigma}_{\mu}^{2} = \sum_{s_t \in C_{\mu}} \left(y_t - \hat{I}_{\mu}\right)^2 / \sum_{s_t \in C_{\mu}} 1 \tag{31d}$$

where C_μ denotes the states of the μth conductance. The denominator corresponds to the number of samples that are idealized into C_μ. Similarly, the transition probability can be estimated by counting the number of transitions occurring from each state,

$$\hat{a}_{ij} = n(i,j) / n(i) \tag{31e}$$

where $n(i)$ is the number of occurrences of state i over the entire sample sequence and $n(i,j)$ the number of occurrences that state j is an immediate successor of state i.

Ideally, the new estimates of the model parameters should agree with those that initiate the idealization. When the model is unknown, however, they may not be equal, in which case the estimates can be used to upgrade the model. This leads to an iterative loop, where an initial model, λ_0, is chosen, and the Viterbi algorithm is used to find an optimal idealization, from which the model parameters are reestimated. The iteration continues until it converges, for example, when the difference of the parameter values in two consecutive iterations becomes less than a preset small tolerance. This is the essence of the segmental k-means method [34].

Experimental testing indicates that the approach has a good tolerance for noise [26]. Reliable idealization can be achieved at a noise-to-signal ratio as high as 2:1. It also provides good estimates on amplitude parameters including the unitary current and the noise variance. The approach, however, shows a poor performance on kinetics parameters. When there are multiple states with the same conductance, it generally cannot resolve the transitions between these states. This occurs because the estimation is based on only a single state sequence. A reliable estimation of kinetics requires the use of the full likelihood approach, which examines all possible state sequences.

The insensitivity of the approach to kinetic parameters, on the other hand, provides ease for selection of models. Data generated by complex models can often be idealized accurately with a two-state model in which each state represents a conductance level. More states become necessary when the channel contains dwell-times that are orders of magnitude different in duration. Under such conditions, introducing a new aggregated state can improve the detection of fast transitions. In practice, an adequate model can be obtained retrospectively. The data can be first idealized with a two-state model. Then the resultant dwell-time distributions can be explored for additional components. Once the number of components is determined, a fully connected and uncoupled model (Scheme II), which attains the maximal complexity for the given number of components, can be used for full idealization.

1.11 Direct Estimation of Kinetics from Single-Channel Recording

The single-channel analysis based on idealized dwell-times has the advantage of being computationally efficient. This occurs because there are fewer dwell-times than samples. The hidden Markov modeling, on the other hand, provides a general paradigm allowing estimation of kinetics directly from single-channel recording. The concept is similar to maximum likelihood fitting of dwell-time sequences. That is, a model is determined to maximize the probability of the observations

$$L(\lambda) = \Pr(y_1 y_2 \ldots y_T \mid \lambda) \tag{32}$$

where λ represents the underlying hidden Markov model and $\mathbf{y} = y_1 y_2 \ldots y_T$ the observed samples. In the absence of noise, the observations reduce to dwell-times. So the maximum likelihood analysis of dwell-times can be considered as a special case of hidden Markov modeling.

Likelihood calculation. As with most applications of maximum likelihood estimation, the core of the problem is how to evaluate the likelihood function and how to search for its maximum [36]. Conceptually, the probability of the observations is determined by the sum of the a posteriori probability in Eq. 30 over all state sequences. This can be equivalently written in the matrix form as

$$L(\lambda) = {}^{\tau}\mathbf{B}_1 \cdot \mathbf{AB}_2 \cdots \mathbf{AB}_T \cdot 1 \tag{33}$$

where $\boldsymbol{\pi} = (\pi_1, \ldots, \pi_N)^{\tau}$ is the starting probability, $\mathbf{A} = [a_{ij}]_{N \times N}$ is the transition probability matrix, and $\mathbf{B}_t = \mathrm{diag}[b_1(y_t), \ldots, b_N(y_t)]$ is a diagonal matrix of the observation distribution for each state at time t. The multiplication of the matrices through samples effectively sums the probability over all possible combinations of states through the entire sequence of samples. Intuitively, the equation says that the likelihood is equal to the initial probability of entering each state, multiplied by the probability of observing the first sample, and then multiplied by the probability to make a transition to the next sample, followed by the probability of observing the second sample, and so on.

The matrix product form of the likelihood function suggests that the likelihood can be computed recursively. This leads to the so-called forward–backward procedure [33]. Let $\alpha_t(i) = \Pr[\Upsilon_1 \ldots \Upsilon_t, s_t = i]$, $1 \leq i \leq N$, $1 \leq t \leq T$ and $\beta_t(j) = \Pr[\Upsilon_{t+1} \ldots \Upsilon_T, s_t = j]$, $1 \leq j \leq N$, $1 \leq t \leq T$ denote the forward and backward variables, respectively, where $\alpha_t(i)$ is the joint probability of the partial observation sequence up to time t assuming the channel is in state i at time t, and $\beta_t(j)$ is the joint probability of the complement observations from the last sample back to time t assuming the channel is in state j at time t. Essentially, they are the elements of the forward and backward partial products of (Eq. 33) up to time t, respectively. Thus the forward and backward variables can be formulated recursively as

$$\alpha_{t+1}(j) = \sum_{i=1}^{N} \alpha_t(i) a_{ij} b_j(y_{t+1}), j = 1 \ldots N, t = 1 \ldots T$$
$$\beta_t(i) = \sum_{j=1}^{N} a_{ij} \beta_{t+1}(j) b_j(y_{t+1}), i = 1 \ldots N, t = T \ldots 1 \quad (34)$$

where the initial conditions $\alpha_0(i) = \pi_i$ and $\beta_T(j) = 1$. These equations imply that the forward and backward variables can be calculated recursively.

From the forward and backward variables, the likelihood function is given by

$$\Pr(\mathbf{Y}) = \sum_{i=1}^{N} \alpha_t(i) \beta_t(i). \quad (35)$$

The equality holds for every $t = 1 \ldots T$. In particular, taking $t = T$, one obtains the likelihood as the sum of the last forward variables over all states. In other words, the forward variables alone suffice for evaluating the likelihood. But as we will see later, the backward variables are needed for reestimation of parameters or calculation of the derivatives of the likelihood function.

Baum-Welch reestimation. The standard hidden Markov modeling relies on Baum's reestimation to estimate model parameters [36]. The algorithm is a precursor of the more general expectation-maximization (EM) approach for maximum likelihood estimation, which states that the maximum likelihood estimates of the model parameters can be constructed iteratively from their expected values based on their current estimates [37]. For a hidden Markov model, there are two fundamental expected probabilities, namely,

$$\xi_t(i,j) = \Pr\left(s_t = i, s_{t+1} = j \mid \lambda^{(n)}, \mathbf{y}\right)$$
$$\gamma_t(i) = \Pr\left(s_t = i \mid \lambda^{(n)}, \mathbf{y}\right)$$

where $\gamma_t(i)$ is the probability of being in state i at time t given the current estimate of the model $\lambda^{(n)}$ and the observations $\mathbf{y}$, and $\xi_t(i,j)$ the probability in state i at time t and state j at time $t+1$, also conditioned on $\lambda^{(n)}$ and $\mathbf{y}$. They can be calculated from the forward and backward probabilities by

$$\xi_t(i,j) = \alpha_t(i) a_{ij} b_i(t) \beta_t(j) / \sum_i \alpha_t(i) \beta_t(i)$$
$$\gamma_t(i) = \alpha_t(i) \beta_t(i) / \sum_i \alpha_t(i) \beta_t(i) \quad (36)$$

where the numerator of $\xi_t(i,j)$ is $\Pr(s_t = i, s_{t+1} = j, \mathbf{y})$, the numerator of $\gamma_t(i)$ is $\Pr(s_t = i, \mathbf{y})$, and the denominator is $\Pr(\mathbf{y})$. The division thus gives the desired probability conditioned on the observation.

The quantities $\xi_t(i,j)$ and $\gamma_t(i)$ can be used to construct the expected values of the model parameters. For example, the summation of $\xi_t(i,j)$ over time t leads to the expected number of transitions from state i to state j, and the summation of $\gamma_t(i)$ over t gives the expected occupancy of state i. The ratio of the two then gives the expected value of the transition probability a_{ij},

$$\hat{a}_{ij} = \sum_{t=1}^{T-1} \xi_t(i,j) \Big/ \sum_{t=1}^{T-1} \gamma_t(i). \tag{37a}$$

The initial probability, the unitary current, and the noise variance can be similarly represented by

$$\hat{\pi}_i = \gamma_1(i) \tag{37b}$$

$$\hat{I}_\mu = \sum_{t=1}^{T-1} \sum_{i \in C_\mu} \gamma_t(i) y_t \Big/ \sum_{t=1}^{T-1} \sum_{i \in C_\mu} \gamma_t(i) \tag{37c}$$

$$\hat{\sigma}_\mu^2 = \sum_{t=1}^{T-1} \sum_{i \in C_\mu} \gamma_t(i) \left(y_t - I_\mu\right)^2 \Big/ \sum_{t=1}^{T-1} \sum_{i \in C_\mu} \gamma_t(i) \tag{37d}$$

where C_μ denotes the states of the μth conductance.

These expected values of the model parameters provide a reestimate of the underlying model. According to Baum's theory, the new estimate always results in an improved likelihood value. Repetition of the reestimation eventually leads to a (local) maximization of the likelihood function.

The reestimation of the parameters given in Eq. 37 should be distinguished from those in Eq. 31. The reestimation in the segmental k-means method is made from the most likely state sequence. The information from a single state sequence is generally inadequate to resolve transitions between aggregated states. The true maximum likelihood estimation overcomes the problem by gleaning information from all state sequences.

Figure 3 illustrates the performance of hidden Markov modeling on noise levels and channel kinetics [31]. For a fixed data length, increasing the noise level reduces the curvature of the likelihood surface. The position of the maximum, however, remains unchanged. In other words, the noise level does not affect the mean values of the estimates but only their confidence limits. Furthermore, the reduced curvature of the likelihood surface at high noise levels can be restored by increasing the data length. This suggests that the algorithm has no intrinsic limit on the noise level provided a sufficient amount of data.

Contrary to the effect of noise, a hard limit appears to exist for kinetic parameters. The likelihood surface is well defined for $k\Delta t < 1$. But when $k\Delta t$ increases up to 2, the upper half of the function becomes virtually flat. The fastest rates that can be extracted reliably are about as fast as the sampling rate. Beyond that, the

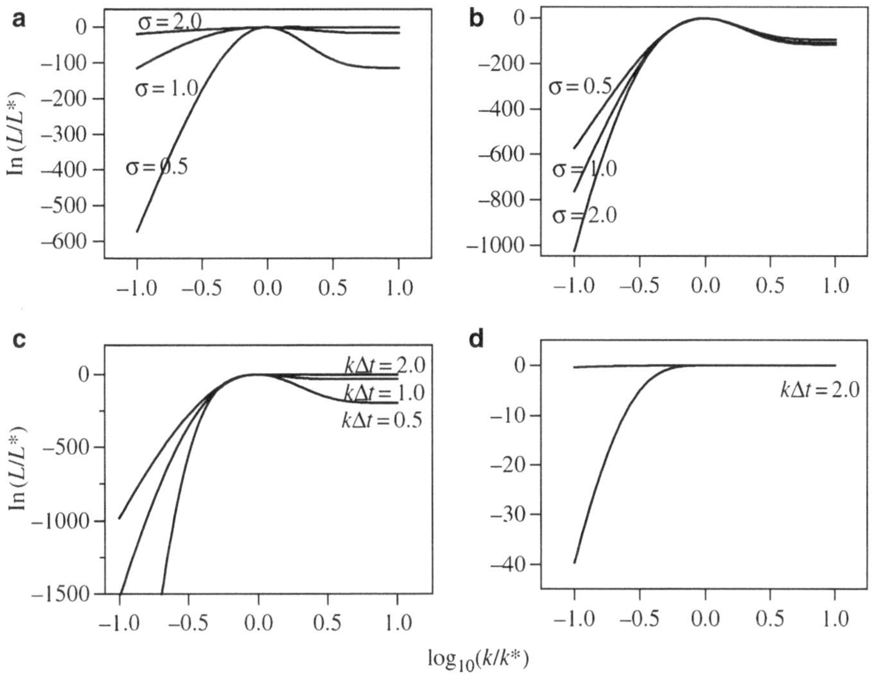

Fig. 3 Dependence of the log likelihood surface on noise and kinetics. (**a**) For a fixed data length, the likelihood surface becomes flatter as the data get noisier. The algorithm reaches the limit at a standard deviation about 2 pA, corresponding to a SNR = 0.5. (**b**) When the noise increases, the data length is also increased so that the log likelihood function maintains about the same curvature at the maxima as in the low noise case. (**c**) As the sampling interval increases, the likelihood surface becomes increasingly asymmetric, where the lower half ($k \leq k^*$) still maintains a good curvature, but the upper half ($k \geq k^*$) becomes flat. The maxima of the likelihood can be reliably identified until $k\Delta t = 2$. (**d**) Two log likelihood functions at $k\Delta t = 2$ with the data length different by a factor of 100. Increasing the data length in this case improves the curvature of the lower half ($k \leq k^*$) of log likelihood function, but has little effect on the flatness of the upper half ($k \geq k^*$). Data were simulated from a two-state model with both rates equal to 100,000 s^{-1} with the unitary currents at 0 and 1 pA respectively. For the noise test, the sampling duration was fixed at $\Delta t = 5 \mu s$, corresponding to $k\Delta t = 0.5$. For the kinetics test, a small amount of noise with $\sigma = 0.1$ pA was added. The curves correspond to the cross sections of the log likelihood surface cut along the direction where the two rates are equal. Republished with permission from ref. 39

algorithm can only tell whether a rate is too slow to fit the data but cannot discriminate its upper limit because all rates faster than the sampling rate give equal likelihood values.

Direct optimization of rate constants. Baum's theory reestimates transition probabilities rather than rate constants of a model. This incurs limitations on allowance for global fitting of data, constraints on rate constants and model topology and so on. One solution to the problem is to convert **A** into **Q** at each iteration and applying constraints to resultant **Q** [28]. A more explicit approach is to treat the maximum likelihood estimation as an optimization problem and resort to optimization techniques to optimize rate constants directly [31].

The derivative of the likelihood function with respect to the model parameters can be calculated analytically. Applying the chain rule to Eq. 33, ones obtains the derivative of the likelihood function to any variable x

$$\frac{\partial L}{\partial x}=\frac{\partial \pi^{\tau}\mathbf{B}_1}{\partial x}\cdot \dagger_1+\cdots+\sum_t \alpha_t^{\tau}\frac{\partial \mathbf{AB}_{t+1}}{\partial x}\beta_{t+1}.$$

Letting x be the initial probability, the transition probability, the current amplitude and the noise standard deviation respectively, then

$$\frac{\partial L}{\partial \pi_i}=\beta_1(i)b_i(1) \tag{38a}$$

$$\frac{\partial L}{\partial a_{ij}}=\sum_t \alpha_t(i)\beta_{t+1}(j)b_j(t+1) \tag{38b}$$

$$\frac{\partial L}{\partial I_k}=\sum_t\sum_{\kappa(i)=k}\alpha_t(i)\beta_t(i)\sigma_k^{-2}(Y_t-I_k) \tag{38c}$$

$$\frac{\partial L}{\partial \sigma_k^2}=\sum_t\sum_{\kappa(i)=k}\alpha_t(i)\beta_t(i)\left[-\frac{1}{2\sigma_k^2}+\frac{(Y_t-I_k)^2}{2\sigma_k^4}\right] \tag{38d}$$

where $\kappa(i)$ denotes the conductance class of state i. The derivative of the transition probability with respect to a rate constant can be obtained from Eq. 23 by setting t to the sampling duration. When the starting probability is chosen as the equilibrium probability at the holding condition, it is also a function of rate constants. Its derivative to rate variables follows

$$\frac{\partial \pi^{\tau}}{\partial x}\mathbf{Q}_{\mathrm{h}}=\pi^{\tau}\frac{\partial \mathbf{Q}_{\mathrm{h}}}{\partial x} \tag{39}$$

which can be solved as linear system equations. Equations38 and 39, along with the derivative of the transition probability to rate constants, constitute the basis for the calculation of the derivative of likelihood with respect to kinetic constants.

Figure 4 compares the convergence of Baum's reestimation and direct optimization of the likelihood function by the variable metric method [31]. The two algorithms proceeded basically in the same direction towards the maximum. Baum's algorithm tends to converge rapidly in the beginning but becomes slow when it gets close to the maximum. This is particularly evident when the likelihood surface is relatively flat in the case of either low signal-to-noise ratio or aggregation of states. The optimization approach, on the other hand, shows little dependence on either the noise level or the aggregation of states. The convergence near the maximum is approximately quadratic. When the likelihood surface is

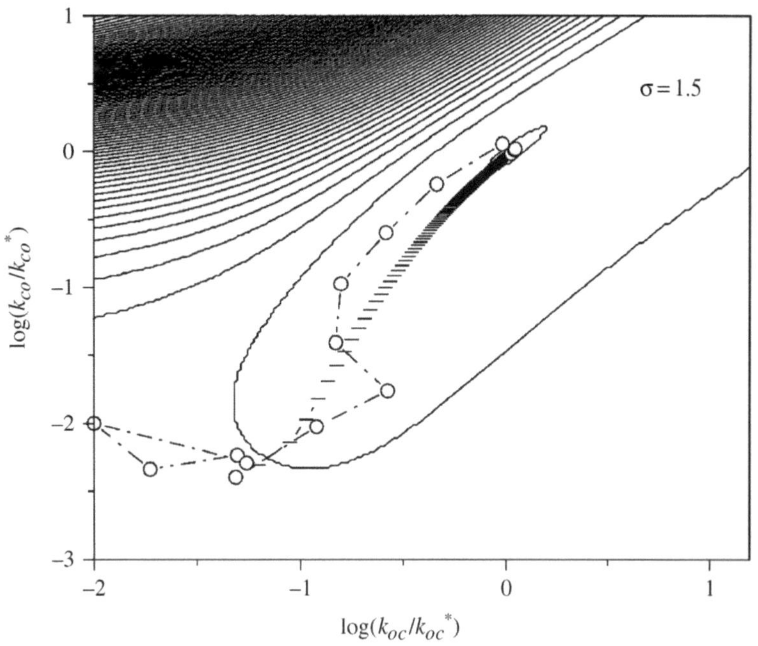

Fig. 4 Comparison of the convergence behavior between Baum's reestimation and the optimization method at a relatively high noise level $\sigma/i = 1.5$. The *circles* represent the optimization approach, and the *bars* correspond to Baum's reestimation. Each *symbol* represents one iteration. Republished with permission from ref. 39

well defined, however, Baum's reestimation can sometimes outperform the optimization approach. The algorithm is thus well suited for estimation of amplitude parameters, where the likelihood function usually attains relatively large curvatures.

1.12 Application Notes

A software suite (QuB) implementing the algorithms described above with user-friendly graphic interface on the 32-bit Windows has been developed (www.qub.buffalo.edu). An implementation with a pClamp-like data visualization interface is also available (www.qinlab.med.buffalo.edu). Besides tools for data analysis, the software also contains modules for data acquisition and supports some cost-effective cards such as those from National Instrument. Examples and tutorials on the practical applications of the algorithms and the usage of the programs are part of the package.

The study of single channel analysis has an extensive history and has resulted in numerous techniques and tools over the years from a number of labs. For advanced kinetics analysis, another free program for the maximum-likelihood dwell-time fitting is HJCFIT [38], which employs the asymptotic approximation for missed event correction. Practical fitting of the 2D dwell-time distribution was first described by Magleby and his coworkers [12]. The commercial TAC software also integrates hidden Markov modeling for single-channel analysis. The method is based on Baum-Welch's reestimation as developed by the Sigworth lab [28–30]. One significant difference between QuB and these other programs is the use of analytically calculated derivatives for optimization of the likelihood function, a feature that is important for fitting complex models and large datasets.

References

1. Colquhoun D, Hawkes AG (1977) Relaxation and fluctions of membrane currents that flow through drug-operated channels. Phil Trans R Soc Lond B 199:231–262
2. McManus OB, Weiss DS, Spivak CE, Blatz AL, Magleby KL (1988) Fractal models are inadequate for the kinetics of four different ion channels. Biophys J 54:859–870
3. McManus OB, Spivak CE, Blatz AL, Weiss DS, Magleby KL (1989) Fractal models, Markov models, and channel kinetics. Biophys J 55:383–385
4. Korn SJ, Horn R (1988) Statistical discrimination of fractal and Markov models of single channel gating. Biophys J 54:871–877
5. Sansom MSP, Ball FG, Kerry CJ, McGee R, Ramsey RL, Usherwood PNR (1989) Markov, fractal, diffusion, and related models of ion channel gating. Biophys J 56:1229–1243
6. Kienker P (1989) Equivalence of aggregated Markov models of ion-channel gating. Proc R Soc Lond B 236:269–309
7. Colquhoun D, Hawkes AG (1981) On the stochastic properties of single ion channels. Proc R Soc Lond B Biol Sci 211:205–235
8. Fredkin DR, Montal M, Rice JA (1985) Identification of aggregated Markovian models: application to the nicotinic acetylcholine receptor. Proceedings of the Berkeley Conference in Honor of Jerzy Neymann and Jack Kiefer, Belmont, CA: Wadsworth, pp 269–289
9. Labarca P, Rice JA, Fredkin DR, Montal M (1985) Kinetic analysis of channel gating: application to the cholinergic receptor channel and the chloride channel from Torpedo California. Proceedings of the Berkeley Conference in Honor of Jerzy Neymann and Jack Kiefer, Belmont, CA: Wadsworth, pp. 469–478
10. Qin F, Li L (2004) Model-based fitting of single-channel dwell-time distributions. Biophys J 86:1657–1671
11. Magleby KL, Pallotta BS (1983) Calcium dependence of open and shut interval distributions from calcium-activated potassium channels in cultured rat muscle. J Physiol 344:585–604
12. Magleby KL, Weiss DS (1990) Identifying kinetic gating mechanisms for ion channels by using two-dimensional distributions of simulated dwell times. Proc R Soc Lond B Biol Sci 241:220–228
13. Magleby KL, Song L (1992) Dependency plots suggest the kinetic structure of ion channels. Proc R Soc Lond B Biol Sci 249:133–142
14. Horn R, Lange K (1983) Estimating kinetic constants from single channel data. Biophys J 43:207–223
15. Ball FG, Sansom MSP (1989) Ion-channel gating mechanisms: model identification and parameter estimation from single channel recordings. Proc R Soc Lond B 236:385–416
16. Qin F, Auerbach A, Sachs F (1996) Estimating single channel kinetic parameters from idealized patch-clamp data containing missed events. Biophys J 70:264–280
17. Colquhoun D, Hawkes AG, Srodzinski K (1996) Joint distributions of apparent open and shut times of single-ion channels and maximum likelihood fitting of mechanisms. Phil Trans R Soc Lond A Math Phys Eng Sci 354:2555–2590
18. Qin F, Auerbach A, Sachs F (1997) Maximum likelihood estimation of aggregated Markov processes. Proc R Soc Lond [Biol] 264:375–383
19. Roux B, Sauve R (1985) A general solution to the time interval omission problem applied to single channel analysis. Biophys J 48:149–158
20. Blatz AL, Magleby KL (1986) Correcting single channel data for missed events. Biophys J 49:967–980
21. Crouzy SC, Sigworth FJ (1990) Yet another approach to the dwell-time omission problem of single-channel analysis. Biophys J 58:731–743
22. Hawkes AG, Jalali A, Colquhoun D (1992) Asymptotic distributions of apparent open times and shut times in a single channel record allowing for the omission of brief events. Phil Trans R Soc Lond B 337:383–404
23. Press WH, Teukolsky SA, Vetterling WT, Flannery BP (1992) Numerical recipes in C. Cambridge University Press, Cambridge

24. Kendall MG, Stuart A (1977) The advanced theory of statistics. Griffin, London
25. Nijenhuis A, Wilf HS (1978) Combinatorial algorithms. Academic Press Inc., New York
26. Qin F (2004) Restoration of single-channel currents using the segmental k-means method based on hidden Markov modeling. Biophys J 86:1488–1501
27. Fredkin DR, Rice JA (1992) Maximum likelihood estimation and identification directly from single-channel recordings. Proc R Soc Lond B 239:125–132
28. Venkataramanan L, Walsh JL, Kuc R, Sigworth FJ (1998) Identification of hidden Markov models for ion channel currents – part I: colored background noise. IEEE Trans Signal Process 46:1901–1915
29. Venkataramanan L, Kuc R, Sigworth FJ (1998) Identification of hidden Markov models for ion channel currents – part II: state-dependent excess noise. IEEE Trans Signal Process 46:1916–1929
30. Venkataramanan L, Kuc R, Sigworth FJ (2000) Identification of hidden Markov models for ion channel currents – part III: bandlimited, sampled data. IEEE Trans Signal Process 48:376–385
31. Qin F, Auerbach A, Sachs F (2000) A direct optimization approach to hidden Markov modeling for single channel kinetics. Biophys J 79:1915–1927
32. Qin F, Auerbach A, Sachs F (2000) Hidden Markov modeling for single channel kinetics with filtering and correlated noise. Biophys J 79:1928–1944
33. Rabiner LR (1989) A tutorial on hidden Markov models and selected applications in speech recognition. Proc IEEE 77:257–286
34. Rabiner LR, Wilpon JG, Juang BH (1986) A segmental k-means training procedure for connected word recognition. AT T Tech J 65: 21–31
35. Forney GD (1973) The Viterbi algorithm. Proc IEEE 61:268–278
36. Baum LE, Petrie T, Soules G, Weiss N (1970) A maximization technique occuring in the statistical analysis of probabilistic functions of Markov chains. Ann Math Stat 41:164–171
37. Dempster AP, Laird NM, Rubin DB (1977) Maximum likelihood from incomplete data via the EM algorithm. J Roy Stat 39:1–38
38. Colquhoun D, Hatton CJ, Hawkes AJ (2003) The quality of maximum likelihood estimates of ion channel rate constants. J Physiol 547: 699–728
39. Qin F (2007) Principles of single-channel kinetic analysis. In: Molnar P, Hickman JJ (eds) Patch-clamp methods and protocols. Series: methods in molecular biology, 1st edn. Humana, Totowa, NJ, pp 195–209

INDEX

Marzia Martina and Stefano Taverna (eds.), *Patch-Clamp Methods and Protocols*, Methods in Molecular Biology, vol. 1183, DOI 10.1007/978-1-4939-1096-0, © Springer Science+Business Media New York 2014

S

T

U

V

W

X

MIX
Papier aus verantwortungsvollen Quellen
Paper from responsible sources
FSC® C105338

If you have any concerns about our products,
you can contact us on
ProductSafety@springernature.com

In case Publisher is established outside the EU,
the EU authorized representative is:
Springer Nature Customer Service Center GmbH
Europaplatz 3, 69115 Heidelberg, Germany

Printed by Libri Plureos GmbH
in Hamburg, Germany